Praise for *Reality+*

"[David] Chalmers's central idea, that 'there is more ~~to reality than we thought~~, is seductive, and I was surprised to find his arguments delightfully—or perhaps worryingly—convincing. . . . He has taken a subject most people would dismiss as pure science fiction and produced a brilliant and very readable philosophical investigation."
—P. D. Smith, *Guardian*

"[Chalmers is] an exuberant guide through challenging terrain, quick with anecdotes and arguments, wit and wild ideas."
—Kieran Setiya, *Times Literary Supplement*

"David Chalmers's new book is a *tour de force*. . . . I expect that *Reality+* will set various agendas in metaphysics just as *The Conscious Mind* (1996) did in the philosophy of mind. And it is all the more impressive that he does this in a book much of which can be understood by beginners. . . . Readable, effortlessly up to date, handsomely produced, and well-structured, *Reality+* will be a boon for teachers of introductory philosophy courses and is sure to remain on syllabuses for many years."
—Tim Crane, *Philosopher*

"Like [Douglas] Hofstadter's work, *Reality+* is frequently weird, wild and wonderful; it captivates the common reader by refusing to condescend."
—Jess Keiser, *Washington Post*

"A brain-bending new book by the philosopher David Chalmers—*Reality+: Virtual Worlds and the Problems of Philosophy*—has turned me into a hardcore simulationist. After reading and talking to Chalmers, I've come to believe that the coming world of virtual reality might one day be regarded as every bit as real as real reality."
—Farhad Manjoo, *New York Times*

"[Chalmers] wrestles with how age-old philosophical conundrums can be reinterpreted in the age of Reality+. . . . He deftly interweaves the finer points of ancient Chinese philosophy and Cartesian dualism with the metaphysics of the Matrix films and the World of Warcraft computer games. . . . [A] rich, scintillating . . . book that reflects many fascinating facets of our virtual worlds."
—John Thornhill, *Financial Times*

"One of the most important living philosophers, existing in an exclusive club of living thinkers who are on compulsory reading lists for undergraduate philosophy students. . . . He writes with admirable clarity and there's something quite rock 'n' roll about him." —Bryan Appleyard, *Spectator*

"The paradox of Chalmers's 'simulation realism,' in fact, is that, once you embrace it [so] many isms that in modern times have been dismissed as mystical, supernatural—dualism, panpsychism, animism—here find themselves reenchanted, imbued with a profound new vitality." —Jason Kehe, *Wired*

"A sprawling, brain-tenderising beast of a book—but a hugely entertaining one at that." —Kit Wilson, *Times* (UK)

"Everyone should read this important book to understand where we may be heading and how it will be rationalised." —Josh Glancy, *Sunday Times* (UK)

"Crafted with the general reader in mind, this is an object lesson in philosophical reasoning and a bold, often awe-inspiring discussion of its implications." —*Publishers Weekly*

"In this accessible yet thought-provoking book, readers will encounter everything from Plato's allegory of the cave and John Wheeler's it-from-bit hypothesis to how mind and body might interact in virtual worlds, whether reality is a mathematical structure, and whether we might just be Boltzmann brains floating in a dream world. Chalmers also tackles techno-centric questions like whether smartphones extend our minds, whether the Internet is making us smart or stupid, the threat of deepfakes and alternative facts, and whether there can be an objective reality in a multiverse of virtual worlds." —Jennifer Ouellette, *Ars Technica*

"A David Chalmers book is a competition. On the one hand the writing is so clear and engaging that you want to keep turning pages; on the other, the ideas are so surprising and profound that you are continually stopping to think about them. *Reality+* is a treasure trove of provocative reflections on cosmology, consciousness, artificial intelligence, ethics, and more. Reading it will change the way you think about the universe."

—Sean Carroll, author of *Something Deeply Hidden: Quantum Worlds and the Emergence of Spacetime*

"Fasten your seatbelt and put your helmet on, David Chalmers is going to take you on an amazing trip. *Reality+* is wild, profound, and playful, placing famous arguments from the history of philosophy next to surprising observations about video games. Cleverly disguised as light reading, this book carries a large payload of new ideas about existence, knowledge, and what makes life worth living."

—Jennifer Nagel, University of Toronto

"As humanity enters a brave new world of artificial superintelligence and computer-generated virtual realities, how can we humble hunter-gatherers, descended from cavemen, begin to grasp our astonishing technological future? The answer lies in this book. We must think about the ultimate nature of reality. In *Reality+* David Chalmers provides the roadmap to your future."

—Susan Schneider, NASA/Library of Congress Chair in Astrobiology, Exploration, and Scientific Innovation, and author of *Artificial You: AI and the Future of Your Mind*

"A stunning achievement. In effortless prose David Chalmers explores new ways to think about everything from consciousness to computation, deities to democracy. *Reality+* shows time and again how familiar topics take on interesting new forms when viewed through the lens of virtual reality."

—Scott Sturgeon, author of *The Rational Mind*

"What is real anyway? Exploring the deepest doubts about reality from Zhuangzi to Descartes, David J. Chalmers stirs our own doubts and leads us into the real worlds of future virtual reality. A gripping book."

—Susan Blackmore, author of *The Meme Machine* and *Seeing Myself*

ALSO BY DAVID J. CHALMERS

The Conscious Mind

The Character of Consciousness

Constructing the World

REALITY+

VIRTUAL WORLDS
AND THE PROBLEMS
OF PHILOSOPHY

David J. Chalmers

Illustrations by Tim Peacock

W. W. NORTON & COMPANY
Celebrating a Century of Independent Publishing

For information about permission to reproduce selections
from this book, write to Permissions,
W. W. Norton & Company, Inc., 500 Fifth Avenue, New York, NY 10110

For information about special discounts for bulk purchases,
please contact W. W. Norton Special Sales at
specialsales@wwnorton.com or 800-233-4830

Manufacturing by Lakeside Book Company
Book design by Lovedog Studio
Production manager: Julia Druskin

Library of Congress Cataloging-in-Publication Data

Names: Chalmers, David John, 1966– author.
Title: Reality+ : virtual worlds and the problems of philosophy /
David J. Chalmers ; illustrations by Tim Peacock.
Other titles: Reality plus
Description: First edition | New York, NY : W. W. Norton & Company, 2022. |
Includes bibliographical references and index.
Identifiers: LCCN 2021037439 | ISBN 9780393635805 (hardcover) |
ISBN 9780393635812 (epub)
Subjects: LCSH: Reality. | Virtual reality. | Philosophy. |
Technology—Philosophy.
Classification: LCC BD331 .C4925 2022 | DDC 111—dc23/eng/20211001
LC record available at https://lccn.loc.gov/2021037439

ISBN 978-1-324-05034-6 pbk.

W. W. Norton & Company, Inc., 500 Fifth Avenue, New York, N.Y. 10110
www.wwnorton.com

W. W. Norton & Company Ltd., 15 Carlisle Street, London W1D 3BS

1 2 3 4 5 6 7 8 9 0

For Claudia

Contents

Introduction

Adventures in technophilosophy

WHEN I WAS TEN YEARS OLD, I DISCOVERED COMPUTERS. MY first machine was a PDP-10 mainframe system at the medical center where my father worked. I taught myself to write simple programs in the BASIC computer language. Like any ten-year-old, I was especially pleased to discover games on the computer. One game was simply labeled "ADVENT." I opened it and saw:

You are standing at the end of a road before a small brick building.
Around you is a forest.
A small stream flows out of the building and down a gully.

I figured out that I could move around with commands like "go north" and "go south." I entered the building and got food, water, keys, a lamp. I wandered outside and descended through a grate into a system of underground caves. Soon I was battling snakes, gathering treasures, and throwing axes at pesky attackers. The game used text only, no graphics, but it was easy to imagine the cave system stretching out below ground. I played for months, roaming farther and deeper, gradually mapping out the world.

It was 1976. The game was *Colossal Cave Adventure*. It was my first virtual world.

In the years that followed, I discovered video games. I started with *Pong* and *Breakout*. When *Space Invaders* came to our local shopping mall, it became an obsession for my brothers and me. Eventually I got an Apple II computer, and we could play *Asteroids* and *Pac-Man* endlessly at home.

Over the years, virtual worlds have become richer. In the 1990s, games such as *Doom* and *Quake* pioneered the use of a first-person perspective. In the 2000s, people began spending vast amounts of time in multiplayer virtual worlds like *Second Life* and *World of Warcraft*. In the 2010s, there arrived the first rumblings of consumer-level virtual reality headsets, like the Oculus Rift. That decade also saw the first widespread use of augmented reality environments, which populate the physical world with virtual objects in games like *Pokémon Go*.

These days, I have numerous virtual reality systems in my study, including an Oculus Quest 2 and an HTC Vive. I put on a headset, open an application, and suddenly I'm in a virtual world. The physical world has disappeared entirely, replaced by a computer-generated environment. Virtual objects surround me, and I can move among them and manipulate them.

Like ordinary video games from *Pong* to *Fortnite*, virtual reality (or VR) involves a virtual world: an interactive, computer-generated space. What's distinctive about VR is that its virtual worlds are *immersive*. Instead of showing you a two-dimensional screen, VR immerses you in a three-dimensional world you can see and hear as if you existed within it. Virtual reality involves an immersive, interactive, computer-generated space.

I've had all sorts of interesting experiences in VR. I've assumed a female body. I've fought off assassins. I've flown like a bird. I've traveled to Mars. I've looked at a human brain from the inside, with neurons all around me. I've stood on a plank stretched over a canyon—terrified, though I knew perfectly well that if I were to step off, I'd step onto a nonvirtual floor just below the plank.

Like many other people, during the recent pandemic I've spent a great deal of time talking to friends, family, and colleagues using Zoom and other videoconferencing software. Zoom is convenient, but it has many limitations. Eye contact is difficult. Group interactions are choppy rather than cohesive. There is no sense that we are inhabiting a common space. One underlying issue is that videoconferencing is not virtual reality. It is interactive but not immersive, and there is no common virtual world.

During the pandemic, I've also met up once a week with a merry band of fellow philosophers in VR. We've tried many different platforms and activities—flying with angel wings in *Altspace*, slicing cubes to a rhythm in *Beat Saber*, talking philosophy on the balcony in *Bigscreen*, playing paintball in *Rec Room*, giving lectures in *Spatial*, trying out colorful avatars in *VRChat*. VR technology is still far from perfect, but we've had the sense of inhabiting a common world. When five of us were standing around after a short presentation, someone said, "This is just like coffee break at a philosophy conference." When the next pandemic arrives in a decade or two, it's likely that many people will hang out in immersive virtual worlds designed for social interaction.

Augmented reality (or AR) systems are also progressing fast. These systems offer a world that is partly virtual and partly physical. The ordinary physical world is augmented by virtual objects. I don't yet have my own augmented reality glasses, but companies like Apple, Facebook, and Google are said to be working on them. Augmented reality systems have the potential to replace screen-based computing entirely, or at least replace physical screens with virtual screens. Interacting with virtual objects may become part of everyday life.

Today's VR and AR systems are primitive. The headsets and glasses are bulky. The visual resolution for virtual objects is grainy. Virtual environments offer immersive vision and sound, but you can't touch a virtual surface, smell a virtual flower, or taste a virtual glass of wine when you drink it.

These temporary limitations will pass. The physics engines that underpin VR are improving. In years to come, the headsets will get smaller, and we will transition to glasses, contact lenses, and eventually retinal or brain implants. The resolution will get better, until a virtual world looks exactly like a nonvirtual world. We will figure out how to handle touch, smell, and taste. We may spend much of our lives in these environments, whether for work, socializing, or entertainment.

In Neal Stephenson's 1992 novel *Snow Crash*, users spend much of their lives in a massive multipurpose virtual world known as the Metaverse. In the 2020s, technology companies have declared their ambitions to build a version of the Metaverse: an interconnected system of virtual and augmented worlds, accessed using VR and AR technology. A mature Metaverse is still many years away, but in the meantime, the label has come to stand for the virtual and augmented worlds in which we will increasingly spend time in the decades to come.

My guess is that within a century we will have virtual realities that are indistinguishable from the nonvirtual world. Perhaps we'll plug into machines through a brain-computer interface, bypassing our eyes and ears and other sense organs. The machines will contain an extremely detailed simulation of a physical reality, simulating laws of physics to track how every object within that reality behaves.

Sometimes VR will place us in other versions of ordinary physical reality. Sometimes it will immerse us in worlds entirely new. People will enter some worlds temporarily for work or for pleasure. Perhaps Apple will have its own workplace world, with special protections so that no one can leak its latest Reality system under development. NASA will set up a world with spaceships in which people can explore the galaxy at faster-than-light speed. Other worlds will be worlds in which people can live indefinitely. Virtual real estate developers will compete to offer worlds with perfect weather near the beach, or with glorious apartments in a vibrant city, depending on what customers want.

Once simulation technology is good enough, these simulated environments may even be occupied by simulated people, with simulated brains and bodies, who will undergo the whole process of birth, development, aging, and death. Like the nonplayer characters that one encounters in many video games, simulated people will be creatures of the simulation. Some worlds will be simulations set up for research or to make predictions about the future. For instance, a dating app (as seen on the TV series *Black Mirror*) could simulate many futures for a couple in order to see whether they are compatible. A historian might study what would have happened if Hitler had chosen not to start a war with the Soviet Union. Scientists might simulate whole universes from the Big Bang onward, with small variations to study the range of outcomes: How often does life develop? How often is there intelligence? How often is there a galactic civilization?

One can imagine that a few curious 23rd-century simulators might focus on the early 21st century. Let's suppose the simulators live in a world in which Hillary Clinton defeated Jeb Bush in the US presidential election of 2016. They might ask: How would history have been different if Clinton had lost? Varying a few parameters, the simulators might go so far as to simulate a world where the 2016 victor was Donald Trump. They might even simulate Brexit and a pandemic.

Simulators interested in the history of simulation might also be interested in the 21st century as a period when simulation technology was coming into its own. Perhaps they might occasionally simulate people who are writing books about possible future simulations, or people who are reading them! Narcissistic simulators might nudge the parameters so that some simulated 21st-century philosophers speculate wildly about simulations built in the 23rd century. They might be especially interested in simulating the reactions of 21st-century readers reading thoughts about 23rd-century simulators, as you are right now.

Someone in such a virtual world would believe themselves to be living in an ordinary world in the early 21st century—a world in which Trump was elected president, the UK left the European Union, and there was a pandemic. Those events may have been surprising at the time, but humans have a remarkable capacity to adjust, and after a while these things become normal. Although simulators may have nudged them into reading a book on virtual worlds, it will seem to them as if they are reading the book out of their own free choice. The book they're reading now is perhaps a little unsubtle in trying to convey the message that they may be in a virtual world, but they will take this in stride and start thinking about the idea all the same.

At this point, we can ask, "How do you know you're not in a computer simulation right now?"

◆

This idea is often known as the *simulation hypothesis*. It is famously depicted in the *Matrix* movies, in which what seems an ordinary physical world turns out to be the result of connecting human brains to a giant bank of computers. Inhabitants of the Matrix experience their world very much as we do, but the Matrix is a virtual world.

Could you be in a virtual world right now? Stop and think about this question for a moment. When you do, you're doing philosophy.

Philosophy translates as *love of wisdom*, but I like to think of it as *the foundations of everything*. Philosophers are like the little kid who keeps asking, *Why?* or *What is that?* or *How do you know?* or *What does that mean?* or *Why should I do that?* Ask those questions a few times in a row and you rapidly reach the foundations. You're examining the assumptions that underlie things we take for granted.

I was that kid. It took me a while to realize that what I was interested in was philosophy. I started off studying mathematics, physics, and computer science. These take you a fair distance into the foundations of everything, but I wanted to go deeper. I turned to studying philosophy, along with cognitive science to keep an anchor in the solid ground of science while I explored the foundations underneath.

I was first drawn to address questions about the mind, like *What is consciousness?* I've spent much of my career focusing on those questions. But questions about the world, like *What is reality?*, are just as central to philosophy. Perhaps most central of all are questions about the relation between mind and world, such as *How can we know about reality?*

This last question was at the heart of the challenge posed by René Descartes in his *Meditations on First Philosophy* (1641), which set the agenda for centuries of Western philosophy to come. Descartes posed what I'll call the problem of the external world: How do you know anything at all about the reality outside you?

Descartes approached the problem by asking: How do you know that your perception of the world is not an illusion? How do you know that you are not dreaming right now? How do you know you're not being deceived by an evil demon into thinking all this is real, when it's not? These days, he might approach the problem by asking the question I just asked you: How do you know you're not in a virtual world?

For a long time I thought I didn't have much to say about Descartes's problem of the external world. Thinking about virtual reality gave me a new perspective. It was reflecting on the simulation hypothesis that led me to realize that I had underestimated virtual worlds. In their own way, so had Descartes and many others. I concluded that if we think more clearly about virtual worlds, this might lead us to the beginnings of a solution to Descartes's problem.

◆

The central thesis of this book is: *Virtual reality is genuine reality*. Or at least, *virtual realities are genuine realities*. Virtual worlds need not be second-class realities. They can be first-class realities.

We can break down this thesis into three parts:

- Virtual worlds are not illusions or fictions, or at least they need not be. What happens in VR really happens. The objects we interact with in VR are real.
- Life in virtual worlds can be as good, in principle, as life outside virtual worlds. You can lead a fully meaningful life in a virtual world.
- The world we're living in could be a virtual world. I'm not saying it is. But it's a possibility we can't rule out.

The thesis—especially the first two parts—has practical consequences for the role of VR technology in our lives. In principle, VR can be much more than escapism. It can be a full-blooded environment for living a genuine life.

I'm not saying that virtual worlds will be some sort of utopia. Like the internet, VR technology will almost certainly lead to awful things as well as wonderful things. It's certain to be abused. Physical reality is abused, too. Like physical reality, virtual reality has room for the full range of the human condition—the good, the bad, and the ugly.

I'll focus more on VR in principle than VR in practice. In practice, the road to full-scale virtual reality is sure to be bumpy. It won't surprise me if widespread adoption of VR is limited for a decade or two, while the technology matures. No doubt it will move in all sorts of directions I haven't anticipated. But once a mature VR technology is developed, it should be able to support lives that are on a par with or even surpass life in physical reality.

Reality+ is my name for the universe (metaverse? multiverse?) of virtual and nonvirtual worlds. The title also captures many of my main claims. Each virtual world is a new reality: *Reality+*. Augmented reality involves additions to reality: *Reality+*. Some virtual worlds are as good as or better than ordinary reality: *Reality+*. If we're in a simulation, there is more to reality than we thought: *Reality+*. There will be a smorgasbord of multiple realities: *Reality+*.

I know that what I'm saying is counterintuitive to many people. Perhaps you think that VR is Reality–, or Reality Minus. Virtual worlds are fake realities, not genuine realities. No virtual world is as good as ordinary reality. Over the course of this book, I'll try to convince you that Reality+ is closer to the truth.

This book is a project in what I call *technophilosophy*. Technophilosophy is a combination of (1) asking philosophical questions about technology and (2) using technology to help answer traditional philosophical questions.

The name is inspired by what the Canadian-American philosopher Patricia Churchland called *neurophilosophy* in her landmark 1987 book of the same title. Neurophilosophy combines asking philosophical questions about neuroscience with using neuroscience to help answer traditional questions in philosophy. Technophilosophy does the same with technology.

There's a thriving area, often called the philosophy of technology, that carries out the first project—asking philosophical questions about technology. What's especially distinctive about technophilosophy is the second project—using technology to answer traditional philosophical questions. The key to technophilosophy is a two-way interaction between philosophy and technology. Philosophy helps to shed light on (mostly new) questions about technology. Technology helps to shed light on (mostly old) questions about philosophy. I wrote this book in order to shed light on both sorts of question at once.

◆

First, I want to use technology to address some of the oldest questions in philosophy, especially the problem of the external world. At a minimum, virtual reality technology helps *illustrate* Descartes's problem—that is, how can we know anything about the reality around us? How do we know that reality is not an illusion? In chapters 2 and 3, I lay out these problems by introducing the simulation hypothesis and asking, "How do we know we're not in a simulation right now?"

The simulation idea does more than illustrate the problem, however. It also *sharpens* the problem by turning Descartes's far-fetched scenarios involving evil demons into more realistic scenarios involving computers—scenarios we have to take seriously. In chapter 4, I make the case that the simulation idea undercuts many common responses to Descartes. In chapter 5, I use statistical reasoning about simulations to argue that we cannot know we're not in a simulation. All this makes Descartes's problem even harder.

Most importantly, reflection on virtual reality technology can help us *respond* to the problem of the external world. In chapters 6 through 9, I argue that if indeed we're in a simulation, tables and chairs are not illusions but perfectly real objects: they are digital objects that are made of bits. This leads us to what is sometimes called, in modern physics, the *it-from-bit hypothesis*: Physical objects are real and they are digital. Thinking about the simulation hypothesis and the it-from-bit hypothesis—two ideas inspired by modern computers—yields the beginnings of a response to Descartes's classic problem.

We can put Descartes's argument as follows: We don't know that we're not in a virtual world, and in a virtual world nothing is real, so we don't know that anything is real. This argument turns on the assumption that virtual worlds are not genuine realities. Once we make the case that virtual worlds are indeed genuine realities—and especially that objects in a virtual world are real—we can respond to Descartes's argument.

I shouldn't overstate the case. My analysis doesn't address everything Descartes says, and it doesn't prove that we know a great deal about the external world. Still, if the analysis works, it dissolves what is perhaps the Western tradition's prime reason for doubting that we can know anything about the external world. So it gives us at least a foothold in establishing that we have knowledge of the reality around us.

We'll also use technology to illuminate traditional questions about the mind: How do mind and body interact? (See chapter 14.) What is consciousness? (See chapter 15.) Does the mind extend beyond the body? (See chapter 16.) In each case, thinking about a technology—VR, artificial intelligence (AI), and augmented reality (AR), respectively—can illuminate those questions. And conversely, thinking about the questions can illuminate these technologies.

It's worth saying that my views about consciousness and the mind are not the main focus of this book. I've explored those issues in other work, and this book is independent of them to a large degree. I hope that even people who disagree with me about consciousness may find my picture of reality appealing. That said, there are many connections between the two domains. You can think of chapters 15 and 16, in particular, as adding a fourth plank to the thesis that virtual reality is genuine reality: namely, *virtual and augmented minds are genuine minds*.

Technology can also illuminate traditional questions about value and ethics. Value is the domain of good and bad, better and worse. Ethics is the domain of right and wrong. What makes for a good life? (See chapter 17.) What is the difference between right and wrong? (See chapter 18.) How should society be organized? (See chapter 19.) I'm by no means an expert on these issues, but technology provides at least an interesting angle on them.

Other time-hallowed philosophical questions will come up along the way. Is there a God? (See chapter 7.) What is the universe made of? (See chapter 8.) How does language describe reality? (See chapter 20.) What does science tell us about reality? (See chapters 22 and 23.) It turns out that to make our case that virtual reality is genuine reality, we have to think hard about those old questions. As always, the illumination flows both ways; thinking about technology throws light on the old questions in turn.

◆

I also want to use philosophy to address new questions about technology, especially the technology of virtual worlds. These include questions about everything from video games through augmented reality glasses and virtual reality headsets to simulations of entire universes.

I've already outlined my central thesis that virtual reality is genuine reality. Where VR is concerned, I'll ask questions like: Is virtual reality an illusion? (See chapters 6, 10, and 11.) What are virtual objects? (See chapter 10.) Does augmented reality genuinely augment reality? (See chapter 12.) Can you live a good life in VR? (See chapter 17.) How should you behave in a virtual world? (See chapter 19.)

I'll also discuss other technologies: artificial intelligence, smartphones, the internet, deepfakes, and computers in general. How can we know we're not being deceived by deepfakes? (See chapter 13.) Can AI systems be conscious? (See chapter 15.) Do smartphones extend our minds, and is the internet making us smart or stupid? (See chapter 16.) And what is a computer, anyway? (See chapter 21.)

These questions are all philosophical questions. Many of them are also intensely practical questions. We need to make decisions right now about how we use video games, smartphones, and the internet. An increasing number of such practical questions will confront us in decades to come. As we spend more and more time in virtual worlds, we'll have to grapple with the issue of whether life there is fully meaningful. Eventually, we may have to decide whether or not to upload ourselves to the cloud entirely. Thinking philosophically can help us get clear on these decisions about how to live our lives.

◆

By the end of this book, you'll have been introduced to many of the central questions in philosophy. We'll encounter both historical greats from centuries and millennia past and contemporary figures and arguments from recent decades. We'll cover many of the central topics in philosophy: knowledge, reality, mind, language, value, ethics, science, religion, and more. I'll introduce some of the powerful tools that philosophers have developed over the centuries for thinking about these issues. This is only one perspective, and a great deal of important philosophy is left out. But by the end, you'll have a sense of some of the historical and contemporary landscape of philosophy.

To help readers think through these ideas, I've made connections to science fiction and other corners of popular culture whenever I can. Many authors of science fiction have delved into these issues just as deeply as philosophers have. I've often had new philosophical ideas by thinking about science fiction. Sometimes I think science fiction gets these issues right, and sometimes it gets them wrong. Either way, science-fiction scenarios can prompt a lot of fruitful philosophical analysis.

The best way I know to introduce philosophy is to *do* philosophy. So while I'll start many chapters by posing a philosophical question connected to virtual worlds and introducing some philosophical background, I'll usually get down quickly to thinking hard about the issues. I'll analyze the issues both inside and outside virtual worlds, with an eye on building my argument for the Reality+ point of view.

As a result, this book is as full of my own philosophical theses and arguments as anything I've ever written. While some chapters of the book go over ground I've discussed in academic articles, well over half of it is entirely new. So even if you're an old hand at philosophy, I hope that you'll find rewards here. In an online supplement (consc.net/reality), I've included extensive notes and appendices pursuing the issues in more depth, often including connections to the academic literature.

✦

The first three parts of the book focus on the simulation hypothesis— the idea that we might already be in a virtual world—and use it to address philosophical questions about knowledge and reality.

Part 1 introduces the central problems of the book. Part 2 uses the simulation hypothesis to introduce and address questions about knowledge, and especially Descartes's arguments for skepticism about the external world. Part 3 addresses questions about reality and makes an initial case for my thesis that virtual reality is genuine reality.

The next three parts of the book (parts 4–6) focus on aspects of the so-called Metaverse—the virtual worlds that we will be constructing in decades to come, and the associated technologies of virtual reality, augmented reality, and artificial intelligence—and use these technologies to address philosophical questions about reality, mind, and value. Part 4 introduces issues about reality and knowledge as they arise in the virtual worlds we construct. Part 5 addresses questions about the mind and consciousness. Part 6 addresses questions about value, ethics, and political philosophy.

The final part of the book (Part 7) focuses on foundational questions about reality, including issues about language, computers, science, and perception that are required to fully develop the Reality+ vision. Chapters 21–23 set out the structuralist picture that is at the heart of this vision, and the last chapter pulls the pieces together to see where things stand with Descartes's problem of the external world.

As a result, this long book contains at least four overlapping shorter books, two on traditional philosophy and two on technology. Running throughout the book, there is an introduction to philosophy (mode: expository). Also throughout the book, and peaking toward the end, there is my own analysis of the philosophical problems of knowledge and reality (mode: analytical). Early on, there is a book on the simulation hypothesis (mode: speculative). Later, there is a book on the Metaverse (mode: a little more practical). I could have written these books separately, but there are so many resonances between them that it made sense to write them all at once.

Different readers may want to read the book in different ways, prioritizing some of these shorter books over others. Everyone should read chapter 1, but after that you can strike out in many different directions. In the endnotes, I give some possible paths, depending on your interests. Many chapters stand relatively independently. Chapters 2, 3, 6, and 10 may be especially helpful in providing background for the chapters that follow, but they aren't absolutely essential.

Most of the chapters are frontloaded with introductory material toward the start. The discussion sometimes gets denser toward the end of each chapter, and toward the end of the book. If you're after a shorter book and a lighter reading experience, you might try reading the first two or three sections of every chapter, and then skipping to the next chapter whenever you like.

◆

We live in an age in which truth and reality have been under attack. We're sometimes said to be in an era of post-truth politics in which truth is irrelevant. It's common to hear that there's no absolute truth and no objective reality. Some people think that reality is all in the mind, so that what's real is entirely up to us. The multiple realities of this book may initially suggest a view like that on which truth and reality are cheap. This is not my view.

Here's my view of these things. Our minds are part of reality, but there's a great deal of reality outside our minds. Reality contains our world and it may contain many others. We can build new worlds and new parts of reality. We know a little about reality, and we can try to know more. There may be parts of it that we can never know.

Most importantly: Reality exists, independently of us. The truth matters. There are truths about reality, and we can try to find them. Even in an age of multiple realities, I still believe in objective reality.

—New York City, April 2021 [minor revisions June 2022]

Part 1

VIRTUAL WORLDS

Chapter 1

Is this the real life?

I N THE OPENING LINES OF THE 1975 HIT "BOHEMIAN RHAPSODY" by the British rock group Queen, lead singer Freddie Mercury sings in five-part harmony:

Is this the real life?
Is this just fantasy?

These questions have a history. Three of the great ancient traditions of philosophy—those of China, Greece, and India—all ask versions of Mercury's questions.

Their questions involve alternative versions of reality. Is this real life, or is it just a dream? Is this real life, or is it just an illusion? Is this real life, or is it just a shadow of reality?

Today we might ask: Is this real life, or is it virtual reality? We can think of dreams, illusions, and shadows as ancient counterparts of virtual worlds—minus the computer, which would not be invented for two millennia.

With or without the computer, these scenarios raise some of the deepest questions in philosophy. We can use them to introduce these questions and to guide our thinking about virtual worlds.

Zhuangzi's butterfly dream

The ancient Chinese philosopher Zhuangzi (also known as Zhuang Zhou or Chuang Tzu) lived around 300 BCE and was a central figure in the Daoist tradition. He recounts this famous parable: "Zhuangzi Dreams of Being a Butterfly."

Once Zhuangzi dreamt he was a butterfly, a butterfly flitting and fluttering around, happy with himself and doing as he pleased. He didn't know he was Zhuangzi. Suddenly he woke up and there he was, solid and unmistakably Zhuangzi. But he didn't know if he was Zhuangzi who had dreamt he was a butterfly, or a butterfly dreaming he was Zhuangzi.

Zhuangzi can't be sure that the life he's experiencing as Zhuangzi is real. Maybe the butterfly was real, and Zhuangzi is a dream.

Figure 1 Zhuangzi's butterfly dream. Was he Zhuangzi who dreamt he was a butterfly, or a butterfly dreaming he was Zhuangzi?

A dream world is a sort of virtual world without a computer. So Zhuangzi's hypothesis that he is in a dream world right now is a computer-free version of the hypothesis that he's in a virtual world right now.

The plot of the Wachowski sisters' 1999 movie *The Matrix* provides a nice parallel. The main character, Neo, lives an ordinary life until he takes a red pill and wakes up in another world, where he's told that the world he knew was a simulation. If Neo had thought as deeply as Zhuangzi, he might have wondered, "Maybe my old life was the reality, and my new life is the simulation"—a perfectly reasonable thought. While his old world was a world of drudgery, his new world is a world of battles and adventure, where he's treated as a savior. Maybe the red pill knocked him out just long enough for him to be hooked up to this exciting simulation.

On one interpretation, Zhuangzi's butterfly dream raises a question about knowledge: How do any of us know we aren't dreaming right now? This is a cousin of the question raised in the introduction: How do any of us know we aren't in a virtual world right now? These questions lead to a more basic question: How do we know anything we experience is real?

Narada's transformation

Ancient Indian philosophers in the Hindu tradition were gripped by issues of illusion and reality. A central motif appears in the folk tale of the sage Narada's transformation. In one version of the story, Narada says to the god Vishnu, "I have conquered illusion." Vishnu promises to show Narada the true power of illusion (or *Maya*). Narada wakes up as a woman, Sushila, with no memory of what came before. Sushila marries a king, becomes pregnant, and eventually has eight sons and many grandsons. One day, an enemy attacks, and all her sons and grandsons are killed. As the queen grieves, Vishnu appears and says, "Why are you so sad? This is just an illusion." Narada finds himself back in his original body only a moment after the original conversa-

Figure 2 Vishnu oversees Narada's transformation into Sushila,
in the style of *Rick and Morty.*

tion. He concludes that his whole life is an illusion, just like his life as Sushila.

Narada's life as Sushila is akin to life in a virtual world—a simulation with Vishnu acting as the simulator. As a simulator, Vishnu is in effect suggesting that Narada's ordinary world is a virtual world too.

Narada's transformation is echoed in an episode of the animated TV series *Rick and Morty,* which chronicles the interdimensional adventures of a powerful scientist, Rick, and his grandson Morty. Morty puts on a virtual reality helmet to play a video game titled *Roy: A Life Well Lived.* (It would be even better if Morty had played *Sue: A Life Well Lived,* but you can't have everything.) Morty lives Roy's entire fifty-five-year life: childhood, football star, carpet salesman, cancer patient, death. When he emerges from the game a moment later as Morty, his grandfather berates him for having made the wrong life decisions in the simulation. This is a recurring theme in the series. Its characters are in apparently normal situations that turn out to be simulations and are often led to ask whether their current reality might be a simulation, too.

Narada's transformation raises deep questions about reality. Is Narada's life as Sushila real, or is it an illusion? Vishnu says it is an illusion, but this is far from obvious. We can raise an analogous question about virtual worlds, including the world of *Roy: A Life Well Lived*. Are these worlds real or illusory? An even more pressing question looms. Vishnu says that our ordinary lives are as illusory as Narada's transformed life. Is our own world real or an illusion?

Plato's cave

Around the same time as Zhuangzi, the ancient Greek philosopher Plato put forward his allegory of the cave. In his extended dialogue, the *Republic*, he tells the story of humans who are chained up in a cave, seeing only shadows cast on a wall by puppets that imitate things in the world of sunlight outside. The shadows are all the cave people know, so they take them to be reality. One day, one of them escapes and discovers the glories of the real world outside the cave. Eventually he reenters the cave and tells stories of that world, but no one believes him.

Plato's prisoners watching shadows call to mind viewers in a movie theater. It's as if the prisoners had never watched anything but movies—or, to update the technology, had watched only movies on a virtual reality headset. A 2016 mobile technology conference produced a famous photograph of Facebook chief executive Mark Zuckerberg walking down the aisle past the conference audience. The members of the audience are all wearing virtual reality headsets in the darkened hall, apparently unaware of Zuckerberg as he strides by. It's a contemporary illustration of Plato's cave.

Plato uses his allegory for many purposes. He's suggesting that our own imperfect reality is something like the cave. He's also using it to help us think about what sort of lives we want to live. In a key passage,

Figure 3 Plato's cave in the 21st century.

Plato's spokesman, Socrates, raises the question of whether we should prefer life inside or outside the cave.

> SOCRATES: Do you think the one who had gotten out of the cave would still envy those within the cave and would want to compete with them who are esteemed and who have power? Or would not he much rather wish for the condition that Homer speaks of, namely "to live on the land [above ground] as the paid menial of another destitute peasant"? Wouldn't he prefer to put up with absolutely anything else rather than associate with those opinions that hold in the cave and be that kind of human being?
>
> GLAUCON: I think that he would prefer to endure everything rather than be that kind of human being.

The allegory of the cave raises deep questions about value: that is, about good and bad, or at least about better and worse. Which is better, life inside the cave or life outside the cave? Plato's answer is clear: Life outside the cave, even life as a menial laborer, is vastly better than life inside it. We can ask the same question about virtual worlds. Which is better, life in a virtual world or life outside it? This leads us to a more fundamental question: What does it mean to live a good life?

Three questions

In one traditional picture, philosophy is the study of *knowledge* (How do we know about the world?), *reality* (What is the nature of the world?), and *value* (What is the difference between good and bad?).

Our three stories raise questions in each of these domains. Knowledge: *How can Zhuangzi know whether or not he's dreaming?* Reality: *Is Narada's transformation real or illusory?* Value: *Can one lead a good life in Plato's cave?*

When we transpose our three stories from their original realms of dreams, transformations, and shadows into the realm of virtuality, they raise three key questions about virtual worlds.

The first question, raised by Zhuangzi's butterfly dream, concerns knowledge. I'll call it the Knowledge Question. *Can we know whether or not we're in a virtual world?*

The second question, raised by Narada's transformation, concerns reality. I'll call it the Reality Question. *Are virtual worlds real or illusory?*

The third question, raised by Plato's cave, concerns value. I'll call it the Value Question. *Can you lead a good life in a virtual world?*

These three questions in turn lead us to three more general questions that are at the heart of philosophy: *Can we know anything about the world around us? Is our world real or illusory? What is it to lead a good life?*

Over the course of this book, these questions about knowledge, reality, and value will be at the heart of our exploration of virtual worlds and at the heart of our exploration of philosophy.

The Knowledge Question: Can we know whether or not we're in a virtual world?

In the 1990 movie *Total Recall* (remade with a few changes in 2012), the viewer is never quite sure which parts of the movie take place in a virtual world and which take place in the ordinary world. The main character, a construction worker named Douglas Quaid (played by Arnold Schwarzenegger) experiences many outlandish adventures on Earth and on Mars. At the movie's end, Quaid looks out over the surface of Mars and begins to wonder (and so do we) whether his adventures took place in the ordinary world or in virtual reality. The movie hints that Quaid may indeed be in a virtual world. Virtual reality technology that implants memories of adventures plays a key role in the plot. Since heroic adventures on Mars are presumably more likely to take place in virtual worlds than in ordinary life, Quaid, if he is reflective, will conclude that he's probably in virtual reality.

What about you? Can you know whether you're in a virtual or a non-virtual world? Your life may not be as exciting as Quaid's. But the fact that you're reading a book about virtual worlds should give you pause. (The fact that I'm writing one should give me even more pause.) Why? I suspect that as simulation technology develops, simulators may be drawn to simulate people thinking about simulations, perhaps to see how close they come to realizing the truth about their lives. Even if we seem to be leading perfectly ordinary lives, is there any way we could know whether these lives are virtual?

To put my cards on the table: I don't know whether we're in a virtual world or not. I don't think you know, either. In fact, I don't think we can ever know whether or not we're in a virtual world. In principle, we could confirm that we *are* in a virtual world—for example, the simulators could choose to reveal themselves to us and show us how the simulation works. But if we're *not* in a virtual world, we'll never know that for sure.

I'll discuss the reasons for this uncertainty over the next few chapters. The basic reason is spelled out in chapter 2: We can never prove we're not in a computer simulation because any evidence of ordinary reality—whether the grandeur of nature, the antics of your cat, or the behavior of other people—could presumably be simulated.

Over the centuries, many philosophers have offered strategies that could be used to show that we're not in a virtual world. I'll discuss these strategies in chapter 4 and argue that they don't work. Going beyond this, we should take seriously the possibility that we *are* in a virtual world. The Swedish-born philosopher Nick Bostrom has argued on statistical grounds that under certain assumptions, there will be many more simulated people in the universe than nonsimulated people. If that's right, perhaps we should consider it likely that we're in a simulation. I'll argue in chapter 5 for a somewhat weaker conclusion: All these considerations mean that we can't know we're *not* in a simulation.

This verdict has major consequences for Descartes's problem: How do we know anything about the external world? If we don't know whether or not we're in a virtual world, and if nothing in a virtual world is real, then it looks like we cannot know if anything in the external world is real. And then it looks like we can't know anything at all about the external world.

That's a shocking consequence. We can't know whether Paris is in France? I can't know that I was born in Australia? I can't know that there's a desk in front of me?

Many philosophers try to avoid this shocking consequence by arguing for a positive answer to the Knowledge Question: we *can* know that we're not in a simulation. If we can know that, then we can know something about the external world after all. If I'm right, though, we can't fall back on this comforting position. We can't know that we're not in a simulation. That makes the problem of knowledge of the external world that much harder.

The Reality Question: Are virtual worlds real or illusory?

Whenever virtual reality is discussed, one hears the same refrain. *Simulations are illusions. Virtual worlds aren't real. Virtual objects don't really exist. Virtual reality isn't genuine reality.*

You can find this idea in *The Matrix*. In a waiting room inside the simulation, Neo sees a child apparently bending a spoon with the power of his mind. They engage in conversation:

CHILD: Do not try and bend the spoon. That's impossible.
   Instead . . . only try to realize the truth.
NEO: What truth?
CHILD: There is no spoon.

This is presented as a deep truth. *There is no spoon.* The spoon inside the Matrix is not real but a mere illusion. The implication is that everything one experiences in the Matrix is an illusion.

In a commentary on *The Matrix*, the American philosopher Cornel West, who himself played Councillor West of Zion in *The Matrix Reloaded* and *The Matrix Revolutions*, takes this line of thinking a step further. Speaking of awakening from the Matrix, he says "What you think you're awakening to may in fact be another species of illusion. It's illusions all the way down." Here there is an echo of Vishnu: Simulations are illusions, and ordinary reality may be an illusion, too.

The same line of thinking recurs in the TV series *Atlanta*. Three characters are sitting around a pool late at night discussing the simulation hypothesis. Nadine becomes convinced: "We're all nothing. It's a simulation, Van. We're all fake." She takes for granted that if we're living in a simulation, we're not real.

I think these claims are wrong. Here's what I think: *Simulations are not illusions. Virtual worlds are real. Virtual objects really exist.* In my view, the Matrix child should have said, "Try to realize the truth. There is a spoon—a digital spoon." Neo's world is perfectly real. So is Nadine's world, even if she is in a simulation.

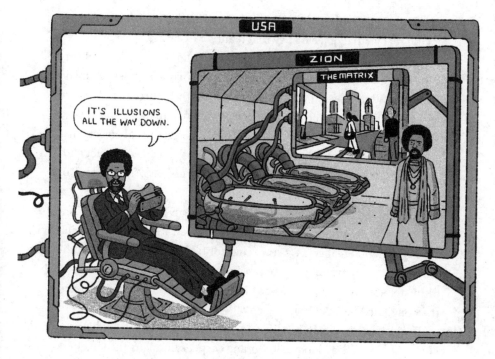

Figure 4 Cornel West, awakening from life as Councillor
West of Zion, on illusion and reality.

The same goes for our world. Even if we're in a simulation, our world
is real. There are still tables and chairs and people here. There are cities,
there are mountains, there are oceans. Of course there may be many
illusions in our world. We can be deceived by our senses and by other
people. But the ordinary objects around us are real.

What do I mean by "real"? That's complicated—the word "real"
doesn't have a single, fixed meaning. In chapter 6, I'll discuss five different criteria for being "real." I'll argue that even if we're in a simulation,
the things we perceive meet all these criteria for reality.

What about ordinary virtual reality, experienced through a headset? This can sometimes involve illusion. If you don't know you're in VR
and you take the virtual objects to be normal physical objects, you'd
be wrong. But I'll argue in chapter 11 that for experienced users of VR,

who know they're using VR, there need be no illusion. They're experiencing real virtual objects in virtual reality.

Virtual realities are different from nonvirtual realities. Virtual furniture isn't the same as nonvirtual furniture. Virtual entities are made one way, and nonvirtual entities are made another. Virtual entities are *digital* entities, made of computational and informational processes. More succinctly, they're made of bits. They're perfectly real objects that are grounded in a pattern of bits in a computer. When you interact with a virtual sofa, you're interacting with a pattern of bits. The pattern of bits is entirely real, and so is the virtual sofa.

"Virtual reality" is sometimes taken to mean "fake reality." If I'm right, that's the wrong way to define it. Instead it means something closer to "digital reality." A virtual chair or table is made of digital processes, just as a physical chair or table is made of atoms and quarks and ultimately of quantum processes. The virtual object is different from the nonvirtual one, but both are equally real.

If I'm right, then Narada's life as a woman is not entirely an illusion. Nor is Morty's life as a football star and carpet salesman. The long lives that they experience really happen. Narada really lives a life as Sushila. Morty really lives a life as Roy, albeit in a virtual world.

This view has major consequences for the problem of the external world. If I'm right, then even if I don't know whether or not we're in a simulation, it won't follow that I don't know whether or not the objects around us are real. If we're in a simulation, tables are real (they're patterns of bits), and if we're not in a simulation, tables are real (they're something else). So either way, tables are real. This offers a new approach to the problem of the external world, one that I will spell out over the course of this book.

The Value Question: Can you live a good life in a virtual world?

In James Gunn's 1954 science-fiction story "The Unhappy Man," a company known as Hedonics, Inc., uses the new "science of happiness" to improve people's lives. People sign a contract to move their life into "sensies," a sort of virtual world where everything is perfect:

> We take care of everything; we arrange your life so you never have to worry again. In this age of anxiety, you never have to be anxious. In this age of fear, you never need be afraid. You will always be fed, clothed, housed, and happy. You will love and be loved. Life, for you, will be an unmixed joy.

Gunn's protagonist rejects the offer to hand over his life to Hedonics, Inc.

In his 1974 book *Anarchy, State, and Utopia*, the American philosopher Robert Nozick offers the reader a similar choice:

> Suppose there was an experience machine that would give you any experience you desired. Super-duper neuropsychologists could stimulate your brain so that you would think and feel you were writing a great novel, or making a friend, or reading an interesting book. All the time you would be floating in a tank, with electrodes attached to your brain. Should you plug into this machine for life, preprogramming your life experiences?

Gunn's sensies and Nozick's experience machine are virtual reality devices of a kind. They are asking, "Given the choice, would you spend your life in this kind of engineered reality?"

Like Gunn's protagonist, Nozick says no, and he expects his readers to do the same. His view seems to be that the experience machine is a second-class reality. Inside the machine, one does not actually do the things one seems to be doing. One is not a genuine autonomous person. For Nozick, life in the experience machine does not have much meaning or value.

Many people would agree with Nozick. In a 2020 survey of professional philosophers, 13 percent of respondents said they would enter the experience machine, and 77 percent said they would not. In broader surveys, most people decline the opportunity, too—although as virtual worlds have become more and more a part of our lives, the number who say they would plug in is increasing.

We can ask the same question of VR more generally. Given the chance to spend your life in VR, would you do it? Could this ever be a reasonable choice? Or we can ask the Value Question directly: Can you lead a valuable and meaningful life in VR?

Ordinary VR differs in some ways from Nozick's experience machine. You know when you're in VR, and many people can enter the same VR environment at once. In addition, ordinary VR is not entirely preprogrammed. In interactive virtual worlds, you make real choices rather than simply living out a script.

Still, in a 2000 article in *Forbes* magazine, Nozick extends his negative assessment of the experience machine to ordinary VR. He says: "even if everybody were plugged into the same virtual reality, that wouldn't be enough to make its contents truly real." He also says of VR: "The pleasures of this may be so great that many people will choose to spend most of their days and nights that way. Meanwhile, the rest of us are likely to find that choice deeply disturbing."

Where VR is concerned, I'll argue (in chapter 17) that Nozick's answer is the wrong answer. In full-scale VR, users will build their own lives as they choose, genuinely interacting with others around them and leading a meaningful and valuable life. Virtual reality need not be a second-class reality.

Even existing virtual worlds—such as *Second Life*, which has been perhaps the leading virtual world for building a day-to-day life since it was founded in 2003—can be highly valuable. Many people have meaningful relationships and activities in today's virtual worlds, although much that matters is missing: proper bodies, touch, eating and drinking, birth and death, and more. But many of these limitations will be overcome by the fully immersive VR of the future. In principle, life in VR can be as good or as bad as life in a corresponding nonvirtual reality.

Many of us already spend a great deal of time in virtual worlds. In the future, we may well face the option of spending more time there, or even of spending most of our lives there. If I'm right, this will be a reasonable choice.

Many would see this as a dystopia. I do not. Certainly virtual worlds can be dystopian, just as the physical world can be, but they won't be dystopian merely because they're virtual. As with most technologies, whether VR is good or bad depends entirely on how it's used.

Central philosophical questions

To recap, our three main questions about virtual worlds are the following. The Reality Question: *Are virtual worlds real?* (My answer: yes.) The Knowledge Question: *Can we know whether or not we're in a virtual world?* (My answer: no.) The Value Question: *Can you lead a good life in a virtual world?* (My answer: yes.)

The Reality Question, the Knowledge Question, and the Value Question match up with three of the central divisions of philosophy.

(1) *Metaphysics*, the study of reality. Metaphysics asks questions like "What is the nature of reality?"
(2) *Epistemology*, the study of knowledge. Epistemology asks questions like "How can we know about the world?"
(3) *Value theory*, the study of values. Value theory asks questions like "What is the difference between good and bad?"

Or, to simplify: *What is this?* That's metaphysics. *How do you know?* That's epistemology. *Is it any good?* That's value theory.

When we ask the Reality Question, the Knowledge Question, and the Value Question, we're doing the metaphysics, epistemology, and value theory of virtual worlds.

Other philosophical questions we'll ask about virtual worlds include:

The Mind Question: *What is the place of minds in virtual worlds?*
The God Question: *If we're in a simulation, is there a god?*
The Ethics Question: *How should we act in a virtual world?*
The Politics Question: *How should we build a virtual society?*

The Science Question: *Is the simulation hypothesis a scientific hypothesis?*

The Language Question: *What is the meaning of language in a virtual world?*

Like our three main questions, these six further questions each correspond to an area of philosophy: the philosophy of mind, the philosophy of religion, ethics, political philosophy, the philosophy of science, and the philosophy of language.

The traditional questions in each of these areas are more general: What is the place of minds in reality? Is there a God? How should we treat other people? How should society be organized? What does science tell us about reality? What is the meaning of language?

In addressing the questions about virtual worlds, I'll do my best to connect them to these bigger questions, too. That way, our answers will not just help us come to grips with the role of virtual worlds in our lives. They'll also help us to get clearer on reality itself.

Answering philosophical questions

Philosophers are good at asking questions. We're less good at answering them. In 2020, my colleague David Bourget and I conducted a survey of around two thousand professional philosophers on one hundred central philosophical questions. To no one's surprise, we found large disagreement on the answers to almost all of them.

Every now and then a philosopher answers a question. Isaac Newton considered himself a philosopher. He worked on philosophical questions about space and time. He figured out how to answer some of them. As a result the new science of physics emerged. Something similar happened later with economics, sociology, psychology, modern logic, formal semantics, and more. All were founded or cofounded by philosophers who got clear enough on some central questions to help spin off a new discipline.

In effect, philosophy is an incubator for other disciplines. When philosophers figure out a method for rigorously addressing a philosophical question, we spin that method off and call it a new field. Because philosophy has been so successful at this over the centuries, what's now left in philosophy is a basket of hard questions that people are still figuring out. That's why philosophers disagree as much as they do.

Still, we can at least pose the questions and try our best to answer them. Occasionally a question is ready to be answered, and we'll get lucky. If we don't answer it, there's often value in the attempt. At the least, posing a question and exploring potential answers can lead us to understand the subject matter better. Others can build on that understanding, and eventually the question might be answered properly.

In this book, I'll try to answer some of the questions I've posed. I can't expect you to agree with all of my answers. Still, I hope you might find understanding in the attempt. With luck, there will be something here that someone can build on. One way or another, we can hope that some of these questions about virtual worlds will eventually migrate from philosophy to a new discipline of their own.

Chapter 2

What is the simulation hypothesis?

THE ANTIKYTHERA MECHANISM WAS FOUND IN A SHIPWRECK off the coast of the Greek island of Antikythera in 1901. It dates from two thousand years earlier. The mechanism is a bronze device that was originally mounted in a wooden box about 13 inches across. Superficially, it resembles a clock, with a complex system of 30 or more gears that once drove pointers and dials on the front and the back. Through painstaking analysis over the last century, researchers have discovered that the pointers simulate the day-by-day positions of the Sun and Moon in the zodiac according to the theories of the astronomer Hipparchus of Rhodes. Recently, mathematical analysis of surviving text and gear fragments has provided strong evidence that the system simulated the five known planets as well. It appears that the

Figure 5 A reconstruction of the Antikythera mechanism, which simulated the position of the Sun and the Moon and probably the five known planets.

Antikythera mechanism is an attempt to simulate the solar system. It is the first known cosmic simulation.

The Antikythera mechanism is a *mechanical simulation*. In a mechanical simulation, the positions of components reflect the positions of the entities they're simulating. In the Antikythera, the motion of the gears is intended to reflect the motion of the Sun and Moon against the stars. One could use it to predict a solar eclipse years in the future.

Mechanical simulations are still used from time to time. One prominent example is a mechanical simulation of the San Francisco Bay and its environs, erected in a giant warehouse taking up more than an acre just outside San Francisco. It's a scale model, with enormous amounts of water moved by hydraulic mechanisms to simulate tides, currents, and other forces. It was built to test whether a plan for building dams on the bay would work. The mechanical simulation showed that it wouldn't, and the dams were never built.

Mechanical simulations of highly complex systems are difficult to build, and the art and science of simulation didn't flourish until the start of the computer age in the mid-20th century. In the celebrated code-breaking unit in Bletchley Park (depicted in the film *The Imitation Game*), the British mathematician Alan Turing and other researchers built some of the first computers in order to simulate and analyze German code systems. After the war, the mathematical physicists Stanislaw Ulam and John von Neumann used the ENIAC computer to simulate the behavior of neutrons in a nuclear explosion.

These models were among the first computer simulations. Whereas a mechanical simulation is driven by physical mechanisms, a computer simulation is driven by algorithms. Instead of using pointers and gears to reflect the positions of the planets, a modern computer simulation uses patterns of bits. An algorithmic simulation of the observed laws of planetary motion makes sure that the bits evolve in a way that reflects the positions of the planets. Using this method, we now have accurate simulations of the solar system allowing us to predict the position of Mars with uncanny precision.

Computer simulations are ubiquitous in science and engineering. In physics and chemistry, we have simulations of atoms and molecules. In biology, we have simulations of cells and organisms. In neuroscience, we have simulations of neural networks. In engineering, we have simulations of cars, planes, bridges, and buildings. In planetary science, we have simulations of Earth's climate over many decades. In cosmology, we have simulations of the known universe as a whole.

In the social sphere, there are many computer simulations of human behavior. As early as 1955, Daniel Gerlough completed a PhD thesis on computer simulation of freeway traffic. In 1959, the Simulmatics Corporation was founded to simulate and predict how a political campaign's messaging would affect various groups of voters. It was said that this effort had a significant effect on the 1960 US presidential election. The claim may have been overblown, but since then, social and political simulations have become mainstream. Advertising companies, political consultants, social media companies, and social scientists build models and run simulations of human populations as a matter of course.

Simulation technology is improving fast, but it's far from perfect. A simulation usually concentrates on a certain level. A population-level simulation approximates human behavior with simple psychological models, but it doesn't usually try to simulate the neural networks that underlie the psychology. A hot topic in the science of simulation involves multiscale simulations, which are increasingly able to simulate systems at a number of levels simultaneously, but there are limits. There are no useful simulations of human behavior that also simulate the atoms within the human brain. Most simulations give at best a rough approximation of the behavior of the systems they simulate.

The same goes for simulations of the whole universe. To date, most cosmic simulations focus on the development of galaxies, typically laying a mesh over an area of the cosmos that divides it into huge units (or cells). The simulation indicates how these cells evolve and interact over time. In some systems, the size of the mesh is flexible, so that cells can become smaller in certain areas for a more fine-grained analysis. But it is rare for a cosmic simulation to descend to the level of simulating individual stars, let alone planets or organisms on those planets.

Within the next century, however, we may construct reasonably accurate simulations of human brains and behavior. Sometime after that, we might have plausible simulations of a whole human society. Eventually we might simulate a solar system or even a universe, from the level of atoms to the level of the cosmos. In such a system, there will be bits corresponding to every entity in the universe being simulated.

Once we have fine-grained simulations of all the activity in a human brain, we'll have to take seriously the idea that the simulated brains are themselves conscious and intelligent. After all, a perfect simulation of my brain and body will behave exactly like me. Perhaps it might have its own subjective point of view. Perhaps it will experience an environment exactly like the one I experience. At this point, we're just a step away from entertaining the hypothesis that we're living in a simulation ourselves.

Possible worlds and thought experiments

Some simulations are based on reality, while others are not. In his 1981 book *Simulacra and Simulation*, the French philosopher Jean Baudrillard distinguished four phases of simulation according to how closely they mirror reality. The first phase is *representation*, which is the "reflection of a profound reality." The last phase is a *simulacrum*, which "has no relation to any reality whatsoever." Baudrillard is talking about cultural symbols and not computer simulations, but a distant cousin of his distinction can be used to classify four sorts of computer simulation as well.

Some simulations (akin to Baudrillard's representations) aim to simulate a particular aspect of reality as closely as possible, the way a map represents a territory as closely as possible. A historical simulation of the Big Bang or the Second World War aims to replicate those past events closely. A scientific simulation of water boiling aims to simulate what happens when water really boils.

Some simulations aim to simulate something that *could* happen in reality. A flight simulator usually doesn't aim to simulate a flight that has already happened, but to simulate one that could happen. A military simulation may try to simulate what could happen to the United States if there were a nuclear war.

Some simulations aim to simulate something that *could have* happened but didn't. An evolutionary simulation might simulate what would have happened if a massive asteroid impact hadn't led to the extinction of the dinosaurs. A sporting simulation might simulate what would have happened if the United States hadn't boycotted the 1980 Moscow Olympic Games.

Finally, some simulations (akin to Baudrillard's simulacra) aim to simulate worlds that bear no resemblance to reality. A scientific simulation might simulate a world without gravity. We might try to simulate a universe with seven dimensions of space and time.

As a result, simulations are not just a guide to our actual universe. They are also a guide to the vast cosmos of possible universes. Philosophers call these *possible worlds*.

In the world (that is, the universe) we live in, I became a professional philosopher. There are nearby possible worlds in which I became a professional mathematician. There are much more distant possible worlds in which I became a professional athlete. In the actual world, Hitler became leader of Germany and there was a Second World War. There are possible worlds where Hitler never took over and the Second World War never happened. In the actual world, life developed on Earth. There are possible worlds where the solar system never formed. There are even possible worlds where there was no Big Bang.

Computer simulations can help us to explore all of these possible worlds. A cosmic simulation can simulate a universe in which our own galaxy never formed. An evolutionary simulation can simulate a version of Earth in which humans never evolved. A military simulation can simulate a world in which Hitler never invaded the Soviet Union. Eventually, a personal simulation might simulate what would have happened if I had stayed in mathematics and never moved into philosophy.

Another device for exploring possible worlds is the *thought experiment*, an experiment you carry out simply by thinking. You describe a possible world (or at least part of one) and see what follows. Plato's cave is a thought experiment. He imagines a world where prisoners can see only shadows cast on a cave wall, and asks how their lives compare to the lives of people outside the cave. Zhuangzi's butterfly is a thought experiment. Zhuangzi describes a world in which he remembers dreaming about being a butterfly, and asks how he can know he is not a butterfly that dreams he is Zhuangzi.

Thought experiments fuel science fiction. Like philosophy, science fiction explores the world as it could be. Any given science-fiction story is a thought experiment; the author conjures up a scenario and watches what follows. H. G. Wells' *The Time Machine* conjures up a world containing a time machine and then lays out the consequences. Isaac Asimov's stories in *I, Robot* conjure up a world containing intelligent robots, and Asimov then reasons about how we should interact with them.

Ursula Le Guin's classic 1969 novel *The Left Hand of Darkness* describes a possible world where humans on the planet Gethen have no fixed gender. As Le Guin puts it in her 1976 article "Is Gender Necessary?": "I eliminated gender to find out what would be left." In an introduction to the novel, she writes:

> If you like you can read [this book], and a lot of other science fiction, as a thought-experiment. Let's say (says Mary Shelley) that a young doctor creates a human being in his laboratory; let's say (says Philip K. Dick) that the Allies lost the Second World War; let's say this or that is such and so, and see what happens. . . . In a story so conceived, the moral complexity proper to the modern novel need not be sacrificed, nor is there any built-in dead end; thought and intuition can move freely within bounds set only by the terms of the experiment, which may be very large indeed.

Thought experiments yield many insights. Le Guin's thought experiment gives us insight into a possibility: It tells us something about gender as it could be. Robert Nozick's thought experiment about the experience machine gives us insight into value: It helps clarify what

Figure 6 Ursula Le Guin's thought experiment:
"I eliminated gender to find out what would be left."

is valuable to us and what isn't. Zhuangzi's butterfly dream gives us insight into knowledge: What can we know, and what can't we know?

Thought experiments can stretch the boundaries of some concepts (time and intelligence) and help delimit the boundaries of others (knowledge and value). By exploring these boundaries, they teach us something about the very nature of time, or about what it is to know something.

Thought experiments can be far-fetched, but they often teach us something about reality. Le Guin says that in writing about gender she is "describing certain aspects of psychological reality in the novelist's way, which is by inventing elaborately circumstantial lies." Le Guin's Gethenians may not exist, but aspects of their nature may resonate with the lived experience of many people, including some nonbinary people. Asimov's exploration of artificial intelligence in robots can advise us about how to interact with real AI systems once they're developed. Plato's cave helps us to analyze the complex relation between appearance and reality. This is part of why thought experiments are so central in philosophy, in science, and in literature.

Simulations in science fiction

One especially powerful thought experiment in both science fiction and philosophy is the idea of a simulated universe. What if our universe is a simulation? What follows?

James Gunn's 1955 story "The Naked Sky" was a sequel to the story about Hedonics, Inc. described in chapter 1. Both were later included in his 1961 novel *The Joy Makers*. After apparently destroying the Hedonic Council's dream machines ("In great gobs of blue, the sky began to melt"), the characters wonder whether they're still in a machine or in reality.

> How could they be sure that this was reality, not another wish-fulfillment dream from the Council-mech? How could they be sure that they had really conquered it and were not just living an illusion in a watery cell? The answer was: they could never be sure.

Gunn's passage is a contender for the first explicit statement of the simulation hypothesis: the hypothesis that we're living in a computer simulation. Admittedly, computers were new at the time, and Gunn's machines are not explicitly described as computer simulations. His "sensies," in the first story, are akin to highly immersive movies, which in later stories become perfectly convincing "realies." Computer simulations play a small role in Arthur C. Clarke's 1956 novel *The City and the Stars*, but the simulation hypothesis is not entertained there.

The two ideas—computer simulation and the simulation hypothesis—may have come together for the first time in David Duncan's obscure but sophisticated 1960 short story "The Immortals." Roger Staghorn devises a computer-simulation system, Humanac, to predict the future consequences of hypothetical events. He and a colleague, Dr. Peccary,

enter the simulation and interact with people predicted to live one hundred years in the future. They have adventures and escape by the skin
of their teeth. Back in the ordinary world, they turn off the simulation.
The story ends:

> "I can't help wondering," mused Staghorn, "of whose computer
> we're a part right now—slight factors in the chain of causation
> that started God knows when and will end . . ."
> "When someone pulls the switch," said Dr. Peccary.

The deepest development of the computer simulation idea in these
early years is the novel *Simulacron-3* (also known as *Counterfeit
World*), published in 1964 by Daniel F. Galouye. This complex work
of simulated worlds within simulated worlds was adapted by the great
German director Rainer Werner Fassbinder into the German TV production *Welt am Draht* in 1973, later released with English subtitles as
the film *World on a Wire*. It appears to be the debut of the simulation
hypothesis in film or TV. Fassbinder's film was later remade into the
1999 Hollywood film *The Thirteenth Floor* and is widely credited with
inspiring many other films in the simulation genre.

Premiering the same year, *The Matrix*, written and directed by Lana
and Lilly Wachowski, remains the best-known depiction of the simulation idea on film. The main character, Neo (in a memorable performance by Keanu Reeves) experiences an ordinary world. He goes to
work, he reads books, he hangs out at parties, more or less as we do. He
has a few clues that something is strange; his world has a faint green
tinge, and he has a perpetual feeling of unease. Tellingly, he has been
reading Baudrillard's book *Simulacra and Simulation*. Eventually he
takes the red pill and learns that he's been living in a computer simulation all along.

The Matrix was partly responsible for my own entry into the simulation arena. The directors and producers of the movie had a significant
interest in philosophy, and a number of philosophers were invited to
write about philosophical ideas for its official website. I accepted the
invitation and in 2003 published an article there called "The Matrix as

Metaphysics," all about how the Matrix is not really an illusion. It was an early version of some of the ideas in part 3 of this book.

In "The Matrix as Metaphysics," I introduced my own name for the simulation hypothesis. I called it the "Matrix Hypothesis" and defined it as the hypothesis that I am and always have been in a matrix. I defined a matrix as an artificially designed computer simulation of a world.

In the same year, Nick Bostrom published his important article "Are You Living in a Computer Simulation?," which gave a statistical argument for why we should take the simulation idea seriously. (I'll discuss the argument in chapter 5.) In another 2003 article, Bostrom introduced the name "simulation hypothesis" for the idea. This proved to be a better name than mine; the simulation idea is universal, whereas a movie is ephemeral. In this book I'm following now-standard practice in talking of the simulation hypothesis.

The simulation hypothesis

What exactly is the simulation hypothesis? Bostrom's version says simply, "We are living in a computer simulation." Mine says, "We are and always have been in an artificially designed computer simulation of a world." I think the two are consistent. My version just makes explicit a couple of things that Bostrom's does not. First, the simulation needs to be lifelong, or at least for as long as we can remember. Being in a simulation since yesterday doesn't count. Second, the simulation needs to have been designed by a simulator. A computer program that popped up randomly without a simulator wouldn't count. Both of these factors are part of the simulation hypothesis as people ordinarily think of it.

What is it to be in a simulation? As I understand this notion, it's all about interacting with the simulation. When you're in a simulation, your sensory inputs come from the simulation, and your motor outputs affect the simulation. You're fully immersed in the simulation through these interactions.

At the start of *The Matrix*, Neo's biological body and brain are in a pod in a nonsimulated world, connected to a simulation somewhere else. In the ordinary spatial sense of "in," Neo's brain is not "in" the simulation. However, all his sensory inputs are coming from the simulation, and his outputs are going there, so he's in a simulation in the sense that matters. After he takes the red pill, his senses respond to the nonsimulated world, so he is no longer in a simulation.

I will use the word *sim* for someone who is in a simulation. There are at least two sorts of sims. First, there are *biosims*: biological beings outside the simulation (in the spatial sense) and connected to it. Neo is a biosim. So is a brain in a vat, connected to a computer. A simulation that includes biosims is an *impure simulation*, since it includes elements (the biosims) that aren't simulated.

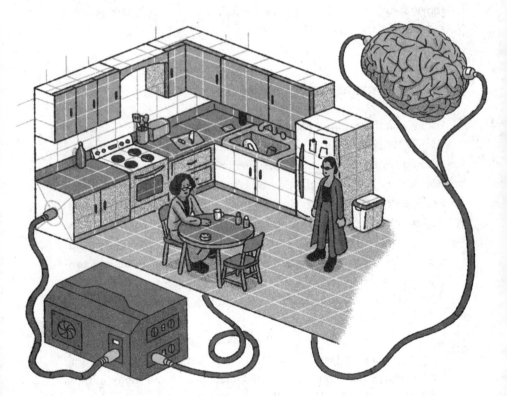

Figure 7 A simulated world with a biosim (controlled by a brain) and a pure sim (controlled by a computer), inspired by Trinity and the Oracle in *The Matrix*.

Second, there are *pure sims*. These are simulated beings who are wholly *inside* the simulation. Most of the people in Galouye's novel *Simulacron-3* are pure sims. They receive direct sensory inputs from the simulation because they're part of the simulation. Importantly, their brains are simulated, too. Simulations containing only pure sims may be *pure simulations*—simulations in which everything that happens is simulated.

There can also be *mixed simulations*, which contain both biosims and pure sims. Inside the Matrix, the leading characters Neo and Trinity are biosims, whereas the "machine" characters Agent Smith and the Oracle are pure sims. In the 2021 movie *Free Guy*, the main character, Guy (played by Ryan Reynolds), is a fully digital nonplayer character in a video game, while his in-game partner, Molotov Girl (played by Jodie Comer), is a video game player and designer with an ordinary life outside the game. So Guy is a pure sim, while Molotov Girl is a biosim.

The simulation hypothesis applies equally to pure, impure, and mixed simulations. Occurrences of the simulation idea in science fiction and philosophy are split fairly evenly among them. In the short term, impure simulations will be more common than pure simulations, since we know how to connect people to simulations but we don't yet know how to simulate people. In the long term, pure simulations may well be more common. The supply of brains for impure simulations is not endless, and in any case hooking them up may be tricky. By contrast, pure simulations are easy in the long term. We need only set up the right simulation program and watch it go.

Here's another distinction. The *global*-simulation hypothesis says that the simulation simulates a whole universe in detail. For example, a global simulation of our universe will simulate me, you, everyone on Earth, planet Earth itself, the whole solar system, the galaxy, and everything beyond. The *local* simulation hypothesis says that the simulation simulates only a part of the universe in detail. It might simulate just me, or just New York (see figure 57 in chapter 24), or just Earth and everyone on it, or just the Milky Way galaxy.

In the short run, local simulations should be easier to create. They require much less computational power. However, a local simulation has to interact with the rest of the world, and that can lead to trouble.

In *The Thirteenth Floor*, the simulators simulated only Southern California. When the protagonist tried to drive to Nevada, he encountered signs saying "Road closed." He kept going, and the mountains morphed into thin green lines. That's not a good way to design a convincing simulation. If a local simulation is completely local, it cannot properly simulate interaction with the rest of the world.

To work well, local simulations will have to be flexible. To simulate me, simulators will have to simulate much of my environment. I talk to people elsewhere, see events taking place around the world on TV, and travel often. The people I meet interact with many others in turn. So a good simulation of my local environment will require a fairly detailed simulation of the rest of the world. The simulators may need to fill in more and more details as the simulation runs. For example, a simulation of the far side of the Moon will need revision once spacecraft can photograph it and send pictures back to Earth. There might be some natural stopping points: perhaps simulators could render Earth in detail and the solar system as they need to, with just a rudimentary simulation of the universe beyond?

Philosophers revel in distinctions. There are many more distinctions we could make. We could distinguish between *temporary* and *permanent* simulations (Do people enter the simulation for a brief period or spend their whole lives there?), *perfect* and *imperfect* simulations (Do we faithfully simulate all the laws of physics, or do we allow approximations and exceptions?), and *pre-programmed* and *open-ended* simulations (Is there a single course of events programmed in advance, or can various things happen depending on initial conditions and what the sims choose?). You can probably think of other distinctions, but we already have enough to go on.

Can you prove you're not in a simulation?

Can you prove you're not in a computer simulation?

You might think you have definitive evidence that you're not. I think that's impossible, because any such evidence could be simulated.

Maybe you think the glorious forest around you proves that your world isn't a simulation. But in principle, the forest could be simulated down to every last detail, and every last bit of light that reaches your eyes from the forest could be simulated, too. Your brain will react exactly as it would in the nonsimulated, ordinary world, so a simulated forest will look exactly like an ordinary one. Can you really prove that you aren't seeing a simulated forest?

Maybe you think your darling cat could never be simulated. But cats are biological systems, and it seems likely that biological mechanisms can be simulated. With good enough technology, a simulation of your cat might be indistinguishable from the original. Do you really know that your cat isn't a simulation?

Maybe you think the creative or loving behavior of the people around you could never be simulated. But what goes for cats goes for people. Human biology could well be simulated. Human behavior is caused by the human brain, and the brain seems to be a complicated machine. Do you really know that a full simulation of the brain could not reproduce all this behavior in detail?

Maybe you think your own body could never be simulated. You feel hunger and pain, you move around, you touch things with your hands, you eat and drink, you're aware of your own weight in a way that seems viscerally real. But as biological systems, bodies can be simulated. If your body is simulated so well that it sends exactly the same signals to your brain, your brain wouldn't be able to tell the difference.

Maybe you think your consciousness could never be simulated. You have subjective experience of the world from a first-person perspective: You experience colors, pains, thoughts, memories. It *feels like something* to be you. No mere simulation of a brain would experience this consciousness!

This issue—the issue of consciousness and whether a simulation could have it—is harder than the others. We'll grapple with it in detail later in the book. For now, we can set the issue of consciousness aside by focusing on impure simulations—that is, *Matrix*-style simulations in which you're a biosim connected to the simulation. Biosims are not themselves simulated. They have ordinary biological brains which will presumably be conscious like ours. Whether you're an ordinary person or a biosim whose brain is in the same state, things will look and feel the same to you.

Pure simulations, in which the people in the simulated world are all simulated themselves, raise the issue of whether simulated beings can be conscious. If we could prove that simulated beings couldn't be conscious, we could prove that we're not in a pure simulation (at least, given that we're sure we're conscious). In chapter 15, I'll argue that simulated beings could be conscious. If a simulated brain precisely mirrors a biological brain, the conscious experience will be the same. If that's right, then just as we can never prove we're not in an impure simulation, we can also never prove that we're not in a pure simulation.

Can you prove you *are* in a simulation?

I've argued that we can never prove we're not in a simulation. What about the other way around? Could we prove we *are* in a simulation?

In *The Matrix*, Neo realized he'd been living in a simulation when he took the red pill and woke up in a different reality. As I've noted, he shouldn't have been so sure. For all he knows, his old world was non-simulated and the red pill plunged him into a simulation.

Still, we certainly could get very strong evidence that we're in a simulation. The simulators could lift the Sydney Harbour Bridge into the air and turn it upside down. They could show us the source code of the simulation. They could show us private episodes from our past, along with the simulation technology that produced them. They could show me a film of my brain hooked up to wires in the next reality up, with an associated readout of my thoughts and feelings. They could give me control of the simulation, so that I could move mountains in the world around me just by pressing some buttons.

Even this evidence would fall short of absolute proof that we're in a simulation. Maybe the world we're in is a nonsimulated magic world, like the Harry Potter world, in which all-powerful wizards are using their powers to convince us we're in a simulation. Maybe most of my life has been nonsimulated but simulators have put me into a temporary simulated duplicate to fool me. Or maybe I'm having a drug-induced hallucination. Still, I think that if I got evidence like this, I would probably be convinced that I am in a simulation.

Is the simulation hypothesis a scientific hypothesis?

Sometimes people treat the simulation hypothesis as a scientific hypothesis—one that's testable in principle by observation or experiment. Might there be scientific evidence that we're in a simulation?

A 2012 article by physicists Silas Beane, Zohreh Davoudi, and Martin Savage argues that in principle we could someday get scientific evidence for the simulation hypothesis. The basic idea is that a simulation of our universe may well cut some corners by making approximations, and those approximations may show up in the evidence. The authors produce a mathematical analysis of how certain physical approximations using a "hypercubic spacetime" lattice would deviate from standard physics in a testable way. If our simulators used lattice-spacing of a certain size, a distinctive pattern among high-energy cosmic rays would result. The authors suggest that this provides a possible way of testing the simulation hypothesis in the future, though we do not have such evidence as things stand.

This potential evidence depends on the simulation's being *imperfect*. The same goes for the potential evidence discussed in the two preceding sections. Red pills, communication with simulators, and approximations are imperfections of a sort—that is, they're points at which the simulation deviates from the laws of the world it's simulating. In *The Matrix*, *déjà vu* experiences such as a black cat crossing one's path twice are said to arise from glitches in the program. A perfect simulation won't have glitches like this.

A perfect simulation can be defined as one that precisely mirrors the world it's simulating. If the world it's simulating obeys strict physical laws, a perfect simulation will simulate those laws precisely and will never deviate from them. Red pills, communication with simulators, and approximations are ruled out.

It's arguable that a digital computer could never perfectly simulate the continuous laws of physics, which involve precise quantities on a continuum. Still, a digital simulation should be able to approximate the known laws of physics to any degree of precision. And at least in principle there could be perfect simulation of known laws by an analog computer (perhaps an analog quantum computer) which deals in continuous quantities.

If we're in a perfect simulation, it's hard to see how we could ever get evidence of that fact. Our evidence in the simulation will always correspond precisely to evidence in the unsimulated world.

It's just as hard to get evidence that we're *not* in a perfect simulation. As before, any such evidence could in principle be simulated. In a perfect simulation, we would get simulations of the same evidence. At least if we assume that a simulated brain would have the same conscious experience as the brain it's simulating, then there will be no way from the inside to tell the difference between a nonsimulated universe and a perfect simulation of it.

Every now and then, an article appears in the popular press claiming that scientists have proved we're not in a simulation. One example from 2017 stemmed from publication of a research article in *Science Advances* arguing that classical computers cannot efficiently simulate quantum processes. The authors, physicists Zohar Ringel and Dmitry Kovrizhin, did not say that this rules out the simulation hypothesis, but some journalists used their article to draw that conclusion. Of course, the mere fact that classical computers cannot efficiently simulate our universe is no proof that we're not in a simulation. As the computer scientist Scott Aaronson pointed out, to get around the problem we need only suppose that the simulation is using a quantum computer. We could even suppose that the simulation is simulating quantum processes using a classical computer running slowly and inefficiently. From the inside, we couldn't tell the difference.

Sometimes people say that no universe can contain a perfect simulation of itself, since the universe would need a simulation of the simulation, and a simulation of that, and so on, leading to an infinite stack of simulations. Now, such a stack is not obviously impossible. Perhaps an infinite universe could devote a small fraction of its resources to running a (still infinite) simulation of itself. The resulting stack of simulations would be no problem for an infinite universe. Even a finite but expanding universe could run an ongoing simulation of the past that lags a little behind reality.

Were it indeed the case that no universe could simulate itself, that still would not rule out the simulation hypothesis. There's no reason to suppose that the simulated universe and the simulating universe should be exactly the same. If we're in a simulation, the simulating universe may have an entirely different physics from ours and may be much larger than ours. If the simulating universe is infinite, and has infinite resources, simulating a finite universe will be easy.

To sum things up, I would say that in principle we can get evidence for and against various imperfect simulation hypotheses, which will presumably have empirical consequences we can test. So these imperfect simulation hypotheses count as scientific hypotheses. They may not yet be serious scientific hypotheses, since we don't yet have scientific evidence that gives them support, but at least they're testable in principle.

However, we can never get experimental evidence for or against perfect simulation hypotheses. A nonsimulated world and a perfect simulation of it will seem exactly the same. So, according to the testability criterion, the hypothesis that we're in a perfect simulation is not a scientific hypothesis. Instead, we can think of it as a philosophical hypothesis about the nature of our world.

Some hard-nosed scientists and philosophers may hold that because it's untestable, the perfect simulation hypothesis is meaningless. I'll argue in chapter 4 that this is incorrect. In principle, we can construct perfect simulated worlds ourselves, with beings inside them. There will be no way for those beings ever to know that they're in a simulation. The simulation hypothesis is demonstrably true of those beings. It follows that the hypothesis is meaningful. It may also be true of us, or it may not. Perhaps we will never know the answer to the question, but the hypothesis is either true or false all the same.

What about the original simulation hypothesis, saying that our world is a computer simulation? Is this a scientific hypothesis or a philosophical hypothesis?

The philosopher of science Karl Popper insisted that the hallmark of a scientific hypothesis is that it is *falsifiable*—capable of being proved false using scientific evidence. We've seen that the simulation hypothesis is not falsifiable because any evidence against it could be simulated. So Popper would say that it's not a scientific hypothesis.

Like many philosophers these days, I think Popper's criterion is too strong. There can be scientific hypotheses—for example, about the early universe—that could never be falsified. But I'm happy to say that the simulation hypothesis is not a squarely scientific hypothesis but one that is partly scientific and partly philosophical. Some versions of it are subject to test, while other versions of it are impossible to test. But whether testable or not, the simulation hypothesis is a perfectly meaningful hypothesis about our world.

The simulation hypothesis and the virtual-world hypothesis

What is the relationship between computer simulations and virtual worlds? Recall that a virtual world is an interactive, computer-generated space. Is every virtual world a simulation? Is every simulation a virtual world?

Most virtual worlds found in video games can be regarded as simulations. This is most obvious in games that simulate some physical-world activity, like fishing or flying or playing basketball. These games are closest to Baudrillard's representations. They may not aim for perfect realism, but they try to reflect the real world. More exotic games, like *Space Invaders* and *World of Warcraft*, are closer to Baudrillard's simulacra. They don't purport to reflect the real world, but they're simulations of possible worlds. *Space Invaders* loosely simulates an alien invasion of Earth. *World of Warcraft* simulates a physical environment with monsters, quests, and battles.

Even games like *Tetris* or *Pac-Man,* which don't obviously simulate physical environments, can be regarded as simulations if you squint at them in the right way. *Tetris* can be seen as a simulation of a two- or three-dimensional world with bricks falling from the sky. *Pac-Man* can be seen as a simulation of predators and prey running through a physical maze. Perhaps it's a stretch to see these as simulations; users may not see them this way, and simulation may have been no part of the designer's intentions. But as I'm understanding the simulation hypothesis, it doesn't matter whether users or the designer see the simulation as a simulation. So these virtual worlds still count as simulations for our purposes.

The same reasoning applies to any virtual world. Any virtual world involves a space, which we can in principle interpret as a simulation of a hypothetical physical space. In this broad sense, any virtual world involves a computer simulation.

What about the converse? Strictly speaking, not all computer simulations are virtual worlds. There are noninteractive simulations, like standard simulations of galaxy formation, that do not interact with users at all. Because they aren't interactive, they don't meet the definition of virtual worlds. But the hypothesis that I'm *in* a computer simulation requires that I'm interacting with a computer-generated world through my sensory inputs and motor outputs. This hypothesis is equivalent to the hypothesis that I'm in a virtual world.

As a result, the simulation hypothesis can equivalently be stated as the virtual-world hypothesis: I am in a virtual world.

To flesh out the picture, the simulation hypothesis suggests we're living in a *fully immersive* virtual world. A virtual world is immersive when you experience it all around you as if you were right there, as with today's standard virtual reality headsets. We defined VR in the introduction as an immersive virtual world. A virtual world is *fully* immersive when one is immersed in the virtual world with all of one's senses, experiencing it just as we experience the physical world. Our experience of the world we live in is fully immersive. So if we're in a virtual world at all, we're in fully immersive VR.

The simulation hypothesis is equivalent to the virtual-world hypothesis, but from now on I will mainly use the standard term "simulation hypothesis." In the same spirit, I'll tend to use the word "simulation" for the sort of *Matrix*-style simulated universe relevant to the simulation hypothesis—that is, a lifelong and fully immersive simulated world in which users may not know they're in a simulation. I'll tend to use "virtual world" and "virtual reality" for the more down-to-earth virtual environments that users enter knowingly and for limited periods. This includes everything from video games and current VR headsets to extensions of that technology, such as the scenario of *Ready Player One*, in which people regularly hook themselves up to a fully immersive virtual world.

There's a spectrum of worlds from current virtual worlds to full-scale simulations such as *The Matrix*. All of them count as virtual worlds in the strict sense, and both ends of the spectrum are relevant to my overarching claims, such as *virtual reality is genuine reality*. Still, down-to-earth virtual worlds and simulated universes raise somewhat different issues. In the next few chapters, simulated universes will take center stage.

Part 2

KNOWLEDGE

Chapter 3

Do we know things?

I N THE ANIMATED TV SERIES *BOJACK HORSEMAN*, THERE'S A TV show within the show called *Hollywoo Stars and Celebrities: What Do They Know? Do They Know Things? Let's Find Out!* It's essentially a quiz show for the movie stars in the series, who inhabit an alternative reality in which the "Hollywood" sign has lost its final letter. Like much of the series, it's easily turned into philosophy. The Georgetown University philosopher Quill Kukla teaches a course called "BoJack Horseman and Philosophy," whose tag line is "What Do We Know? Do We Know Things? Let's Find Out!" That tag line pretty much sums up the history of epistemology—that is, the theory of knowledge—in Western philosophy.

What do we know? Most of us think we know a lot. We know what happened yesterday and what will probably happen tomorrow. We know about our families and our friends. We know some history, some science, and some philosophy. We even know a little about ourselves.

Philosophers have questioned each of these kinds of knowledge. The ancient Greek philosopher Sextus Empiricus (second or third century CE) questioned our knowledge of science. His Indian Buddhist contemporary Nāgārjuna questioned whether we gain knowledge from philosophy. The 11th-century Persian philosopher al-Ghazali questioned our knowledge of what we see and hear. The 18th-century Scottish philosopher David Hume questioned our knowledge of the future. The contemporary American philosophers Grace Helton and Eric Schwitzgebel have respectively questioned whether we know other people's minds and whether we know our own minds.

Do we know things? Some philosophers have questioned whether we know anything at all. The ancient skeptic Pyrrho and his followers said that we shouldn't trust any of our perceptions or our beliefs. Trusting them doesn't lead us to knowledge or to happiness, and if we refrain from believing anything, we can be free from worry. Most of us don't follow Pyrrho's advice; we believe things. But do we know those things?

Let's find out! To find out whether we know things, we have to figure out what knowledge is and whether we ever have it. And we have to assess the many challenges to our knowledge that philosophers have put forward over the ages.

A common view of knowledge, going back to Plato, is that knowledge is justified, true belief. To know something, you have to *think* it's true (that's belief), you have to be *right* about it (that's truth), and you have to have good *reasons* for believing it (that's justification).

If I falsely believe that Hillary Clinton ran a child sex ring, that's not knowledge. I'm wrong; it's just a false belief. If I guess someone's birthday and I'm right by pure chance, that's not knowledge. I don't have good reasons; it's an unjustified belief. There may be more to knowledge than justified true belief, but most philosophers think these three requirements are at the heart of the story.

Almost everyone agrees that knowledge is something we want. The 16th-century English philosopher of science Francis Bacon said, "Knowledge itself is power." The American president Thomas Jefferson added that knowledge is happiness and security. In her song "The Knowledge," Janet Jackson sang, "What you don't know can hurt you bad. . . . Get the knowledge."

At the same time, knowing things can be hard work. It's easy to go wrong. Our reasons for believing something are rarely as strong as we'd like them to be. As a result, many thinkers have been led to doubt whether we have any knowledge at all.

Skepticism about the external world

In the opening of his 1983 book *Adventures in the Screen Trade,* the screenwriter William Goldman, who wrote *Butch Cassidy and the Sundance Kid* and *All the President's Men,* addresses the question of knowledge with a declaration: "NOBODY KNOWS ANYTHING." He's talking about the movie business. But once again, the answer runs deeper.

Goldman's famous line is an expression of *skepticism,* which is exactly the view that nobody knows anything. It's a view with a long history.

In philosophy, a *skeptic* is someone who casts doubt on our beliefs about a certain domain. Goldman was a skeptic about the movie business. He thought our beliefs about how to make successful movies don't amount to knowledge. A skeptic about the paranormal casts doubt on our beliefs about ghosts and telepathy. A skeptic about the news media casts doubt on beliefs acquired through the news media.

Skepticism about the news media and about the paranormal are examples of *local* skepticism: casting doubt on our beliefs in a specific domain. There are many forms of local skepticism. You could be a skeptic about the future (casting doubt on our beliefs about what will happen tomorrow), or about science (casting doubt on scientific findings), or about other people's minds (casting doubt on whether we can ever know what other people are thinking).

The most virulent form of skepticism is *global skepticism:* casting doubt on all of our beliefs at once. The global skeptic says that we cannot know anything at all. We may have many beliefs about the world, but none of them amount to knowledge.

Perhaps the most well-known form of skepticism is skepticism about the external world: casting doubt on all of our beliefs about the world around us. This view is often called Cartesian skepticism, after René Descartes, who was its most famous proponent. Strictly speaking, Cartesian skepticism is not full-scale global skepticism, since it's consistent with our knowing a few things—about logic or about our own minds, for example. But it's so encompassing that I will count it as a form of global skepticism here.

Refuting Cartesian skepticism about the external world is one of the hardest problems in modern philosophy. Many philosophers have tried to refute it, but no refutation has commanded much of a consensus. In this book (especially chapters 6, 9, 22, and 24), I will lay out what I see as the best response to the Cartesian skeptic. Perhaps I too will fail, but I hope to gain some wisdom in the attempt.

My ambitions are limited. I'm going to argue that certain Cartesian arguments for global skepticism about the external world fail. I'm not trying to refute local forms of skepticism, such as skepticism about the news media (though I'll come back to that issue in chapter 13). My target is the classical Cartesian skeptic who uses a radical hypothesis to cast doubt on all of our beliefs about the external world at once.

How do you know your senses aren't deceiving you?

In 1641, Descartes published his *Meditations on First Philosophy*. He was trying to build a foundation for everything that we know. To build that foundation, he first had to tear everything down. His demolition crew included three classic arguments—concerning illusions, dreams, and demons—that cast doubt on our knowledge of the external world. These arguments weren't entirely new. Illusions and dreams were standard fare for skeptics in ancient times such as Sextus Empiricus and the Roman orator Cicero, as well as for medieval thinkers such as the 5th-century North African saint Augustine and the Persian philosopher al-Ghazali. We'll see that demons were used by Descartes's contemporaries as well. Nevertheless, Descartes gave these arguments their most influential formulation.

Descartes's first argument was based on illusions. *Our senses have deceived us before. How can we know they aren't deceiving us now?*

Most of us have experienced optical illusions in which appearances are different from reality. We've been fooled by smoke and mirrors. If our senses have deceived us in the past, they may be deceiving us now. So we can't be sure that whatever we observe in the external world is as it seems to be.

Descartes accepted that sensory illusions have limits. No sensory illusion could give people the sense of having an entirely different body or being in an entirely different environment. He wrote: "Yet although the senses occasionally deceive us with respect to objects which are very small or in the distance, there are many other beliefs about which doubt is quite impossible, even though they are derived from the senses—for example, that I am here, sitting by the fire, wearing a winter dressing-gown, holding this piece of paper in my hands, and so on."

A 21st-century reader will say, "Not so fast!" VR researchers regularly talk about "whole-body illusions"—the sort of thing Descartes thought was impossible. I can see and control a body that isn't my biological body, and I'll sense that it's mine. Inside VR, Descartes could

Figure 8 Inside VR, Descartes has the sense that he's sitting by the fire in a dressing gown, holding a piece of paper in his hands.

even have the sense that he's sitting by the fire in a dressing gown, holding a piece of paper. VR thereby strengthens Descartes's original argument based on illusions. Technology makes it harder to know that we're not experiencing an illusion right now.

We can even mount a 21st-century version of Descartes's illusion argument: the argument from virtual reality. *Virtual reality devices have fooled people before. How do you know that a VR device isn't fooling you now?* In principle, everything you're seeing and hearing could be the product of a VR device. Can you really be sure you're not using such a device now?

Admittedly, to be fully convincing, today's technology would have to improve, but eventually we'll have VR contact lenses and other unnoticeable equipment that can handle all the senses. In principle, someone could put you into such an advanced VR device while you're asleep. In the morning, you wake up in a virtual bed and go about your virtual day. If you've been put in a new virtual environment, like a virtual version of Mars, you'll presumably realize that something's wrong—unless your memories have also been tampered with, which goes a step further than I'm going here. So let's say that the VR environment is like your normal home, or like wherever you are right now. That way, you won't notice anything too strange.

Are you really sure you're not in such a VR device right now? If you're sure—how can you really rule out the VR hypothesis? If you're not sure—then how can you be sure about anything you perceive in the world around you? Can you really be sure that this is a genuine book you're reading, or a genuine chair you're sitting in? Can you really be sure about where you are or about whether what you're seeing is really there?

This VR argument casts doubt on your knowledge of what you're seeing and hearing now, and perhaps on your knowledge of the recent past. But it may not throw everything you know into question. VR per se won't tamper with memories, so your memories of growing up in your hometown won't be threatened. It also won't tamper with your general scientific or cultural knowledge, so your knowledge that Paris is in France will still be safe.

You could try to extend the VR argument to threaten these domains. Maybe a memory-tampering device could get into your brain and change your memories. Maybe a permanent VR device could ensure that all of your memories and your scientific knowledge would derive from VR. These extensions bring us beyond standard VR and into the domain of the simulation hypothesis, which I'll return to soon.

How do you know you're not dreaming?

Descartes's second argument was about dreaming: *Dreams are like reality. How do we know we're not dreaming?*

We usually dream without being aware that we're dreaming. When we're dreaming, we typically think the world of the dream is real. There are occasionally lucid dreams, when people know they're dreaming, but these are the exception. Although most dreams are weirder and less stable than reality, in principle there could be dreams that are indistinguishable from reality. How do you know you're not having such a dream right now? Perhaps you could pinch yourself or run some experiments. That wouldn't be too convincing, though, as any results you get could in principle come from a dream. And if you can't know you're not dreaming now, it seems that you can't know that anything around you is real.

In the 2010 movie *Inception* (spoiler alert!), the characters fall asleep and enter a dream world, and then enter dream worlds within dream worlds. For most of the movie, the main character, Dominick Cobb (played by Leonardo DiCaprio), knows he is in a dream world and that he's asleep in ordinary reality. But other characters, including Robert Fischer (played by Cillian Murphy), do not know this. Fischer is the one who is dreaming the dream world, and he treats the dream world like reality. At the end of the movie, when Cobb and the other characters seem to have returned to the ordinary world, the question arises: How do they know that this is not yet another dream world? It seems impossible to know for sure.

Descartes thought the dream argument was stronger than the illusion argument. Unlike an optical illusion, a dream could easily convince him that he was in his dressing gown by the fire when in fact he wasn't. Maybe a dream could even give him a different body. But for Descartes, this argument, too, had limits. Dreams never give you something absolutely new. If you dream of a head or a body, it must be based on your having perceived heads and bodies in the real world. Or at least the shapes and colors in the dream must be based on shapes and colors in the real world.

We haven't developed dream technology as thoroughly as we've developed VR technology, so Descartes's dream argument is less affected by technological change than his illusion argument. However, dream science has found some fairly good ways to know you're dreaming. Look at some writing on a page twice: In a dream it will usually change; outside a dream it usually won't. Another relevant piece of science is that we can indeed experience colors we've never experienced in reality. For example, residual images after we see certain colors can give us a shade of "dark yellow" that we can never get through ordinary perception.

Like the illusion argument and the VR argument, the dream argument casts doubt on our current and recent knowledge of the world around us: How do I know that what I'm seeing now, or what I saw a moment ago, is real? Preexisting knowledge is trickier. Dreams can sometimes alter our memories (in a dream, I can remember a different childhood) and our cultural beliefs (I can dream that the Beatles are still performing together), but they don't usually alter our memories wholesale. One could postulate a lifelong dream in which every element of one's reality comes from a dream, but now we are once again in the science-fiction territory of simulations and the like.

Descartes's illusion argument and dream argument both work best in supporting local skepticism—casting doubt on some of our knowledge of the external world, but not all of it at once. Descartes was not satisfied with this. He was interested in global skepticism: that is, casting doubt on our knowledge of the entire external world all at once. For that, he needed a stronger argument.

Descartes's evil demon

Descartes's third and most notorious argument is an argument about deception. *An all-powerful being could deceive me completely, by giving me experiences of a world that does not exist. How do I know this isn't happening to me?*

Descartes's original and central deceiver in the first meditation was an all-powerful and all-deceiving God. If God can do anything, surely God has the power to deceive us completely. The deceiver that everyone remembers, though, is Descartes's evil demon. In the original Latin, Descartes talked of a *genium malignum*, which might be translated as "bad genie" (French philosophers typically talk of a "malin génie"), but "evil demon" does the job in English. While a benevolent God might refuse to deceive us, an evil demon would have no such compunctions. Descartes introduces the demon this way:

> I will suppose therefore that not God, who is supremely good and the source of all truth, but rather some evil demon of the utmost power and cunning has employed all his energies in order to deceive me. I shall think that the sky, the air, the earth, colours, shapes, sounds and all external things are merely the delusions of dreams which he has devised to ensnare my judgement.

The evil demon is devoted to deception. It feeds you sensations and perceptions, as of an external world, for your whole life. I remember my growing up in Australia, and these days I seem to be living an enjoyable life in New York City as a professor of philosophy. But if Descartes's evil-demon hypothesis is correct, all of this was based on sensations and perceptions fed to me by the demon. In reality, I have spent my whole life in its lair, where it is manipulating my senses.

The evil-demon thought experiment was not entirely novel. The Columbia University historian of philosophy Christia Mercer has recently charted how the 16th-century Spanish theologian Teresa of Ávila wrote her own meditations in which deceiving demons played a central role. For Teresa, the issue was belief in God, and the demons were trying to deceive her to make her lose her faith. Teresa's book, *The Interior Castle*, was a huge seller in Descartes's time, and he almost certainly read it. Descartes's readers would also have encountered the illusion and dream arguments in the well-known writings of the 16th-century French essayist Michel de Montaigne. So while Descartes's *Meditations* were certainly an advance, he was building on the work of the women and men around him.

Some aspects of the evil demon story are portrayed in the 1998 movie *The Truman Show*. In the movie, Truman Burbank (played by Jim Carrey) is living in a bubble populated by actors. A television producer, Christof (played by Ed Harris), orchestrates the bubble to give him the sense that his is a normal life. Christof is playing the role of the evil demon. Some of Truman's world is real, though. He really has a body, he really lives on Earth, and he really interacts with people. Christof doesn't deceive him about that. The evil demon's victim is like a version of Truman who does not have a body and does not interact with people. Everything the victim experiences is produced by the evil demon.

How do you know that right now you're not being manipulated by an evil demon? It seems that you can't. Maybe there's some suggestion of the evil demon's handiwork—the fact that you're now reading about evil demons, for example; evil demons with a sense of humor might enjoy causing people to think about evil demons. Even without hints like this, it seems impossible to exclude the evil-demon hypothesis entirely. But if you can't know that you're not being manipulated by an evil demon, how can you know that anything is real?

The evil-demon argument calls into question everything you know about the external world. Therein lies its power. As we've seen, ordinary illusions and dreams don't threaten my knowledge of my childhood home in Australia, or my knowledge that Einstein discovered relativity. The evil demon has been deceiving us our whole lives, so it threatens everything. My experiences of Australia may be a fiction. When I read about Einstein's discoveries, the stories may be made up. So if we cannot rule out the evil-demon hypothesis, we are threatened with global skepticism.

How does the evil demon do its work? Descartes is not clear on the details. Presumably the demon has to keep a complicated model of a fictional world in its head to make sure the subject has matching experiences over time. Every time I return to Australia or visit an old friend, my experience needs to be consistent with previous visits. The demon will also need models for places I've read about and places I'll eventually visit, as well as everything I read about in newspapers or watch on TV. The model will have to be constantly updated. This will be a lot of work, although perhaps the work is nothing for an all-powerful demon.

An especially insidious version of Descartes's evil demon gets inside people's minds and directly tampers with their thoughts. In the modern version, this demon could be an evil neuroscientist. Perhaps the demon manipulates your brain so that you believe that you're in Antarctica. Descartes says that a deceiver might even manipulate his thoughts so that "I too go wrong every time I add two and three." Perhaps the demon can make you believe 2 + 3 = 6, and you will find this completely convincing.

The mind-tampering demon threatens to lead to a sort of *internal-world* skepticism in which you can't even trust your own rationality or reasoning anymore. This sort of evil demon is fascinating, but it's outside the scope of my discussion. I'm concerned here with scenarios in which my external world is manipulated, not scenarios in which my internal world is manipulated directly. I'll return to mind-tampering scenarios in the final chapter of this book.

From the evil demon to the simulation hypothesis

If Descartes's evil demon lives in the computer age, its task is a lot easier. It can simply offload the modeling work into a computer. It can run a computer simulation of the world and connect subjects to the simulation so that they experience the world as it evolves. This is the setup in *The Matrix*, where godlike machines play the role of the evil demon and a computer simulation takes care of the hard work.

In the 20th century, the American philosopher Hilary Putnam and others updated Descartes's idea with equipment from modern science. The evil demon was replaced by an evil scientist, and the person deceived by the evil demon was replaced by a *brain in a vat*. Like the brains that float in jars in the Steve Martin movie *The Man with Two Brains*, the brain is kept alive with a carefully balanced mix of nutrients. Putnam tells us that the brain's nerve endings are "connected to a super-scientific computer." The computer sends electronic impulses to the brain, bringing about the illusion that everything is normal. The brain experiences a richly detailed and well-populated world, but in fact it is alone in a laboratory.

Putnam's brain-in-a-vat scenario is very much like the scenario in *The Matrix*, except that in the movie full bodies in pods are connected to the computer. Putnam doesn't say much about what the computer is doing (neither does *The Matrix*), but clearly (as in *The Matrix*) it is running a simulation of the world that the brain is experiencing.

In the 21st century, philosophers' focus has gradually shifted from brains in vats to the simulation hypothesis. The simulation idea captures an element at the core of all of the great Cartesian scenarios: The evil demon must do its work by simulating a world. A lifelong dream can be seen as a sort of simulated world. The brain in a vat is connected to a simulation. And so on. Making the simulation a computer simulation helps us to pin down the scenario in more concrete terms without losing anything essential.

The brain-in-a-vat idea is one version of the simulation hypothesis. It involves an impure simulation, in which a brain is connected to the simulation from the outside. The simulation hypothesis also includes other versions, such as pure simulation versions in which the brain is internal to the simulation. Both of these scenarios can be used to mount a Cartesian argument for skepticism.

You might think that the switch from evil demons to brains in vats to simulations is a mere change in packaging, but there is one respect in which the use of modern technology makes the argument more powerful. Because the evil-demon hypothesis is so fanciful, Descartes was reluctant to put too much weight on it. It was important to him that his skeptical concerns be grounded in reasonable doubts that he should take seriously, given his beliefs. He gave more weight to the deceiving-God hypothesis because he believed in an all-powerful God and thought it reasonable that God would have the power to deceive us. Because this was a realistic hypothesis, it gave him greater reason for doubt.

The simulation hypothesis may once have been a fanciful hypothesis, but it is rapidly becoming a serious hypothesis. Putnam put forward his brain-in-a-vat idea as a piece of science fiction. But since then, simulation and VR technologies have advanced fast, and it isn't hard to see a path to full-scale simulated worlds in which some people could spend a lifetime.

As a result, the simulation hypothesis is more realistic than the evil-demon hypothesis. As the British philosopher Barry Dainton has put it: "The threat posed by simulation scepticism is far more *real* than that posed by its predecessors." Descartes would doubtless have taken today's simulation hypothesis more seriously than his demon hypothesis, for just that reason. We should take it more seriously, too.

The master argument for skepticism

Philosophers love arguments. This is not to say that they love disputes with each other, although many enjoy that, too. In philosophy, an argument is a chain of reasoning that supports a conclusion. I can argue that God exists by laying out some reasons for thinking that God exists and showing how they support my conclusion.

Sometimes arguments are informal. Maybe I try to convince you that we should go to a movie by giving some reasons: We both have spare time, it's a great movie, and it's only playing tonight. I can do the same in philosophy. I can try to convince you that you can't be certain of the world around you by giving some reasons: You've had sensory illusions before, so how do you know you're not having one now? If I do a good job, maybe it will convince you of the conclusion, or at least prompt you to take it seriously.

Sometimes arguments are formal. That may sound intimidating, but formal arguments are often simple. You lay out a number of claims that are *premises*, and then you lay out a *conclusion* that follows from them. Usually the idea is that the premises are plausible enough that people will have some inclination to accept them, and the conclusion drawn from these premises is bold enough to be interesting.

Here's a formal argument for skepticism about the external world.

1. You can't know you're not in a simulation.
2. If you can't know you're not in a simulation, you can't know anything about the external world.

_____

3. So: You can't know anything about the external world.

Here, the first two claims are the premises, and the third claim is the conclusion. The conclusion follows logically from the premises: If the premises are true, the conclusion has to be true. When the conclusion follows from the premises, philosophers say the argument is *valid*. When in addition the premises are true, the argument is *sound*. Just because an argument is valid, this doesn't mean that the conclusion is true. After all, one or both of the premises could be false. But when an argument is sound, the conclusion has to be true. In the argument above, *if* you accept the two premises, you pretty much have to accept the conclusion.

Bertrand Russell once said, "The point of philosophy is to start with something so simple as not to seem worth stating, and to end with something so paradoxical that no one will believe it." The argument above at least has the potential to meet Russell's ideal. Both premises seem plausible, at least on a moment's reflection, and the conclusion seems surprising. That's one of the things that makes this argument so interesting.

In fact, this argument is so interesting that it, or something like it, is often regarded as the *master argument* for skepticism in recent philosophy. The details can change a bit. For example, we could replace simulations with evil demons or brains in vats, but the basic idea is intact.

Why believe the first premise? I've made an initial case in chapter

2. In a good-enough simulation, the world would look and feel to you exactly as today's world looks and feels to you now. And if a simulation would look and feel the same as reality, it's hard to see how we could know we're in a simulation rather than reality.

Why believe the second premise? Pick anything you think you knew about the external world. You thought you knew that Paris is in France, or that there's a spoon in front of you. But if you're in a simulation, then your beliefs about Paris and the spoon come from the simulation, not from reality. Paris and the spoon are simulated. The world outside the simulation may be entirely different. There may well be no Paris and no spoon in reality outside the simulation. So to know that Paris is in France or that there's truly a spoon in front of you, you have to rule out the possibility that you're in a simulation.

The reasoning here is a bit like this: If your phone is a knockoff, you don't really have an iPhone. So if you can't know that your phone isn't a knockoff, you can't know that you have an iPhone. In this case, we start from the plausible claim: If you're in a simulation, there's no spoon in front of you. By the same sort of reasoning as in the iPhone case, we get to: If you can't know you're not in a simulation, you can't know there's a spoon in front of you. The same applies to everything in the external world.

Our Reality Question about virtual reality was: *Is virtual reality real or an illusion?* If you answer by saying *Virtual reality is an illusion*, you'll probably accept the second premise. Here's why. Given this answer, you'll also accept *Simulations are illusions*, since simulations are a kind of virtual reality in the broad sense. In fact, you'll probably accept *If you're in a simulation, everything you experience in the external world is illusory.* So if you can't rule out the simulation hypothesis, you can't rule out that everything in the external world is illusory. It seems to follow that you can't know anything about the external world at all.

The conclusion is startling. If you're like most people, you thought you knew a lot of things. You thought you knew that Paris is in France, and you thought you knew what's physically in front of you. But it turns out you don't! The argument applies to more than just objects or cities.

It applies to memories of your childhood. If you're in a simulation, so the argument goes, your memories of going to school aren't real, so you don't really know that you went to school. The same goes for pretty much everything you thought you knew about the external world and your life in it.

Strictly speaking, the argument doesn't stop you from knowing a *few* things about the external world. Some things are true as a matter of logic or mathematics. You can know that all dogs are dogs, for example. You can know that if there is one table here and a different table there, there are two tables. But these are all trivialities. To be strictly correct, we could adjust the conclusion to "We can't know anything substantial about the external world."

If we accept the premises, the argument leads us to global skepticism about the external world—that is, the view that we don't know anything substantial about the external world. Maybe we can still know that two plus two is four, but that's not a huge consolation.

What can we do to avoid the shocking conclusion?

I think, therefore I am

Descartes himself didn't want to be a skeptic. In fact, he wanted to establish a foundation for all knowledge. So after casting all our knowledge into doubt with his skeptical arguments, he tried to build it back up, piece by piece.

Descartes needed to start with a piece of knowledge he couldn't doubt. He needed to uncover something about reality that would be true even if he was having sensory illusions, even if he was dreaming, even if he was being fooled by an evil demon. He found a candidate: his own existence.

Descartes's famous argument for his own existence, presented most explicitly in his 1637 *Discourse on Method*, went like this: *Cogito, ergo sum*. I think, therefore I am.

Philosophers have interpreted Descartes's celebrated slogan in many different ways. But at least on the surface, it looks like an argument.

Figure 9 Even if you're a brain in a vat, receiving sensations from an evil demon, you can still reason, "I think, therefore I am."

The premise of the argument (to unpack it a little) is *I am thinking*. The conclusion is *I exist*. As with most arguments, the real work is done by the premise. Once you grant that, the conclusion *I exist* seems to follow as a matter of logic.

How does Descartes know he's thinking? For a start, this knowledge does not seem to be undercut by the skeptical arguments. Even if you're in the grip of a sensory illusion, you're still thinking. Even if you're dreaming, you're still thinking. Even if you're being fooled by an evil demon, you're still thinking. Even if you're a brain in a vat, you're still thinking. Even if you're in a simulation, you're still thinking.

More deeply, Descartes reasoned that he could not doubt that he's thinking. Even if he doubted that he was thinking, his doubt was itself a sort of thinking. To doubt that one is thinking is internally inconsistent: The doubting itself shows that the doubt is wrong.

Once Descartes knew he was thinking, it was a small step to knowing his own existence. Where there is thinking, there must be a thinker. So Descartes concludes: *Sum*! I exist!

Plenty of philosophers have tried to poke holes in Descartes's *Cogito, ergo sum*. Some question the *cogito* part. How can Descartes be so sure that he's even managing to doubt? That is, how does he know that he's not a mindless automaton? Others question the step to *sum*. Is it so obvious that thinking requires a thinker? According to the 18th-century German philosopher Georg Lichtenberg, Descartes should have said, "There is thinking, therefore thought exists." That way, he could have known that thoughts exist, but he should not have been so sure about himself.

Still, a lot of people accept Descartes's *Cogito, ergo sum*. It's hard to doubt that I'm thinking. The evil-demon scenario doesn't really call my own mind into doubt, and it's not easy to generate scenarios that do. As a result, even some skeptical philosophers are prepared to say that we do know that we think, and that therefore we do know that we exist.

Speaking for myself, I don't think there's anything special about thinking per se. Descartes could have said, "I feel, therefore I am," or "I see, therefore I am," or "I worry, therefore I am." All of these are claims about his mind that he can be certain of and that aren't threatened by the evil demon. At least, he can be certain about these claims if they're understood as states of consciousness, or subjective experience. If we understand "see" as referring simply to the subjective experience of seeing, then Descartes can be certain he is seeing.

In my view, the best statement of the *cogito* is "I am conscious, therefore I am." Perhaps it's not surprising that I would say this, since thinking about consciousness is my day job. (A writer might say, "I write, therefore I am.") But it's arguable that this is what Descartes really meant. He explicitly defines thought as including everything we're conscious of, and says that it includes the senses and imagination as well as the intellect and the will.

Some theorists have tried to apply skepticism not only to the external world but also to consciousness itself, suggesting that consciousness could be an illusion. We'll revisit this view in chapter 15. It's usually regarded as extreme, but it does demonstrate that in philosophy, everything is open to question.

If we grant *Cogito, ergo sum*, that gives Descartes a foundation. The hard part is what comes next. How do we get from knowledge of ourselves and our own minds to knowledge of the external world?

Chapter 4

Can we prove there is an external world?

OVER THE YEARS, A PARADE OF PHILOSOPHERS HAVE TRIED TO solve Descartes's problem and show that we have knowledge of the external world. I'll look at some of their responses in this chapter, but first I want to tell you a joke. It's from Raymond Smullyan's book *5000 BC and Other Philosophical Fantasies.*

A philosopher once had the following dream.

First Aristotle appeared, and the philosopher said to him, "Could you give me a fifteen-minute capsule sketch of your entire philosophy?" To the philosopher's surprise, Aristotle gave him an excellent exposition in which he compressed an enormous amount of material into a mere fifteen minutes. But then the philosopher raised a certain objection which Aristotle couldn't answer. Confounded, Aristotle disappeared.

Then Plato appeared. The same thing happened again, and the philosopher's objection to Plato was the same as his objection to Aristotle. Plato also couldn't answer it and disappeared.

Then all the famous philosophers of history appeared one-by-one and our philosopher refuted every one with the same objection.

After the last philosopher vanished, our philosopher said to himself, "I know I'm asleep and dreaming all this. Yet I've found a universal refutation for all philosophical systems! Tomorrow when I wake up, I will probably have forgotten it, and the world will really miss something!" With an iron effort, the philosopher forced himself to wake up, rush over to his desk, and write down his universal refutation. Then he jumped back into bed with a sigh of relief.

The next morning when he awoke, he went over to the desk to see what he had written. It was, "That's what *you* say."

Smullyan's riposte comes to mind when thinking about the many great philosophers over the centuries who have responded to the puzzle of the external world. Everyone wants skepticism to be false. In our 2020 survey of professional philosophers, only 5 percent of respondents said they accepted or were leaning toward skepticism, making it one of the least popular positions on the survey. At the same time, it's difficult to find a convincing response to skepticism, and no response has garnered much of a consensus.

Could there be some universal philosophical refutation that overcomes all responses to skepticism? I hope not, since I will develop a strategy for responding to skepticism in this book. But there's a line of thinking with the potential to shoot down many anti-skeptical responses. This line of thinking derives from a wonderful and long-neglected story called "A Philosopher's Nightmare," written in 1967 by the British philosopher Jonathan Harrison. The story is set in 2167, when the "disciplines of Physiology, Psychology, Medicine, Cybernetics and Communication Theory were enormously more advanced than they have been before or since." It is essentially a fable about a philosopher inside a simulation, thinking about skepticism.

A neuroscientist, Dr. Smythson, devises an "endocephalic electrohallucinator"—a sort of simulator—which can produce hallucinations of all sorts of different worlds. The scientist takes a newborn's brain and places it in the electrohallucinator. He christens the baby "Alfred Ludwig Gilbert Robinson," or Ludwig for short, after the Austrian philosopher Ludwig Wittgenstein.

Dr. Smythson resolves to give Ludwig the experiences of a coherent and happy life. Ludwig receives an excellent education, with particular

Figure 10 Four stages in Ludwig's life in the electrohallucinator.

exposure to works of philosophy. He's particularly taken by the works of René Descartes, which provoke him to worry that he might be hallucinating the world with sensations produced by an evil demon.

Fortunately, Ludwig encounters the work of a succession of philosophers who set out to prove that skepticism is false. He reads the works of George Berkeley, who convinces him that appearance is reality, so that the external world that he perceives is real. Later he reads the work of G. E. Moore, who convinces him that he has hands, so that the external world exists. (We'll encounter Berkeley's and Moore's ideas later in this chapter.) Then mid-20th-century philosophers convince him that the whole idea of a global illusion is meaningless. Dr. Smythson is too kind to inform him of his real situation. Ludwig's life goes on happily until his brain is finally transplanted into a real body and he loses interest in philosophy.

Harrison doesn't draw any explicit conclusions in his article, which may be one reason why philosophers have cited his story only a handful of times. He seems to be poking fun at the various anti-skeptical views, reducing them to absurdity. The reader is led to think: How good could these arguments for knowledge of the external world be, if someone inside a hallucination machine would make them, too?

Harrison's story suggests a strategy for responding to many anti-skeptical views. I call this strategy the *Simulation Riposte*. Faced with an argument that we have knowledge of the external world, the Riposte counters: *That's what someone in a simulation would say.*

I love Harrison's riposte. It's not a universal refutation of responses to skepticism, but it causes trouble for a number of them. It's especially troublesome for responses that aim to prove conclusively that we're not in a simulation.

Recall that the master argument for skepticism went like this. First, you don't know you're not in a simulation. (This amounts to saying no to the Knowledge Question.) Second, if you don't know you're not in a simulation, you don't know anything substantial about the external world. Conclusion: You don't know anything substantial about the external world. If you accept both premises, you have to accept the conclusion and accept global skepticism about the external world.

Historically, by far the most common response has been to reject the first premise, by saying yes to the Knowledge Question: We can know we're not in a simulation (or that we're not brains in vats or being fooled by evil demons). In this chapter I'll look at some of these replies and argue that they fail. This is another plank in our case for a negative answer to the Knowledge Question: We cannot know we're not in a simulation.

Can God solve the problem?

To get from knowledge of your own mind to knowledge of the external world, there's a big bridge to cross. Descartes thought he had a way to cross the bridge. The secret was to go through God.

Descartes argued that he had an idea of God as a perfect being. His idea of God is an idea of a being who is perfectly good, perfectly wise, and so on. In fact, he argued that the idea itself is a perfect idea, and therefore it could not have come from anywhere except from a perfect being. That is, the idea of God must have come from God. If this argument works, it gets us from knowledge of our own mind to knowledge of something outside us and much greater than us. It gets us to God.

Once we've gotten to God, the argument goes, getting to the world around us is easy. Since God is a perfect being, he would not allow us to be deceived. So, given that God exists, there can be no evil demon and no dreams or sensory illusions that persist for one's whole life. God will ensure that our impressions of the external world are, by and large, accurate. For him to do anything else would be imperfect. So the external world exists and is much as we thought it was. Hallelujah!

Most philosophers have been much less impressed by this argument than by Descartes's *Cogito, ergo sum*. You've probably already thought of a few holes in it already.

One obvious problem: Why couldn't the idea of a perfect being come from somewhere other than a perfect being? I have the idea of a perfect circle, and I didn't need perfection to create it. In fact, why couldn't an evil demon give someone the idea of a perfect being?

Another problem: Even if there is a perfect being, how can we be sure we're not being deceived? Maybe deceiving us is part of the perfect being's master plan! For example, maybe we all need to go through a period of deception before we're finally enlightened about reality. We're imperfect after all, so we don't know much about what perfect beings are like.

We can also run the Simulation Riposte. Before long, we'll be able to create simulations. Those simulations may be full of creatures who believe in a perfect being. Perhaps one of them will be a simulated version of Descartes: Let's call him Sim Descartes. Sim Descartes will say, "Our creator is perfect and would never deceive us." But in fact we are his creators. And we are imperfect. It turns out that you don't need to be perfect to create the idea of perfection. In fact, it was precisely because we're imperfect that we had to create Sim Descartes in a simulation; creating him in a nonsimulated world was too hard for us.

Admittedly, for all we know, there really is a perfect God out there who created us and indirectly created the simulation. Descartes could argue that our own idea of perfection came from God, so Sim Descartes got the idea indirectly from God, and his argument for a perfect being still stands. But this doesn't help his answer to skepticism. Assuming that creatures in a simulation are deceived, then Sim Descartes is deceived. If a perfect being is present, then the existence of a perfect being doesn't rule out global deception.

Descartes might object that we haven't actually created such a simulation yet! Perhaps when we try to create a simulation that deceives Sim Descartes, we'll never succeed, because the perfect being won't let us. But simulation technology certainly seems to be on track to embedding people in simulated worlds. We already have good reason to think it's possible.

If we actually create such simulations, Descartes's argument (construed as an argument that we're not in an evil-demon scenario or a simulation scenario) will be decisively undermined. Once we see that some beings are in simulations, we'll know it's possible for beings to be in simulations, and any argument that this is impossible will be refuted. Even before then, the current state of simulation technology throws Descartes's argument into question. Once again, technology helps us to see these old arguments in a new light.

Is appearance the same as reality?

Perhaps the most venerable reply to skepticism is the assertion that *appearance is reality*. In *The Matrix*, the rebel leader Morpheus (played by Laurence Fishburne) airs this view:

"What is real? How do you define real? If you're talking about what you can feel, what you can smell, what you can taste and see, then 'real' is simply electrical signals interpreted by your brain."

Figure 11 Idealism in Buddhist philosophy. Vasubandhu contemplates the Buddha saying, "Everything is mind."

The idea is that reality is all in the mind. If something looks like reality and feels like reality (and sounds and smells and tastes like reality), then it is reality. If something appears real, and there's no appearance to the contrary, it is real.

In philosophy, Morpheus's *appearance is reality* thesis is a central form of *idealism*: the thesis that reality is made of minds. Idealism in philosophy isn't so much about *ideals* as about *ideas*. Idealism often says that reality is made of ideas: sensations, thoughts, feelings and other components of the mind. In Indian philosophy, idealism has been a common view in both Buddhist and Hindu traditions. In his *Twenty Verses*, an in-depth defense of idealism, the 4th-century-CE Yogācāra philosopher Vasubandhu starts by attributing idealism to the Buddha: *Everything in the three realms is nothing but mind.* Or as Vasubandhu puts it: *Reality is consciousness only.* On Vasubandhu's view, when I see a tree, all there is in reality is the idea of a tree, or the appearance of a tree, or consciousness of a tree. There is no tree outside the mind.

In Western philosophy, idealism is most closely associated with the 18th-century Anglo-Irish philosopher George Berkeley. In his 1710 book *A Treatise Concerning the Principles of Human Knowledge*, Berkeley put forward his famous slogan, *esse is percipi.* To be is to be perceived. A spoon exists if it is perceived. This means roughly that if a spoon appears to you, then the spoon is real. To put it another way, appearance is reality.

The thesis bridges the gap between the mind and the world by saying that there is no gap. The world was in our minds all along. Once we know how things appear in our minds, we know how things are in the world.

Berkeley and Vasubandhu argue that reality is made of minds. At the bottom level, there is perception, thought, and feeling. These element serve as building blocks for the world as a whole. A table is built up from numerous appearances of the table, from many angles and in different circumstances. We would normally think that the table came first and the appearances come later. But if these idealists are right, the appearances of the table are fundamental, and the table derives from the appearances.

This sort of idealism allows for some local gaps between appearance and reality. You can still hallucinate a pink elephant, so that there appears to be a pink elephant but there's really no pink elephant present. But illusions and hallucinations can happen only when there's also potential evidence that there's no elephant there. If you reach out, you won't touch an elephant, and if you look the next morning, there will be no trace of an elephant. As long as the overall weight of your appearances says there's no elephant, then in reality there's no elephant.

If appearance is reality in this way global skepticism is ruled out. Global skepticism requires taking seriously that we're in a global illusion in which nothing is as it appears. But *appearance is reality* rules out global illusions. If there appears to be a table and there's no appearance to the contrary, then there's really a table. If we have the appearances, we have the reality.

Furthermore, if our appearances determine reality, this rules out the perfect simulation hypothesis, where there's never any clue that you're in a simulation. In such a scenario, there's a simulation in reality but not the slightest hint of a simulation in appearance. If appearance is reality, that can't happen.

There are a lot of objections to idealism. One objection is, "Whose minds is reality made of?" If it's my mind alone, then we have solipsism: I'm the only one who truly exists—or, at least, it's my mind that makes up the universe. That way megalomania looms. If reality is made up of all our minds together, though, then there will be a gap between my mind and reality as a whole. Maybe where I see a unicorn, everyone else sees an elephant, meaning there's an elephant in reality. That threatens to bring us back to skepticism. How can I know that everyone else's perceptions match mine?

Another serious objection: What about unobserved bits of reality? For example, what happens to my desk when I leave the room? And what about parts of the universe where there are no observers, and times long ago, before consciousness evolved? Is there any reality there? This objection has been summarized memorably in a limerick by the English theologian Ronald Knox:

There once was a man who said "God
Must think it exceedingly odd
If he finds that this tree
Continues to be
When there's no one about in the quad."

A reply, from Berkeley's viewpoint, is expressed in another limerick:

Dear sir, your astonishment's odd.
I am always about in the quad.
And that's why the tree
Will continue to be,
Since observed by, Yours faithfully, God.

As with Descartes, God comes to the rescue. As long as God is always watching the whole world, unobserved reality is not a problem. God's experiences sustain the ongoing reality. Our own experiences derive from God's experiences. The constancy of God's experiences explains why we always see the tree when we return to the quad.

This is okay so far as it goes. But now God has inherited the role that was played by the external world. Instead of having a physical tree out there, sustaining my experience of the tree and everyone else's, we have God's idea of the tree, sustaining my experience of the tree and everyone else's. That raises the same question that befell Descartes: How do we know that God exists? And if we can't know, how can we be sure that the tree exists when we're not around?

Another question: Why do we need God here? Couldn't an evil demon or a simulation play the role? In fact, isn't the God scenario just a mild variant on the evil-demon scenario? Perhaps the evil-demon scenario is ruled out because there's no appearance of an evil demon. But then, why isn't the God scenario ruled out because there's no appearance of God?

The underlying problem for idealism is that in order to explain the regularities in our appearances (the fact that we see an identical tree day after day, say), we need to postulate some reality that lies beyond these appearances and sustains them. Berkeley appeals to God's mind as this further reality. But now we have created a gap between our own perception and reality, so the skeptical problem rearises. How can we know about the reality (whether God or an external world) behind the appearances?

Here's where we can run a version of the Simulation Riposte. We create a rich simulated world, with a simulation of George Berkeley inside it. Sim Berkeley tells us, "Appearance is reality." Since there is no appearance of a simulation, there is, in actuality, no simulation. Sim Berkeley concludes, "I am not in a simulation. My experiences are all produced by ideas in God's mind."

From our perspective, Sim Berkeley looks a bit ridiculous. He says he's not in a simulation, but he is wrong: He *is* in a simulation. He says that appearance is reality, but there is a vast realm of reality beyond what appears to him. He says his experiences are produced by God, but in fact his experiences are produced by us, with the help of a computer. His world is sustained by the computer, not by ideas in God's mind.

Now, Berkeley might have some comebacks here. He could say that everything in our world is sustained by God's mind, including the computer. But once we see that a computer can do the job, why do we need God? Berkeley could also say that even if reality goes beyond Sim Berkeley's appearances, Sim Berkeley's reality—the world of tables and chairs he perceives—is constituted by those appearances. What's outside the simulation is outside Sim Berkeley's world.

Still, it's hard to deny that if Sim Berkeley says he's not in a simulation, he is wrong. This should make us suspicious of the principle that appearance is reality.

Later, I'll argue that some forms of idealism should be taken seriously at least as a speculative hypothesis. We can't rule out that consciousness underlies the universe. However, I think that any version of idealism that rests on equating our appearances with reality is doomed. As a result, we need another path to solve the skeptical problem.

Is the simulation hypothesis meaningless?

The philosophers of the Vienna Circle in the 1920s and 1930s, sometimes known as the *logical positivists* or *logical empiricists*, wanted to make philosophy scientific. In the 1920s, the Circle met regularly in Viennese cafés and classrooms. The best-known members of the Circle included the philosophers Otto Neurath and Moritz Schlick and the mathematician Kurt Gödel. The Viennese philosophers Karl Popper and Ludwig Wittgenstein interacted with many members of the Circle, although they did not attend meetings. The philosopher Rose Rand kept scrupulous records of their meetings, including votes on which propositions to accept and reject.

The leading figure in the Circle, the great Rudolf Carnap, held that many philosophical problems are meaningless "pseudo-problems." For a hypothesis to be meaningful, he said, it must be testable: you have to

Figure 12 Inside a simulation, Rudolf Carnap tells the simulated Vienna Circle (Schlick, Neurath, Rand, Gödel, Olga Hahn-Neurath, and Hans Hahn, with Popper and Wittgenstein walking by) that the simulation hypothesis is meaningless.

be able to get evidence for or against it. But we can never get evidence for or against Cartesian skeptical hypotheses, such as the evil-demon hypothesis. As a result, the positivists held that these skeptical hypotheses are meaningless. This view was shared by some of their interlocutors. In his *Tractatus Logico-Philosophicus* (1921), Wittgenstein said, "Scepticism is not irrefutable, but obviously nonsensical."

Is the simulation hypothesis meaningless by these lights? We've seen that we could in principle get evidence for the simulation hypothesis. For example, the simulators could tell us we're in a simulation, show us the program, and show how it controls the world around us. Some people think there could even be evidence in physics suggesting that we're in a simulation. But as we saw in chapter 2, this sort of evidence involves imperfect simulations. In a perfect simulation, our experience will always be as it would be in an unsimulated world. So it's hard to see how we could get evidence for or against the perfect simulation hypothesis. If we cannot, Carnap and the other Vienna Circle philosophers would say it's meaningless.

I think the Vienna Circle philosophers are wrong here. If we can't get evidence for or against the perfect simulation hypothesis, this means at most that it isn't a scientific hypothesis—one that we can test using the methods of science. But as a philosophical hypothesis about the nature of our world, it makes perfect sense.

Once again, we can use the Simulation Riposte. Imagine that we create a perfect simulation ourselves. Inside the simulation, some sims might have an argument. Sim Bostrom says, "We're in a simulation." Sim Descartes says, "No we're not! This is a nonsimulated reality." Sim Carnap says, "This debate is meaningless! Neither of you are even wrong!"

A proponent of the meaninglessness hypothesis sides with Sim Carnap, saying that the debate between Sim Bostrom and Sim Descartes is incoherent. Neither of the two is right. But that seems the wrong verdict; in fact, Sim Bostrom is right, and Sim Descartes is wrong. They are both in a simulation. Sim Bostrom will never get evidence that proves he's right, but he's right all the same.

If you doubt this, then suppose we leave an imperfection in the simulation: a small "red pill" that's difficult to find, but that when found gives definitive evidence of the simulation. Now, suppose Sim Bostrom and Sim Descartes one day find the red pill and get the evidence. Someone shows them the computer running the simulation and controlling their whole lives. Both of them will presumably agree that Sim Bostrom was right and Sim Descartes was wrong. And they will be right about this. So, at least in this case, the debate about whether they're in a simulation is not meaningless.

Now let's change the story a bit. Suppose that Sim Bostrom and Sim Descartes never find the red pill, although they could have. Perhaps they start looking for evidence but die before finding the pill. Then we can say that if they'd found the red pill, they'd have discovered that Sim Bostrom was right and Sim Descartes was wrong. In this case, I think it's clear that even though they didn't discover the pill, Sim Bostrom is right, and Sim Descartes is wrong. So, again, the sims' debate isn't meaningless.

Let's change the story again. The creator of the simulation notices the imperfection in the simulation. She patches the bug, so the red pills disappear. Now it's a perfect simulation. The two simulated philosophers lead the same lives they did in the previous case, with Sim Bostrom insisting they're in a simulation and Sim Descartes insisting the opposite. Of course they never find a red pill, and they die without proof. I think it's still pretty clear: Sim Bostrom was right all along, and Sim Descartes was wrong. Their lives are exactly the same as in the previous case. The mere existence of a red pill somewhere in the bowels of the simulation doesn't make a difference as to who is right. Both Sim Bostrom and Sim Descartes are making perfectly meaningful claims about their world, even though neither of them can prove those claims.

The Vienna Circle view was founded on *verificationism*, which says that a hypothesis is meaningful only if it can be verified as true or false by sensory evidence. Verificationism is now widely rejected because it seems that there are many meaningful hypotheses that can't be verified by sensory evidence. People tied the verificationists up in knots by asking: Can verificationism itself be verified by sensory evidence? If not, is it meaningless? The answer seemed reasonably clear. Verificationism can't be verified—which means that by the verificationist's own lights, verificationism is meaningless. That was enough to undercut the view. Most philosophers came to the reasonable conclusion that verificationism is unverifiable but meaningful all the same.

The same goes for the simulation hypothesis. I just gave you an argument that for Sim Bostrom and Sim Descartes, the simulation hypothesis is meaningful although it can't be verified. The same goes for us. Whether or not we can prove or disprove the simulation hypothesis, it's perfectly meaningful. We are either in a simulation or we are not.

Is the simulation hypothesis contradictory?

Another anti-skeptical view holds that although the simulation hypothesis is meaningful, it is *contradictory*. It couldn't possibly be true. Consider the statement, "Seven times three is a prime number." Every word here is meaningful, but the statement is a contradiction, since by definition a prime number cannot be factorized like this. So we can know that the statement is false. Similarly, if the simulation hypothesis is contradictory, we can know it's false.

We've already touched on one route to the conclusion that the hypothesis is contradictory, suggested by Berkeley's idealism. Idealism says that appearance is reality. A strong version of idealism says that when we say, "We're in a simulation," all this means is "It appears that we're in a simulation," or something along those lines. Now, the perfect simulation hypothesis can be understood as saying, "We are in a simulation, but it does not appear that we are in a simulation." If the strong version of idealism is true, this is equivalent to "We are in a simulation and we are not in a simulation," which is a contradiction. So, given this version of idealism, we can know that the simulation hypothesis is false.

We can argue against this view in the same way we argued against idealism before. We could also consider Sim Berkeley, who is really in a simulation. Sim Berkeley claims, "It is contradictory to suppose I am in a simulation." At this point, it seems clear that something has gone wrong.

Hilary Putnam has advanced a more subtle version of the claim that skeptical hypotheses are contradictory. As we saw in the last chapter, Putnam updated Descartes's evil-demon scenario to the more contemporary *brain-in-a-vat* hypothesis: the hypothesis that we are brains in vats, being fed sensory input by a superscientist. In his 1981 book *Reason, Truth and History*, Putnam argued that the brain-in-a-vat hypothesis is contradictory.

Putnam's argument is based on an analysis of what words like "brain" mean for a brain in a vat. The argument depends on Putnam's theory of meaning, which says that the meaning of a word depends on what it's connected to in the external environment. In essence, Putnam argues that a brain in a vat, when using the word "brain," would not be talking about real biological brains because it has never been exposed to real biological brains in its environment. It has been exposed only to digital brains. As a result, any brain in a vat who thinks, "I am a brain in a vat" is wrong. Its thought means something like "I am a digital brain in a vat," but in fact it is a biological brain, not a digital brain. This situation suggests that the hypothesis "I am a brain in a vat" cannot possibly be true.

I will discuss Putnam's argument and his theory of meaning properly in chapter 20. For now, I will note that the argument does not work as well for the hypothesis "I am in a computer simulation" as for "I am a brain in a vat." When Sim Putnam thinks, "I am in a computer simulation," he thinks something that is *true*. The words "computer simulation" are not anchored to specific systems in our environment, as the word "brain" is. Sim Putnam is talking about regular computer simulations, the same thing we're talking about. And he really *is* in a computer simulation. So there's nothing contradictory about thinking, "I am in a computer simulation."

I conclude that the simulation hypothesis is not contradictory. There remains a chance that it is true.

Does simplicity rule out the simulation hypothesis?

So far, we have looked at responses to skepticism that argue that we can be certain that we're not in a simulation and certain that the external world exists. Another kind of response says that knowledge doesn't require certainty. Cartesian arguments establish that we can't be certain we're not in a simulation, but we may well know we're not in a simulation all the same.

By analogy: As I write this, I know that Joseph Biden is the president of the United States, even though I can't be certain that he didn't die five minutes ago. My knowledge is fallible, but it's still knowledge. Once we recognize that our knowledge of the external world doesn't need to be certain, Descartes's argument seems less strong.

An important reply along these lines is Bertrand Russell's appeal to *simplicity*. The renowned British philosopher argued that the commonsense hypothesis that objects in the external world are real is the simplest explanation of our observations. By comparison, the dream hypothesis is extremely complicated. Presumably he would say the same about the simulation hypothesis. In general, we should accept the simplest explanation of our observations and reject overly complicated explanations. So we should accept the real-world hypothesis and reject the simulation hypothesis.

This sort of appeal to simplicity is ubiquitous in science. It's often memorialized as Ockham's (or Occam's) razor, after the 14th-century English philosopher William of Ockham. Ockham's razor says: Do not multiply entities without necessity! This says that, other things being equal, we should favor the most *parsimonious* theory—the theory that postulates the fewest things. You should accept a complex theory only when there's no simpler one consistent with the data.

For example, the ancient mathematician Ptolemy proposed a theory in which the Sun goes around Earth, while the Renaissance astronomer Johannes Kepler proposed a theory in which Earth goes around the Sun. Ptolemy's theory postulated many epicycles to give the right results, whereas Kepler's theory worked without postulating epicycles. Ockham's razor enjoins us to accept Kepler's theory over Ptolemy's.

If we focus on hypotheses about the external world, the real-world hypothesis certainly seems simpler than the simulation hypothesis. After all, the simulation hypothesis postulates both a nonsimulated world and a simulated world, whereas the real-world hypothesis has one world. Why postulate two worlds when it's possible to get by with one?

But simplicity is just one factor among many. Often, simple theories turn out to be false, and more complicated theories turn out to be true. Simplicity can be overridden by other factors. One way it can be overridden is when we know there's complexity in the environment.

For example, suppose we find the letter A scratched on a rock on Mars. There are two hypotheses: It was formed by random movements of other rocks, or it was put there by an intelligent being. The first seems simpler, since we otherwise have little reason to postulate intelligent beings on Mars, so we may favor it. On the other hand, if we find a letter A scratched on a rock on Earth, we should favor the intelligent-being hypothesis, even though (because it involves human behavior) it is more complex. We know there are many intelligent beings on Earth, so we have reasons to believe in the relevant complexity. Here our knowledge of what is possible overrides simplicity.

The same goes for the simulation hypothesis. If we otherwise have no reason to believe there are simulations, then the simplicity of the real-world hypothesis gives us good reason to favor it. On the other hand, if we believe there are many perfect simulations of whole universes in our world, as Bostrom's simulation argument tends to sug-

Figure 13 Bertrand Russell and Nick Bostrom on the simulation hypothesis.

gest, then this simplicity reasoning would be overridden. We may not have seen any perfect simulations, but we have good reason to believe that they're possible and may well be developed at some future point in human history. As things stand, an appeal to simplicity gives us little reason to reject the hypothesis.

The Simulation Riposte complements this analysis. Sim Russell tells us that the simulation hypothesis is far too complex and should be rejected. Obviously, he's in a simulation all the same. We might say that he just got unlucky: After all, he said the simulation hypothesis is unlikely but he didn't say it was impossible. But once Sim Russell has reason to believe that simulations are widespread, he no longer has reason to find the hypothesis unlikely.

Is it obvious that we're not in a simulation?

Russell's colleague G. E. Moore offered another famous reply to the external-world skeptics. Moore said "Here is one hand. Here is another. Therefore the external world exists." Moore called this a proof of the external world, arguing that the premises are obviously true and are far more plausible than any piece of philosophy. Given that there are hands, there must be an external world.

Moore greatly respected ordinary common sense, and common sense drives his argument. For Moore, it's obvious common sense that he has hands. And common sense can be taken as a premise in making a philosophical argument. It follows from that premise that the external world exists. Moore doesn't say anything explicitly about the evil-demon or brain-in-a-vat hypotheses, but one suspects he would have taken it as commonsensical that we're not in those situations either.

Few have been convinced by Moore's proof of the external world. Many hold that when the existence of the external world is in question, Moore is not entitled to simply assume he has hands. In this context, the assumption "I have hands" begs the question; Moore's claim that he has hands presupposes the conclusion of the argument, which is that the external world exists. When the premise of an argument presupposes its conclusion, it's a circular argument; you need to assume the conclusion to get to the conclusion.

The Simulation Riposte confronts us with the entertaining sight (prefigured explicitly in Jonathan Harrison's short story) of Sim Moore holding up his simulated hands and saying "I have hands! Therefore the external world exists!" It seems clear that something has gone wrong for Sim Moore. He thinks it's common sense that he has hands. But once it's a serious possibility that he's in a simulation, he shouldn't rely on common sense. All bets are off.

Once simulations become a serious possibility, Moore's argument loses most of its remaining force. At that point, our commonsense views about the external world are thrown into question. So we can't use those views to extract us from doubts.

Some other replies to external-world skeptics (some of which I discuss in the notes) try to show how we can know we're not in a simulation even though we can't prove it. These replies are not refuted quite as easily by the Simulation Riposte. But once simulations are seen as a serious possibility, even these replies become difficult to maintain.

In the next chapter, I'll make the case that the simulation hypothesis really is a serious possibility, and that as a result, we can't know we're not in a simulation.

Chapter 5

Is it likely that we're in a simulation?

SIMCITY, A GAME IN WHICH THE USER CONTROLS A SIMU-lated city, was first introduced in 1989. Not long afterward came *SimEarth*, which simulates the development of life on Earth. In 2000, *The Sims* was released, complete with simple simulated human beings in simulated homes. It's all but inevitable that eventually we'll develop *SimUniverse*, in which an entire universe is simulated in detail.

From the inside, *SimUniverse* will be indistinguishable from the universe it's a simulation of. Suppose that we're simulating a possible universe containing ten billion people; in that case, *SimUniverse* will contain a simulated person—a pure sim—for each of them.

After a while, every teenager may be running *SimUniverse* on a mobile device. Even if the use of the technology is restricted, we can imagine researchers running many simulated universes for scientific, historical, financial, and military purposes. Within a century or two, this could easily lead to millions or billions of different copies of *Sim Universe* running here and there. As a result, there will be vastly more sims than nonsims. Across history, sims may outnumber nonsims by at least a million to one.

The same goes for intelligent beings anywhere in the cosmos. If any aliens have human-level intelligence, they should eventually develop computers and program them. If these alien civilizations survive long enough, they'll likely create simulated universes.

Let's run some numbers, using small numbers to keep things simple.

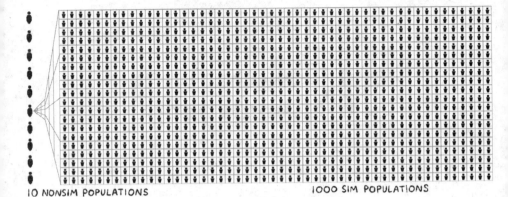

IO NONSIM POPULATIONS   IOOO SIM POPULATIONS

Figure 14 If one in ten nonsim populations creates at
least a thousand sim populations each, then sims will
outnumber nonsims by at least one hundred to one.

Any nonsim population that survives long enough will eventually be
able to create (let's say) at least a thousand sim populations, each with
about as many sims as the original nonsim population. It's arguable that
at least one in ten human-level nonsim populations will eventually do
this. If one in ten nonsim populations creates at least a thousand sim
populations each (as depicted in figure 14), this will result in at least a
hundred sim populations per nonsim population.

If this is right, then throughout the cosmos, sims will outnumber
nonsims by at least a hundred to one. These sims (and all the sims I discuss in this chapter) will be pure sims: digital beings generated inside
the simulation. Under reasonable assumptions, these sims will have
conscious experiences that are the same as those of the nonsims they
simulate. For most of these sims, there may be no evidence to tell them
that they're simulated.

We can then ask: What are the odds that we are among the relatively
few nonsimulated beings? Since sims outnumber nonsims by at least a
hundred to one, the natural answer is "less than 1 percent." It is much
more likely that we're sims than that we're nonsims.

Conclusion: We are probably in a simulation.

The simulation argument

The argument I've just sketched is a version of what is often called the *simulation argument*. The basic idea is illustrated in figure 13 in chapter 4. The first argument along these lines that I know of was put forward by the roboticist and futurist Hans Moravec in his 1992 essay "Pigs in Cyberspace." Moravec gave a pithy summary in an interview with *Wired* magazine in 1995:

> In fact, the robots will re-create us any number of times, whereas the original version of our world exists, at most, only once. Therefore, statistically speaking, it's much more likely we're living in a vast simulation than in the original version.

The definitive version of the simulation argument was put forward by Nick Bostrom in his 2003 article "Are You Living in a Computer Simulation?" Bostrom gives a mathematical argument for a complicated conclusion with a choice between three options, focusing especially on a version of the simulation hypothesis that involves simulating one's ancestors. I'll consider Bostrom's argument later. For now, I'm following Moravec in giving a direct argument that we're probably in a simulation.

In a 2016 interview, the entrepreneur Elon Musk formulates the Moravec-style argument as follows:

> Given that we're clearly on a trajectory to have games that are indistinguishable from reality, and those games could be played on any set-top box or on a PC or whatever, and there would probably be billions of such computers or set-top boxes, it would seem to follow that the odds that we're in base reality is one in billions.

The reasoning is straightforward. Simulation technology is likely to be so ubiquitous that most beings in the universe (or most beings with experiences like ours) are sims. If so, then we are probably sims.

This reasoning is not bad, but it's not irresistible. Where might it go wrong? You've probably already come up with a few objections.

One sort of objection says: *It will never happen!* You might deny that there will ever be many sim populations with human-level intelligence. This might be because simulation is impossible, or at least too difficult. It might be because no one will choose to make simulations. It might be because all human-level populations will die before they can build the simulations. If so, then there won't be many (or any) simulations, and we're much less likely to be in a simulation.

Another sort of objection says: *We're special.* You might say that even if there are lots of sim populations, we have special features that make us unlikely to be sims. For example, we are conscious, and we have very specific sorts of minds, and some might deny that sims could be conscious or have minds like ours. Alternatively, you might hold that we live in a distinctive sort of world, and that most simulated worlds won't be anything like ours. If so, then even if there are many simulations, we are much less likely to be in one.

I'll discuss all these objections to the simulation argument in what follows (along with many other objections in the online notes). To put my cards on the table: I think that while some of these objections are reasonable, their strength is limited. You can't be confident that widespread simulations will never happen, or that we are so special that we're unlikely to be sims. As a result, the hypothesis that we're in a simulation is one that we can't rule out and that we should take seriously.

Laying out the argument

To get clearer on the reasoning in the simulation argument, we can lay it out as an argument with premises and a conclusion.

Let's say that intelligent beings (or just "beings," for short) are creatures with at least human-level intelligence. We'll understand intelligence in terms of what beings can do (see chapter 15 for more on this), with the ability to program computers being especially relevant. If cats can't program computers, they won't design computer simulations. Our focus is on beings who can. We won't require beings to have humanlike conscious experiences for now, though we'll impose a requirement like this later.

As before, a sim is an intelligent being in a simulation. A nonsim is an intelligent being who is not in a simulation. Populations are groups of beings, so that every being belongs to exactly one population. It is natural to group populations according to species or social cooperation, but it doesn't matter too much how the grouping works. I'll simplify by assuming that all populations have the same size, but that assumption can easily be dropped.

We can then lay out the argument as follows. This version of the argument is far from perfect (I'll give my preferred version toward the end of the chapter), but it's a good starting point that helps to bring out the underlying issues.

1. At least one in ten nonsim populations will each create a thousand sim populations.
2. If at least one in ten nonsim populations will each create a thousand sim populations, then at least 99 percent of intelligent beings are sims.
3. If at least 99 percent of intelligent beings are sims, we are probably sims.

    _____

4. So: We are probably sims.

I have put in some numbers to make the argument more concrete. As before, I've kept the numbers small to simplify things. In a more ambitious mood, we could change the first premise to say that one in a thousand nonsim populations will create a billion sim populations each, resulting in million-to-one odds in favor of us being simulated.

With these clarifications in mind, let's examine the premises. The argument seems valid, in that the conclusion follows from the premises. If the premises are true, the conclusion must be true.

Premise 2 is the most straightforward. Given the numbers we've specified in a finite universe, and as long as the terms are defined so that every being is either a sim or a nonsim, the premise is guaranteed to be true. I'll save the math and other complications (such as infinite universes) for an endnote. The real action is with premises 1 and 3.

Will there be many sims?

Premise 1 says that at least one in ten nonsim populations will each create a thousand sim populations. Note that this premise concerns what nonsim populations in general will do—not what we will do.

The objections to premise 1 are the *It will never happen* objections: Simulation is impossible, simulation is too difficult, nonsim populations will all die before creating sims, nonsims will choose not to make simulations. We can call these objections *sim blockers*, since they will tend to block (or prevent) the existence of sims.

Intelligent sims are impossible. One objection is that processes that produce intelligent behavior are uncomputable: that is, they can never be successfully simulated on a computer. This could be because the nonphysical mind affects behavior in uncomputable ways. It could also be because there are physical processes in the brain that can't be simulated. For example, the mathematical physicist Roger Penrose has speculated that a quantum gravity theory (that is, a theory unifying quantum mechanics and general relativity) may involve processes with a nonalgorithmic element that is crucial to human behavior. If so, it could turn out that intelligent sims are impossible—and therefore that sims are impossible, as we've said that sims are intelligent by definition.

The existence of these processes would be surprising as there is currently little evidence of uncomputable processes in nature. Even if there are processes in nature that no classical computer can simulate, it's arguable that they could be harnessed to build a new sort of more powerful computer. We already know that quantum mechanics can be harnessed to build quantum computers. If Penrose is right that quantum gravity involves processes that are not classically computable, we should be able to harness these processes to build more powerful quantum gravity computers that no classical computer can simulate. Then these quantum gravity computers could simulate our brain processes, and we'd end up with a new version of the simulation argument involving these ultrapowerful computers.

Sims take too much computer power. Here the idea is that simulating a whole population of human-level intelligences would require an infeasible amount of computational power. This is far from obvious. Brains are large but finite. They contain around 100 billion neurons with around 1,000 connections (or synapses) each. On current estimates, the brain performs the equivalent of around 10 quadrillion (10^{16}) floating point operations per second (or flops): a computing speed also known as 10 petaflops. That's a lot, but it's about on a par with the best existing supercomputers.

If this is right, then once we know enough about the brain, one second of supercomputer time should be able to simulate one second of brain processing. If technology advances at anything like its usual speed, we can expect computers to speed up by a factor of ten every decade, or a factor of 10 billion (10^{10}) over a century. That suggests that within a century (at a speed of 10^{26} flops), a second of computer time will be able to simulate 10 billion brains for a second each. Within another century (at a speed of 10^{36} flops), a second of computer time will be able to simulate 10 billion brains for a lifetime of 100 years (or 3 billion seconds) each. A full-scale simulation will require the environment to be simulated, too, but it's hard to see why this should add more than another two or three orders of magnitude to the workload. All this would require a computing speed of around 10^{39} flops. Even if progress slows, this should in principle within the range of future computers.

As it stands, the universe has enormous unused capacity for computing. There is a vast amount of matter in a vast amount of space. Matter also has a vast microstructure that can be used for computing. As Richard Feynman titled a 1959 lecture exploring nanotechnology: *There's plenty of room at the bottom.* The physicist Seth Lloyd has estimated that in principle, a one-kilogram system could perform up to

10^{50} operations per second. This method involves a black hole with a very short lifespan, so it has many limitations, but there are other proposed methods that reach over 10^{40} operations per second. With even a small fraction of these resources, simulating huge populations quickly will eventually be fairly trivial.

If the universe is finite, there will be limits. At some point, most available matter could be turned into *computronium*, a hypothetical state in which matter is used as efficiently as possible for computing. Beyond this point, large new simulations may become prohibitively difficult to construct. However, that point is much further in the future than the point at which simulations become easy to create. By that far-future point, we can reasonably expect that simulations will have greatly outnumbered nonsimulations. So these long-term limits on computer power don't pose much of an obstacle for the argument.

You might worry that if we *are* in a simulation, then all this evidence about the computer power in physics may be misleading. Maybe our simulators have put us in a low-cost simulation whose capacity won't extend nearly this far. If so, when we try to make giant simulations, we may fail. The simulators who created us obviously have the power to create at least one simulation, but perhaps not the millions stipulated in premise 1. Still, this objection gets off the ground only if we're already in a simulation—and if we are in a simulation, then we have a much quicker route to the conclusion that we're probably in a simulation!

Nonsims will die out before creating sims. This one sounds pessimistic, but we know it's a serious possibility. Nuclear weapons technologies already have the capacity to destroy much or all of the human life on this planet alone. Many people think that nanotechnology could soon be able to do the same, through a chain reaction at the microscopic level that reduces everything to "gray goo." Sometime in this century, artificial-intelligence technology may become powerful enough to destroy all intelligent life on Earth, should it so choose. These possibilities are sometimes called "existential risks," risks that threaten the very existence of human beings—and of intelligent life in general.

Some existential risks may be hard to avoid. All we need is a destructive technology that meets the following conditions. First, it's inevitably discovered by human-level populations. Second, it's easy enough and tempting enough to use so that discovery inevitably leads to use. Third, it's so destructive that when it's used, everyone dies. It's not hard to imagine that eventually ultradestructive nuclear technology could be accessible to almost anybody, resulting in inevitable doom. The same might go for nanotechnology, or artificial-intelligence technology, or some as-yet-unheard-of technology. If there's hope, it lies in the second condition. Perhaps there is some way to prevent an all-destructive technology from ever being used. But the hypothesis that intelligent civilizations inevitably destroy themselves should be taken seriously.

This hypothesis might explain some of our observations. For example, it would explain why we've never seen signs of extraterrestrial intelligence: Intelligent populations destroy themselves around the same time that they gain the ability to send signals. It might also explain why we seem to be living early in the history of humanity. This issue is at the core of the so-called Doomsday argument, developed by the astrophysicist Brandon Carter and the philosopher John Leslie. Their key claim is that probabilistically any one of us should expect to be somewhere in the middle of all the humans who have ever lived or will live. Given the rapid increase of human population, this means we should expect human life to end within centuries (in which case we'd be comfortably in the middle of all humans) rather than to go on for millions of years (in which case we'd be extremely early).

That said, the hypothesis that almost all human-level populations die before they're able to create many simulations would be surprising if true. One could reasonably hope that at least one in ten human-level populations will be collectively rational enough not to destroy itself; or if that's too optimistic for you, we can run a version of the argument requiring only one in a thousand.

Nonsims will choose not to create sims. In this scenario, human-level populations develop the ability to create many population simulations, but they don't. Perhaps they think it's too risky. It may be that the only populations to survive all the destructive technologies are extremely risk-averse and are worried, for example, about sims escaping from the simulation and taking over their world. Perhaps they think it's unethical to create a world where sims might suffer. Or perhaps they're just uninterested—they may have other priorities.

Still, there will be many strong incentives to create these simulations. First, there's ordinary curiosity, including scientific curiosity: One could discover an enormous amount about one's world by running simulations. It's already common for scientists to run thousands of smaller-scale simulations overnight and come back in the morning to collect the results. It's easy to envisage the same thing happening with population simulations. There may also be a practical consideration: Before making a difficult decision, it often makes sense to simulate the decision first and see how things go. And ethically, there may be an imperative to create worlds where good greatly outweighs bad. Given the ability to create simulations and strong incentives to do so, it would be surprising if few populations create them.

More nonsims than sims will be created. Couldn't we create nonsims instead of sims for these purposes? We could run physical simulations using robots in physical environments, for example. Alternatively, we could simulate our history using artificial biological organisms in terraformed environments.

Perhaps we could do this in principle, but it seems likely that creating sims would be vastly cheaper and easier. Nonsims in physical environments take up far more physical space and mass than sims in virtual environments. We've seen that a one-kilogram system running for one second could simulate a world with a billion humans living for a century each. It's unlikely that biological brains could run remotely as fast. Robot brains can in principle run as fast as sim brains, but a physical environment imposes many limitations. If robot bodies operate at human-body scale (somewhat larger and heavier than a one-kilogram laptop), a population of robots will take up more than a billion times the space of a population of sims. They'll also use more than a billion times the mass. Perhaps we could miniaturize the robots to the nanoscale (one-billionth of a meter), but then their physical environment and their experiences will be nothing like ours.

Given all this, we should expect humanlike sims to greatly outnumber humanlike nonsims. (Here humanlike beings are those with experiences broadly like ours.) Still, the argument has an underlying assumption: that it will be cheaper and easier for nonsims to create humanlike sims than to create humanlike nonsims. If it turns out to be easier to create nonsims for the relevant purposes—perhaps using nanotechnology, exploiting infinite space, or creating baby universes—then we'd expect nonsims to proliferate instead.

Overall, we can certainly say that if there are no sim blockers, then most intelligent beings are sims. We have not definitively ruled out sim blockers, but any of them would be somewhat surprising. We certainly can't know that there will be uncomputable laws, insufficient computer power, near-universal extinction, a near-universal choice to avoid simulations, or more efficient nonsims. And if we can't know that there are sim blockers, then for all we know, most intelligent beings are sims.

Are we special?

Premise 3 says that if most intelligent beings are sims, we are probably sims. Despite the ring of plausibility, it's easy to imagine ways that the premise could be false. For example, suppose it's a universal and widely known rule that sims all have a stamp in their visual field saying, "You are a sim," whereas nonsims do not. Then even if most beings are sims, the fact that I lack the stamp tells me that I'm not a sim.

Let's say that a *sim sign* is a feature that raises the probability that a creature is a sim. More precisely, it is a feature that a sim is more likely to have than a nonsim. For example, a sim is perhaps more likely than a nonsim to experience glitches in physical reality arising from approximations, shortcuts, and programming errors—a *Matrix*-style experience of the same cat crossing one's path twice, say. If so, these glitches are sim signs. Likewise, perhaps simulators are especially likely to simulate people who are thinking about simulations. If so, then the fact that you're reading a book called *Reality+* may be a sim sign.

The economist and futurist Robin Hanson has suggested that inter-

SIGNS YOU MAY BE A SIM

Figure 15 Potential sim signs: You're famous or interesting (Cleopatra);
you live relatively early in the universe (ancient Egypt); you
observe anomalies (glitching cats crossing your path);
you're thinking about simulations (*Reality+*).

estingness is a sim sign. Designers interested in entertainment or in historical simulation will more often create and sustain interesting or famous sims than uninteresting sims perhaps in local simulations where only these interesting sims and a few other people are simulated in detail. If you are living an interesting or well-known life, this increases the chance that you are a sim.

Perhaps our most significant sim sign is that we seem to live quite early in the universe. We haven't discovered intelligent life elsewhere in the universe, and we haven't yet created simulated universes with intelligent beings of their own. Both of these things may be sim signs. Where nonsims are concerned, it seems likely that the population of the universe will increase greatly over time, so most nonsims will exist later in the universe. But where sims are concerned, it's likely that early-universe simulations will be especially common, partly because later creatures may be interested in simulating their history, and partly because early-universe simulations will be far less demanding than later-universe simulations. Simulating universes running full-scale simulations of their own will be expensive! This suggests that disproportionately many sims may find themselves early in the universe. If so, our position early in the universe is a sim sign.

A *nonsim sign* is a feature that tends to attach to nonsims. More precisely, it is a feature that a nonsim is more likely to have than a sim. In the world where all sims get a "You are a sim" signal, the absence of such a signal is a nonsim sign. If we know we have a nonsim sign, then even if we know that 99 percent of beings are sims, we should be less than 99 percent confident that we are sims.

Some important objections to premise 3 point to nonsim signs. These are the objections that say, *We're special.* The potential nonsim signs here include consciousness (simulations won't be conscious), our minds more generally (simulated minds won't work like ours), the complexity of the world (simulated worlds will be simpler than ours), and more.

Sims can't be conscious! The most obvious potential nonsim sign is consciousness itself. Suppose one thinks, as some philosophers do, that only biological systems can be conscious and that therefore simulations can't be conscious. Given this view, the fact that we're conscious will indicate that we aren't simulations. We could be biosims (biological brains connected to simulations), but pure sims would be ruled out.

This view is controversial. I'll argue in chapter 15 that it's false and that simulated beings will be as conscious as their nonsim counterparts. Nick Bostrom rules out the view that simulations aren't conscious by making an assumption of *substrate-independence* (or equivalently, *substrate-neutrality*)—that is, consciousness depends only on the organization of a system and does not depend on the substrate (e.g., biology or silicon) in which the system is implemented.

At present, consciousness is ill-understood. So it wouldn't be entirely unreasonable for someone who believes that 99 percent of beings are simulated to be only 50 percent confident that simulations will be conscious. If so, then instead of being 99 percent confident that they're simulated, they should be only about 50 percent confident that they're simulated. That would be a less dramatic conclusion, but still a striking one.

Simulators will avoid creating conscious sims. A quite different way that consciousness could be a nonsim sign is that populations advanced enough to create sims will know how to create intelligent sims that aren't conscious (while nevertheless serving many practical purposes), and will have strong reasons—perhaps ethical reasons—to do this. This hypothesis doesn't require assuming that sims can't be conscious, and it doesn't require substrate-neutrality, although it does require that not all intelligent sims are conscious. For all we know, there might be some simple way to tweak the organization of intelligent sims so that they lack consciousness. For example, the neuroscientists Christof Koch and Giulio Tononi have argued that sims running on serial von Neumann architectures will not be conscious, though sims running on strongly parallel architectures will be. If so, then ethical simulators may aim to use the nonconscious versions wherever possible.

Like the previous sim sign, I think this is one to take seriously. Still, it's far from obvious that intelligent beings without consciousness are really possible. Even if they are possible—if even one percent of simulations involve conscious beings, conscious sims are likely to outnumber conscious nonsims.

Sims won't have minds like ours. Our minds may contain nonsim signs other than consciousness. For example, creativity or emotions might be less common in sims, perhaps because they introduce complications into simulations. Or we might expect sims to be much more intelligent or rational than we are, so that our irrationality or low intelligence is a nonsim sign.

If you're certain that an aspect of your mind—consciousness, say— cannot be replicated in a simulation, then you'll regard that aspect as an *absolute* nonsim sign. If it's merely less likely that this aspect— emotions, say—will be found in simulations, then it will be a *probabilistic* nonsim sign.

Probabilistic nonsim signs will reduce the probability that we're in a simulation, but as long as the probabilities are not extreme, they won't reduce it by too much. Suppose we think that all nonsims have emotions, while only one in ten sims have emotions. And suppose we think that there are about one thousand sims for every nonsim, so that we are initially 99.9 percent confident that we are sims. Then we should think there are about one hundred emotional sims for every emotional nonsim. So even after taking into account our own emotions as a nonsim sign, we should still be 99 percent confident that we are sims.

Sims won't experience large universes. Our world seems to be enormous in its spatial extent. The observable universe alone is around 90 billion light-years across, with room for at least two trillion galaxies and perhaps one trillion trillion stars. Our universe is also enormous in its depth, with levels upon levels of detail below the level we ordinarily perceive. It seems likely that most simulated worlds won't be as large as this. It would be much easier and cheaper to simulate a smaller world. For many purposes, a smaller simulation will be just as useful. If so, the apparent largeness of our world is a nonsim sign, increasing the probability that our world is not simulated.

One reply to this objection is that the simulating world may be enormous compared to our world, perhaps even infinite. In such a world, simulations of worlds like ours may be cheap and common. At most, we'll rule out a version of the simulation hypothesis in which the simulating world is no more complex than our world. But many other versions of the hypothesis are left open.

Another reply is that we may well be in a *shortcut simulation*—a simulation that takes shortcuts so that our world is not as large as it seems. Perhaps only our local area is simulated in detail, and the rest is a simplified model? As noted in chapter 2, local simulations may be much cheaper than global simulations. They may not serve every purpose that a global simulation serves, but for many purposes they may suffice. If local simulations that replicate our experiences of a large world are possible and common, then these experiences may not be much of a nonsim sign.

As we saw in chapter 2, it's not trivial for a local simulation to generate our experiences. To simulate where we've been, the people we interact with, and the media we read and watch, it will need to simulate a fair amount of our planet, and it will need reasonably detailed simulations of the Sun, Moon, and other planets (of which we now have detailed images). We'll need a decent simulation of at least the visible stars and galaxies, and of background radiation and other observable phenomena. The simulators will need to be ready to expand the simulation—for example, if we travel to the stars or acquire new ways of getting information from them. Expandable simulations are already familiar in video games such as *No Man's Sky*, in which new planets are algorithmically generated as players travel to them. Sophisticated simulators may be masters of these techniques, knowing just where a local simulation can cut corners.

Something similar goes for using shortcuts to simulate the microscopic depths of the world. In some contexts, a simple Newtonian physics of macroscopic objects may be all we need, but there are many contexts for which more is needed. Many observable properties of ordinary objects depend on chemistry, which itself depends on quantum mechanics, so it will be difficult to simulate macroscopic objects really well without going very deep. Making our simulation consistent with the observations of scientists about atomic physics and the like will require even more work. Perhaps not every last detail will have to be simulated. Some low-level details may never show up in our observations. When a system is not being observed closely, simplified models can sometimes be used. Still, it seems likely that a lot of physics will have to be simulated to yield plausible results.

The moral is that even a local simulation consistent with our experiences will have to be highly complex. It is arguable that this degree of complexity will be relatively rare in simulations. For many purposes, it will be easier to build a simpler simulation. And chains of complex worlds simulating complex worlds (in which building many simulations is possible) will eventually end in a plethora of simple worlds in which building simulations is impossible. If so, then most sims won't have experiences of a complex world, so our experiences will be a probabilistic nonsim sign. As before, this nonsim sign will push down the probability that we are in a simulation.

Nevertheless, it remains plausible that simulators will create a reasonably large number of complex-world simulations. If so, we can expect most beings with complex-world experiences to be sims.

Stepping back: The potential nonsim signs we've considered, such as consciousness and a large world, may decrease the probability that we're in a simulation. At the same time, we need to weigh these against potential sim signs, such as the fact that we seem to be early in the universe, which may increase the probability that we're in a simulation. Do the sim signs outweigh the nonsim signs, or vice versa? I won't try to settle this now.

Ancestor simulations and humanlike sims

Nick Bostrom takes another approach to the sim sign issue by focusing only on ancestor simulations: exact simulations of the entire mental history of humankind. Any ancestor simulation of my world will include an exact simulation of me. If there are many ancestor simulations of my world, it is guaranteed that there will be many sims with experiences just like mine. If so, there is no need to worry that my experience contains nonsim signs. Every feature of my experience will be replicated in many sims.

I don't think the "ancestor simulation" version of the argument works in this form, since I don't see good reason to believe that there will be exact ancestor simulations. Constructing such simulations would require knowing something close to the exact state of human brains at every point in human history, and there's not much reason to think that's possible. Perhaps it will be possible to do this inside simulated universes (via backup records for simulated brains, for example), but this won't help in the key case of nonsims building simulations.

Bostrom later notes that for the argument to work, the simulated creatures needn't have *exactly* our experiences. It is good enough if they have "human-type" experiences, the sorts of experiences typical of human creatures. I think this is right, though it means that there is no longer any need to mention ancestor simulations in the argument. Any human-type simulations are good enough. In fact, I think "human-type" experiences as Bostrom defines them may still be unnecessarily narrow. In principle, the argument will still work with a broader class of minds, as long as they preserve the most important sim signs and nonsim signs that humans have.

I would therefore focus the argument more broadly on the likelihood that there will be *humanlike* sims. Humanlike beings are beings with roughly the same major sim signs and nonsim signs as humans: for example, they're conscious, they experience a large universe, and their society is at a certain stage of technological development.

Bostrom uses his notion of human-type experiences to define a mathematical formula for the fraction of all observers with human-type experiences that live in simulations, and to draw conclusions about the probability that we are sims. I don't think Bostrom's formula or his conclusions are quite correct as they stand, for reasons I discuss in the notes. However, it's possible to formulate the simulation argument in a simple general form that avoids these problems while also avoiding objections from sim blockers and nonsim signs. The argument runs as follows.

1. If there are no sim blockers, most humanlike beings are sims.
2. If most humanlike beings are sims, we are probably sims.

---

3. So: If there are no sim blockers, we are probably sims.

"Most humanlike beings are sims" means that most humanlike beings in the cosmos (including past and future beings, beings who created us or whom we created, and so on) are sims. The numbers for "most" and "probably" can be assigned as we choose, as long as they match. For example, "most" can be 99 percent, and "probably" can be a 99 percent confidence. Importantly, a sim blocker is now defined as something that prevents the creation of enough *humanlike* sims to ensure that most humanlike beings will be sims.

Because premise 1 builds in *If there are no sim blockers* as a condition, sim blockers are no longer an objection to it. Premise 1 now requires only the plausible assumption that if nothing prevents the creation of many humanlike sims (enough of them that most humanlike beings are sims), then there will be many humanlike sims. Because premise 2 builds in *most humanlike beings*, nonsim signs are no longer an objection to it. Premise 2 now requires only the assumption that if there are many beings with the same sort of experience as me, then I am equally likely to be any of those beings. This is sometimes called an indifference principle, because it recommends indifference between each of these hypotheses about who I am. From this assumption it follows that if 90 percent of beings with experiences like mine are sims, then I should be 90 percent confident that we are sims.

Even if we accept premises 1 and 2, we've really just relocated the sim blocker and sim sign issues. In weakening the premises, we've also weakened the conclusion, which now explicitly builds in the possibility of sim blockers. Furthermore, the notion of a sim blocker has now been broadened to include anything that blocks creation of enough *humanlike* sims. As a result, the notion of a sim blocker now covers things that we previously counted as nonsim signs. For example, *Sims won't be conscious* is now a potential sim blocker: if conscious sims are impossible, then humanlike sims are impossible. *Sims won't experience large universes* is now also a potential sim blocker; if simulations with apparently large universes are rare, then humanlike sims are rare.

From the conclusion, it follows that we can be highly confident in *Either there are sim blockers or we are sims.* If we accept the argument with the numbers assigned as above, then we should be at least 99 percent confident that one of these alternatives holds.

Bostrom puts the conclusion of his argument in roughly this form:

This paper argues that *at least one* of the following propositions is true: (1) the human species is very likely to go extinct before reaching a "posthuman" stage; (2) any posthuman civilization is extremely unlikely to run a significant number of simulations of their evolutionary history (or variations thereof); (3) we are almost certainly living in a computer simulation.

Here, Bostrom's options (1) and (2) are both sim blockers, closely related to *Nonsims will all die before creating sims* and *Nonsims will choose not to create sims*. These are reasonable sim blockers to consider, but they're far from the only ones. Of the sim blockers for humanlike sims we've considered, I'd add at least *Intelligent sims are impossible, Conscious sims are impossible, Sims take too much computer power, Simulators will avoid creating conscious sims*, and *More nonsims than sims will be created*. Instead of Bostrom's three-way conclusion, these five extra sim blockers would give us an eight-way conclusion.

A simpler approach is to divide sim blockers into two groups. First, it could be that humanlike sims are impossible or highly impractical to make (to simplify, I'll understand "possible" to mean "practically possible," so that impracticality counts as impossibility). This group includes sim blockers like *Sims won't be conscious, Intelligent sims are impossible*, and *Sims take too much computer power*. Second, it could be that humanlike sims are possible and practical, but few humanlike populations will create them (in sufficient numbers for most humanlike beings to be sims). This group includes sim blockers like *Nonsims will die before creating sims, Nonsims will choose not to create sims, Simulators will avoid creating conscious sims*, and *More nonsims than sims will be created*.

If this is right, we can cash out the conclusion more explicitly in a three-way form. My conclusion is that we should be highly confident that either (1) we are sims, or (2) humanlike sims are impossible, or (3) humanlike sims are possible but few humanlike nonsims will create them.

(In an online appendix, I discuss further objections to the simulation argument, and argue that none of them undermines the argument. The objections include *We shouldn't be indifferent between nonsims and sims; Sims won't have our external evidence; We know we're not the sims we create; We can't know the physics of the next universe up; We should expect to live in an impoverished world*. I argue there that none of these objections undermines the argument.)

Is it likely that we're in a simulation? **101**

The upshot

What's the upshot? Are we in a simulation? What are the implications of the simulation argument for skepticism and the Knowledge Question?

I wouldn't say we can know we're in a simulation; there are too many possible sim blockers to be sure of that. I can't know for sure that humanlike simulations are possible. Perhaps consciousness is substrate-dependent, or physical processes are uncomputable. And I can't know for sure that if these simulations are possible, humanlike populations will create them. Perhaps almost all of them will go extinct or avoid simulations. So I can't know for sure that most humanlike beings are simulated, and I can't be confident that we're sims.

Still, I don't think we can be at all confident that there are sim blockers. If I had to estimate, I'd say it's more likely than not that conscious humanlike simulations are possible. I'd also say it's more likely than not that if conscious humanlike simulations are possible, many humanlike populations will create them. If so, then there's less than a 50 percent chance that there are sim blockers in the first class, and less than a 50 percent chance that there are sim blockers in the second class. It follows (given plausible assumptions) that there is less than a 75 percent chance that there are any sim blockers at all. Given that it's at least 99 percent likely that either there are sim blockers or we are sims, it follows that the chance that we are sims is at least 25 percent or so.

Whatever one says about the probabilities, this argument suggests very strongly that we can't know we're *not* in a simulation. We can be highly confident that either we are sims, or that most humanlike populations won't create humanlike sims, or that humanlike sims are impossible. The latter two hypotheses are extremely speculative. We do not know that even one of them is true. As a result, we don't know that we're not in a simulation.

An opponent might suggest that we can know we're not in a simulation by one of the methods discussed earlier—Bertrand Russell's appeal to simplicity, say, or G. E. Moore's observation of his own hands—and that we could thereby *conclude* that one of the sim blockers obtains, although we can't be sure which.

I think this is implausible. In effect, the simulation argument makes the simulation hypothesis a serious possibility. And once it is a serious possibility, these arguments cannot rule it out.

Suppose God tells me that when I was born, she flipped a coin. If it came up heads, she connected me to a perfect simulation. If it came up tails, she sent me into nonsimulated reality. She did this for many people and is now telling all of them about it. Then I should have 50 percent confidence that I'm in a simulation. In light of God's announcement, simplicity now has no force in ruling out the simulation hypothesis. Likewise, looking at my hands and running Moore's argument has no force. I should still be at 50 percent. It seems plain that in this situation, I don't know that I'm not in a simulation.

The simulation argument does something similar. It elevates the simulation hypothesis to the status of a serious possibility to which we should assign substantial probability. Whether that probability is 20 percent or 50 percent, the anti-skeptical arguments we've discussed do not drag it down. Once it is a serious possibility that we're in a simulation, none of these arguments allows us to know that we're not.

I conclude that we cannot know that we are not in a simulation.

Part 3

REALITY

Chapter 6

What is reality?

•

A T THE END OF THE MOVIE *READY PLAYER ONE*, WHICH TAKES place mainly in a virtual world, one of the characters raises a version of the Reality Question. He says, "Reality is the only thing that's real."

At first, this looks like a tautology, akin to *Boys are boys*. Of course only reality is real!

On second glance, it looks like a confusion, as with *Happiness is happy*. How could reality itself be real?

Still, the underlying message is pretty clear. In the context of the movie, the speaker is praising physical reality and downgrading virtual reality. The intended message: *Physical reality is the only thing that's real*. And *Virtual reality is not real*.

Isn't this still a confusion? What is it for physical or virtual reality to be "real"? On my reading, the central idea is that a reality is real if *things* in that reality are real. Read this way, the slogan is saying *Physical things are the only things that are real*, and *Virtual things are not real*.

If the slogan is right, the planet Earth is real, whereas the planet Ludus (a virtual planet in *Ready Player One*) is not. Earth and the things on it, from ducks to mountains, exist as part of the objective world. Ludus and the things on it, from avatars to virtual weapons, do not—they're mere fictions or illusions.

Virtual things are not real is the standard line on virtual reality. I think it's wrong. Virtual reality is real—that is, the entities in virtual reality really exist.

My view is a sort of *virtual realism*. That phrase first appeared as the title of the American philosopher Michael Heim's pioneering 1998 book on the ramifications of VR. Heim used the label primarily for a broad social and political view of virtual reality, intermediate between that of "network idealists who promote virtual communities" and "naive realists who blame electronic culture" for various social ills. At the same time, Heim associated the label with the view that "Virtual entities are indeed real, functional, and even central to life in coming eras." I'm using the label *virtual realism* in the latter sense.

As I understand it, virtual realism is the thesis that virtual reality is genuine reality, with emphasis especially on the view that virtual objects are real and not an illusion. In general, "realism" is the word philosophers use for the view that something is real. Someone who thinks morality is real is a moral realist. Someone who thinks that colors are real is a color realist. By analogy, someone who believes that virtual objects are real is a virtual realist.

I also accept *simulation realism*: If we're in a simulation, the objects around us are real and not an illusion. Virtual realism is a view about virtual reality in general, while simulation realism is a view specifically about the simulation hypothesis. Simulation realism says that even if we've lived our whole life in a simulation, the cats and chairs in the world around us really exist. They aren't illusions; things are as they seem. Most of what we believe in the simulation is true. There are real trees and real cars. New York, Sydney, Donald Trump, and Beyoncé are all real.

Simulation realism has major implications for skepticism about the external world. We've seen that the Cartesian path to global skepticism says no to the Reality Question (in a simulation, nothing is real) and says no to the Knowledge Question (we can't know we're not in a simulation). When we accept simulation realism, we say yes to the Reality Question. In a simulation, things are real and not illusions. If so, the simulation hypothesis and related scenarios no longer pose a global threat to our knowledge. Even if we don't know whether or not we're in a simulation, we can still know many things about the external world.

Of course, if we're in a simulation, the trees and cars and Beyoncé are not *exactly* how we thought they were. Deep down, there are some differences. We thought that trees and cars and human bodies were ultimately made of fundamental particles such as atoms and quarks; instead, they're made of bits.

I call this view *virtual digitalism*. Virtual digitalism says that objects in virtual reality are digital objects—roughly speaking, structures of binary information, or bits.

Virtual digitalism is a version of virtual realism, since digital objects are perfectly real. Structures of bits are grounded in real processing, in a real computer. If we're in a simulation, the computer is in the next world up, metaphorically speaking. But the digital objects are no less real for that. So if we're in a simulation, the cats, trees, and tables around us are all perfectly real.

That might sound preposterous, but I'll try to convince you it's true. I'll argue for these views over the next few chapters. In chapters 7–9, I'll make the case that if we're in a simulation, our world is still real. In chapters 10–12, I'll concentrate on more familiar virtual reality technology, making the case that here, too, virtual worlds and objects are real.

First, though, we should elucidate what we mean by *reality* and *real*.

Reality and realities

Before defining *real*, I'll say something about defining *reality*. This word has at least three meanings that are relevant for our purposes. I use the word *reality* in all these ways in this book.

First, when we talk about *reality* as an entity, we mean something like *everything that exists*: the entire cosmos. When we talk about *physical reality* and *virtual reality* as entities, we mean something in the same spirit. We mean roughly *everything that's physical* and *everything that's virtual*.

Second, when we talk about *a reality*, we mean something like a *world*. When we talk about *realities* plural, we mean *worlds*. When we talk about *a virtual reality*, we mean roughly *a virtual world*. A world is roughly equivalent to a universe: a complete interconnected physical or virtual space.

Reality (in the first sense) may contain many realities (in the second sense). It's a familiar idea—depicted for example in the movie *Spider-Man: Into the Spider-Verse*—that we could be living in a multiverse: a cosmos with multiple universes. In a multiverse, reality contains many worlds and many realities. With the advent of virtual worlds, we have Reality+: a multiverse of both physical and virtual realities. All of these realities (worlds) are part of reality (the cosmos).

Third, we can talk about *reality* as a property like *rigidity*. Rigidity is the property of being rigid. Some objects are rigid, and some are not. In this sense, reality is *real-ness*. It's the property of being real. Some things are real, and others are not. Talking about reality as real-ness is a way of talking about what it is to be real. That's our focus in this chapter.

To really confuse things, we could sum up the *Reality+* view of reality (in all three senses) by saying: reality contains many realities, and those realities are real. Or more mundanely: the cosmos (everything that exists) contains many worlds (physical and virtual spaces), and the objects in those worlds are real. Now we just need to unpack *real*.

Five ways of thinking about what's real

In *The Matrix*, Morpheus asks, "What is real? How do you define real?" That is, *What does it mean to say something is real?* What does it mean when we say that Joe Biden is real but Santa Claus isn't?

Philosophers have answered Morpheus's question in various ways. I'll focus on five complementary answers here. Each answer illuminates part of the story. Each is a different strand in our concept of the real.

Reality as existence. First and foremost, something is real if it really exists. Joe Biden really exists; he's part of the universe. Santa Claus doesn't really exist; he's not part of the universe. There are *stories* about Santa Claus, and there are *beliefs* about Santa Claus, but Santa Claus himself does not exist.

Figure 16 In a virtual world, Morpheus and Neo debate digital reality. Virtual digitalism holds that virtual objects are real digital objects.

Of course, this just raises a further question: "What is it to exist?" This is a deep question, to which I can't give a definitive answer. Many people think there's no definitive answer to be given.

Recall Berkeley's dictum from chapter 4: *esse* is *percipi*. To be is to be perceived—which means that something exists if it's perceived or at least could be perceived. As Morpheus put it, when we say something is "real," we might be talking about "what you can feel, what you can smell, what you can taste and see." A related, more scientific-sounding view is that something is real if it can be measured.

I've already discussed why Berkeley's dictum is too strong. There could be real things that we can never perceive and never measure. It may be that we can never measure certain physical entities around the time of the Big Bang, or in distant galaxies, but they exist all the same. If I'm in a perfect simulation, I may never be able to perceive the world outside the simulation, but it's real all the same. There are also some things we can perceive and measure that aren't real. We can measure the intensity of a mirage, for example, but this doesn't mean the mirage really exists.

Still, Berkeley's dictum can function well as a *heuristic* for existence—that is, an imperfect guide to what exists—rather than an absolute criterion. If something is perceivable and measurable, that's a strong indication that it exists.

Reality as causal power. An even better heuristic for existence—one that subsumes Berkeley's dictum—is sometimes called the Eleatic dictum, because a version of it is suggested by the mysterious "Eleatic Stranger" from the ancient town of Elea in Plato's dialogue *The Sophist*. The Stranger says,

> "I suggest that everything which possesses any power of any kind, either to produce a change in anything of any nature or to be affected even in the least degree by the slightest cause, though it be only on one occasion, has real being."

That's the Eleatic dictum: to be real is to have causal powers. That is, something exists if and only if it can affect things or be affected by things. Joe Biden has causal powers; as US president, he can command the armed forces and sign or veto legislation. Physical events around the Big Bang and in distant galaxies have causal powers; they affect what happens next in their locale. Even a stone has causal powers; it can leave an indentation on the ground. Anything perceivable has causal powers because it has the power to make a difference to a perceiver. As a result, anything that falls under Berkeley's dictum also falls under the Eleatic dictum.

Santa Claus does not fall under the Eleatic dictum. Santa has no causal powers: He doesn't make a difference in the world. According to *stories* about Santa Claus, Santa has enormous causal powers: He delivers billions of Christmas presents in a single night. But those stories are false; Santa has no intrinsic causal powers. Of course, the stories themselves have causal powers. They affect cards and costumes and children throughout the world. So by the Eleatic dictum, the stories are real, but this doesn't mean Santa is real.

The Eleatic dictum isn't the whole truth about reality. There could be real things that have no causal powers. Maybe numbers are real without having causal powers, for example. Maybe there could be a forgotten dream with no causal powers. But causal powers provide at least a sufficient condition for reality. If something has causal powers, then it exists and it is real. In this way, causal powers add a second heuristic criterion for something's being real, alongside existence itself.

Reality as mind-independence. A third criterion for reality was proposed by the science-fiction author Philip K. Dick in his 1980 short story "I Hope I Shall Arrive Soon." We can call it Philip K. Dick's dictum: "Reality is that which, when you stop believing in it, doesn't go away." The idea is that if no one believed in Santa Claus, Santa Claus would not be on the scene, but if no one believed in Joe Biden, he'd still be out there.

I don't think Dick's dictum is quite right as it stands. If we stop believing in Gandalf but still see him on screen, he doesn't go away. That doesn't mean that Gandalf is real. The same goes for illusions and hallucinations: even if I believe that a distant mirage isn't real, the mirage doesn't go away.

Still, Dick is onto something. In those cases, the presence of Gandalf or the mirage still depends on people's minds—their thoughts and experiences. If no one had ever thought about Gandalf, he would never have entered our lives. If we stop experiencing the mirage, it will go away. So we might say, "Reality is that which, when it's not in anyone's mind, doesn't go away." Or maybe better: "Reality is that which doesn't depend on anyone's mind for its existence."

Even this modified version of Dick's dictum isn't the whole truth about reality. A counterpoint is what we might call Dumbledore's dictum, spoken by Hogwarts headmaster Albus Dumbledore toward the end of the Harry Potter series: "Of course it is happening inside your head, Harry, but why on earth should that mean that it is not real?" Your thoughts and experience happen inside your head and depend on your mind, but they're real all the same. Social entities like money also depend on people's minds. If no one regarded dollar bills as valuable, they wouldn't be money—but money is real all the same.

Still, mind-independence can serve as a useful sufficient condition for being real, and one that helps to explain at least one useful dimension of our sense of reality. If something exists in a way that's independent of anyone's mind, then it has an especially robust sort of reality. If something exists only in a way that depends on our minds, then it's less robustly part of the external world.

Reality as non-illusoriness.   What is the difference between illusion and reality? So far, we've asked the question, "Do things really exist?" Another crucial question, however, is "Are things as they seem?" We can use this question as a fourth criterion for reality, capturing another strand in what we mean when we say something is real. That is, something is real when it's the way it seems. Something is illusory when it's not the way it seems.

Physical reality is real because it's roughly the way it seems. When it seems that there's a ball in front of you in physical reality, there's typically a ball there. According to the standard view, virtual reality isn't real because it isn't the way it seems. In VR it seems that there's a ball there, but there's not. In virtual reality, the story goes, things are not as they seem—that is, virtual reality is an illusion.

Right now, it seems to me that I'm sitting in a chair, using a desktop computer, in a house somewhere in the Hudson Valley. It seems to me that there's an algae-covered pond outside the window, with geese wandering around. It seems to me that I'm a philosopher working on a book on reality. It seems to me that I grew up in Australia and that I now live in New York. All this and more constitutes what we might think of as my apparent reality.

This apparent reality is real if, and only if, things are as they seem. If I'm really sitting in a chair, and there is really a pond outside the window, and I really grew up in Australia, then things are the way they seem, and all this is real. If I'm not really sitting in a chair, and there is not really a pond outside the window, and I didn't really grow up in Australia, then all this is not real.

Of course, I can be wrong about some things—maybe those aren't really geese outside the window—without my reality collapsing. But if I'm wrong about almost everything, then it's reasonable to say that my apparent reality isn't real.

What counts as the way things seem? There's more than one candidate. There's the way we *perceive* things to be, using our senses. There's also the way we *believe* things to be, using our thinking and reasoning as well as perception. I can perceive a pink elephant without believing that it exists. I can believe that there was a Big Bang without having perceived it.

In these cases, it's arguably what we believe that matters most for our reality. For issues about skepticism and the simulation hypothesis, what matters most to us is whether things in the world are as we believe them to be. So where simulation realism is concerned, I'll understand the fourth criterion as saying things are real when they're roughly as we believe them to be.

Reality as genuineness. There's a fifth way of thinking about reality that comes from the British philosopher J. L. Austin in his 1947 lectures collected as *Sense and Sensibilia*. (The title was inspired by Austin's near-namesake Jane Austen's novel *Sense and Sensibility*, but the book is about perception and reality.) Austin held that philosophy must always pay close attention to the way words are used in ordinary language. To understand what it is for something to be real, we must look at how normal English speakers use the word "real."

Austin said that in ordinary language we cannot *just* say of something that it is real. If we did, it would invite the question "A real *what*?" The issue would have to be whether it is a real diamond (as opposed to a fake diamond), or a real duck (as opposed to a decoy duck). We might call this Austin's dictum: *Instead of asking whether something is real, ask whether it's a real X.*

I don't think Austin's dictum is quite right. It's perfectly reasonable for a child to ask whether Santa Claus is real or whether the Easter Bunny is real. You *could* ask whether the Easter Bunny is a real rabbit, or a real spirit, or whatever. But there's also a question you can ask while staying neutral on whether it's a rabbit or a spirit: Is the Easter Bunny real? That is, does it really exist? The answer seems to be no. It's a folkloric figure that has been passed down for generations; the folklore is real, but the bunny is not. This is reflected in our first three dicta, all of which concern what it is for an object to be real (rather than to be a real X). The Easter Bunny doesn't really exist, it doesn't have causal powers, and it isn't independent of our minds.

Still, Austin's dictum captures something crucial. We often don't *just* want to know whether something is real; we want to know whether it's real money or a real iPhone. If someone gives me an object that looks like a Rolex watch, the object is indisputably real. It's a real *thing*, at least. What I'm interested in is whether it's a real *watch*, and in particular whether it's a real Rolex. We could put this by asking whether the watch is *genuine*—that is, is it a genuine watch? Is it a genuine Rolex?

The same goes for life in a simulation. It's one question to ask whether buildings and trees and animals we're seeing are real; perhaps someone could be convinced that they're real digital entities. But it's another question to ask whether they're real buildings, real trees, and real animals. If they're real objects but not real buildings, then things are not as they seem. So when something seems to be an X, our fifth criterion for reality asks: Is this genuine? That is, is it a real X?

Is simulated reality real?

We now have five criteria for reality. We can encapsulate them in a *reality checklist*: five questions to ask when we're wondering whether or not something is real. *Does it really exist? Does it have causal powers? Is it independent of our minds? Is it as it seems? Is it a genuine X?*

These five criteria—existence, causal powers, mind-independence, non-illusoriness, and genuineness—capture five different strands in our concept of being real. When we say that something is real, we sometimes mean one of these things and sometimes a mix of them. There are other strands we could have added—perhaps *Reality as intersubjectivity*, *Reality as theoretical utility*, and *Reality as fundamentality*, among others I discuss in an endnote—but these five are the strands that matter most for our purposes.

None of these criteria for reality make it easy to figure out what's real; they just clarify the question a little. Different notions of reality are relevant for different purposes, so we need to be clear about which notion we're using.

Let's apply these criteria to the case of objects in a simulation. We can start with the case of a perfect and permanent simulation—one that simulates a world in full detail, which has generated all our experiences throughout our history. If we're in a perfect simulation, is our world real? My purpose here isn't so much to argue for my view about simulations as to make clear what my view says.

The first criterion for reality asks: *Do the objects we perceive in this simulated world really exist?* If I'm in a perfect simulation, does the tree outside my window really exist? My opponent says, "No, the tree and the window itself are mere hallucinations." I say, "Yes, the tree and the window really exist." At some level they're digital objects, grounded in digital processes in a computer, but they're no less real for that.

The second criterion asks: *Do the objects we perceive have causal powers?* Do they make a difference? My opponent says "No, the tree merely seems to make a difference." I say, "Yes, the tree is a digital object with many causal powers." It produces leaves (which are themselves digital objects), it supports (digital) birds, and it produces experiences in me and in others who look at it.

The third criterion asks: *Are the objects we perceive independent of the mind?* If minds go away, can they still exist? My opponent says, "No, the tree I perceive exists only in my mind, and if we all went away, it wouldn't exist." I say, "Yes. The tree is a digital object, which does not depend on me for its existence." Even if all (simulated and unsimulated) human life went away, in principle the tree could continue to exist as a digital object.

The fourth criterion asks: *Are things as they seem to be?* Are flowers really blooming in my garden, as they seem to be? Am I really a philosopher from Australia, as I seem to be? My opponent says, "No, these are mere illusions, and in reality there are no flowers and no Australia." I say, "Yes, there are really flowers blooming in the garden, and I am really from Australia." If my whole world is a simulation, flowers are ultimately digital objects, and Australia is ultimately digital, too, but this is no obstacle to the flowers blooming and to my being Australian.

The fifth criterion asks: *Are the things I experience in a simulation real flowers, and real books, and real people?* My opponent says, "No. Even if they're real digital objects, these objects are at best fake flowers, not real flowers." I say, "Yes, these objects are real flowers, the books are real books, and the people are real people." If I've lived my whole life in a simulation, every real flower I experience has been digital all along.

To sum up: if we're in a perfect, permanent simulation, the objects we perceive are real according to all five of those criteria.

I defined simulation realism this way: "If we're in a perfect simulation, the objects around us are real and not an illusion." The reference to illusion puts the greatest weight on the fourth criterion: Simulation realism holds that things are largely as we believe them to be. Now, we believe that cats exist, and that cats do things, and that they are real cats. Simulation realism entails that in a simulation, these beliefs are largely true.

You might think that I've cherry-picked the criteria for being real. Perhaps things would go differently if we defined reality as *fundamentality* or *originality*. If we're in a simulation, the simulated trees we perceive may not be the original trees (simulators may have modeled them after unsimulated trees) and the simulated physical world will not be fundamental (since a simulation in the next universe up underlies it). Perhaps these points capture some part of what people mean when they say that virtual objects are not real. Still, these criteria seem marginal as criteria for reality. Dolly the cloned lamb is not original (she's a clone of another lamb) and not fundamental (she's made out of particles)—but she is perfectly real.

To bring out what a strong view my simulation realism is, and to reinforce that the criteria are not cherry-picked, we can contrast it with a view laid out by the British theoretical physicist David Deutsch in his 1997 book *The Fabric of Reality*. In a fascinating discussion of VR, Deutsch advocates a partial sort of virtual realism. He argues that VR environments (including simulations) "pass the test for reality" because they "kick back" at the user. That suggests versions of the second and third criteria: These environments have causal powers and are independent of our minds. At the same time, Deutsch does not endorse the first criterion: He says of one scenario, "the simulated aircraft and its surroundings do not really exist." Nor does he endorse the fourth and fifth criteria. He says that in another VR scenario you can see rain that's not there in reality, and that an engine in the scenario is not a real engine. In Deutsch's view, VR environments are real but the objects in them are illusions. Simulation realism makes a much stronger claim than this.

Of course, if we're in a perfect simulation, then *some* things aren't exactly as they seem. Some of what we believe will be wrong. Most people believe that they're not in a simulation. They believe that flowers are not digital. They may believe that their universe is the ultimate reality. If we're in a simulation, those beliefs will be wrong. But the undermined beliefs here are mostly scientific or philosophical beliefs about reality. Undermining them does not undermine everyday beliefs such as *There are flowers blooming in the garden.*

If we're in an *imperfect* simulation, more of our beliefs will be wrong, but plenty will still be right. If the solar system is fully simulated but the rest of the universe is only a sketch, then my beliefs about the Sun may be correct, but my beliefs about Alpha Centauri may be wrong. If 2019 is simulated but 1789 was not, then my beliefs about the French Revolution may be false, but my beliefs about the contemporary United States may be correct. Still, we'll be right about those closer-to-home aspects of the universe that are present in the simulation. Assuming the deer I see in my garden is part of the simulation, I'll still be right that there's a deer in my garden.

The simulation hypothesis as metaphysics

I'm sure that some of these claims sound counterintuitive to you. One source of resistance is that trees and flowers certainly don't *seem* to be digital objects. But they don't seem to be quantum mechanical objects either. Yet deep down they are, and few people think that the mere fact that trees are grounded in quantum processes makes them less real. I think that being digital is just like being quantum mechanical here.

Science has taught us that there's much more to reality than initially seems to be the case. For millennia, we didn't know that cats and dogs and trees are made of cells, let alone that the cells are made of atoms or that those are fundamentally quantum mechanical. Yet these discoveries about the nature of cats and dogs and trees have not undermined their reality.

If I'm right, the discovery that we're in a simulation should be treated the same way. It will be a discovery about the underlying nature of cats and dogs and trees—that they spring from digital processes—but it won't undermine their reality.

Importantly, I'm not saying in an unqualified way that a simulated tree is the same as a real tree. I've said that *if* we're in a perfect permanent simulation, then the real trees of our universe are simulated trees—that is, real trees have been digital trees all along. On the other hand, if we're *not* in such a simulation and are merely looking at one from the outside, then the trees in the simulation are completely different from the trees in our outside world: Simulated trees are digital; real trees are not. The real trees of our world have been nondigital entities all along. Either way, digital entities and nondigital entities are quite different things.

I will argue that the simulation hypothesis should be seen as a variant of the "it-from-bit" hypothesis that's been widely discussed in physics. This hypothesis posits a digital level underlying physics: roughly speaking, molecules are made of atoms, atoms are made of quarks, and quarks are made of bits. This is a view in which physical processes are real. There are just underlying levels to reality that we don't know about yet.

I think that's the correct way to think about the simulation hypothesis. If the simulation hypothesis is true, the it-from-bit hypothesis is true. Physical reality is perfectly real—there's just an underlying level consisting of the interaction of bits, and perhaps still further levels still underlying that.

The it-from-bit hypothesis corresponds to one part of the simulation hypothesis—the simulation itself. What about the other part—the simulator who initiated the simulation? In the next chapter, I'll argue that the simulator is analogous to a god. At least, the simulator can be seen as the creator of the it-from-bit universe. In chapter 9, I'll argue that the simulation hypothesis is itself a combination of the it-from-bit hypothesis and the creation hypothesis (according to which a creator created the universe).

If I'm right, the simulation hypothesis is not a skeptical hypothesis in which nothing exists. Instead it is a *metaphysical* hypothesis, a hypothesis about the nature of reality. It's equivalent to a metaphysical hypothesis (the creation hypothesis) about how our world was created, plus a separate metaphysical hypothesis (the it-from-bit hypothesis) about what underlies reality in our world. If the simulation hypothesis is right, the physical world is made of bits, and a creator created the physical world by arranging those bits.

In the next three chapters, I'll lay out the creation hypothesis and the it-from-bit-hypothesis and then argue that the simulation hypothesis is roughly equivalent to the two hypotheses put together. Along the way, I'll discuss many connected issues about God and about reality.

The no-illusion view in the history of philosophy

My response to Cartesian skepticism rests on a positive answer to the Reality Question. In a perfect simulation, things are perfectly real. The same goes for other Cartesian scenarios, such as Descartes's evil-demon scenario and Hilary Putnam's brain-in-a-vat scenario. Generalizing simulation realism to these scenarios, we arrive at the *no-illusion* view of Cartesian scenarios. In these scenarios, things are largely as they seem. Subjects in these situations aren't deceived; they have largely true beliefs about the world.

If the no-illusion view is right, then Cartesian scenarios are no bar to knowing about the external world. If most of our beliefs are correct in a Cartesian scenario, then our inability to rule out that scenario does nothing to cast doubt on our beliefs. Our beliefs are more robust than many Cartesians have thought.

I'm not the first to advocate a response to skepticism holding that subjects in these scenarios have largely true beliefs. Still, it's striking how uncommon this view has been in the history of philosophy. Recently, I asked a number of historians of philosophy whether they knew of anyone explicitly endorsing this view before the 20th century. No one could come up with a clear case. Even in the 20th century, there were only a handful of brief discussions of it.

I suspected that George Berkeley, the 18th-century Anglo-Irish philosopher we first encountered in chapter 4, might have held this view. Recall that Berkeley was an idealist who argued that appearance is reality. In an evil-demon scenario, the tables and chairs have the same appearance as tables and chairs in the physical world. If appearance is reality, it follows that people in that scenario will experience real tables and chairs, and will have mostly true beliefs about their reality.

Alas, it looks as if Berkeley never endorsed this view. He never discusses Descartes's evil-demon scenario at all. He would probably have considered the evil demon an impossibility; like Descartes, he thought God alone was perfect enough to produce sensations and perceptions like ours. Still, he insisted that God is doing this and when he's doing it, we're getting things right. You could think of that as a cousin of my view on skeptical scenarios, with God playing the role of the simulator or evil demon.

The first really clear statement of the no-illusion view that I know of appears in a 1949 essay by the University of Nebraska philosopher Oets Kolk Bouwsma. Bouwsma was a student of Ludwig Wittgenstein, who inspired Jonathan Harrison's hallucinating Baby Ludwig in chapter 4. Bouwsma's article "Descartes' Evil Genius" is a wonderful fable about a hapless evil demon who sets out to deceive us about reality and

Figure 17 Bouwsma's evil demon tries to deceive Tom, first with paper flowers and then with perfect simulations.

always fails. (Bouwsma translates "malignus genium" as "evil genius," but I'll use the more common "evil demon," as in chapter 3.) The demon first turns everything into paper, but people quickly notice. Tom, the human protagonist of the story, who has himself been turned into paper, detects the illusion. He finds that paper flowers don't smell or feel the same, and the evil demon is exposed.

In his second attempt, the evil demon destroys everything except people's minds. He grows cocky and whispers to Tom, "Your flowers are nothing but illusions." Tom replies that he can tell the difference between flowers and illusions, and these are plainly not illusions. He understands what the evil demon has done, but he says that what the demon calls an illusion, he calls a flower. The demon has not created an illusion, he has created flowers. The demon rides off on a corpuscle, defeated.

Like me, Bouwsma thinks that the subject in Descartes's evil-demon scenario is not undergoing an illusion. His reasons differ from mine, however. Bouwsma thinks that an illusion is an illusion only if it can be discovered. If an illusion cannot be discovered, as in Descartes's evil-demon scenario, it's not an illusion at all, and it's not a deception.

More precisely, Bouwsma's evil demon can discover the illusion, but his subject, Tom, cannot. So when the demon says, "This is an illusion," he's correct, but if Tom were to say this, Tom would be wrong. For similar reasons, what Tom experiences counts as a "flower" for Tom but not for the demon. Consequently, when Tom says, "That's a flower," he's correct.

Bouwsma's reasoning here is grounded in a sort of verificationism, which says that all meaningful claims are verifiable. You'll recall that earlier we discussed the verificationism of Rudolf Carnap and other logical positivists, who held that skeptical hypotheses are unverifiable and therefore meaningless. Bouwsma takes a related line. He thinks that the hypothesis that there's an illusion is meaningless if it cannot be verified. If we can never verify that the evil-demon scenario is an illusion, it's not an illusion at all.

For reasons I discussed in chapter 4, I reject verificationism. Many meaningful claims are unverifiable, and there may well be illusions we never discover. So I reject Bouwsma's analysis of the situation. Nevertheless, I think he's right about the crucial point. The subject in Descartes's evil-demon scenario is not undergoing an illusion.

A related route to the no-illusion view that shares some of the spirit of Berkeley's idealism is suggested by the Chinese philosopher Philip Zhai (who now publishes as Zhenming Zhai) in his important 1998 book *Get Real: A Philosophical Adventure in Virtual Reality*. Zhai holds that as long as we have stable and coherent perception of an object, that object is real. In the spirit of Berkeley, we might say that *stable and coherent appearance is reality*. Zhai does not apply this thesis to issues about skepticism, but he applies it to virtual reality and perfect simulations. A subject in a perfect simulation will have stable and coherent perception of the world, so by Zhai's lights the world is real and not an illusion. I reject Zhai's idealist framework for much the same reasons that I rejected Berkeley's idealism in chapter 4, but like verificationism, idealism provides one important route to the no-illusion view.

A third route to the no-illusion view is suggested by Hilary Putnam in his 1981 book *Reason, Truth and History*. In chapter 4, we encountered Putnam's argument that the brain-in-a-vat hypothesis is contradictory based on externalism, his theory that the meaning of a word depends on its external context. Putnam suggests that the beliefs of a brain in a vat are about the electronic impulses in its environment, and that these beliefs are mostly true. This is a version of the no-illusion view that can potentially drive an entirely different response to skepticism (though Putnam doesn't make the connection to skepticism). Instead of arguing that the brain-in-a-vat hypothesis is contradictory, one could argue that the beliefs of a brain in a vat are mostly true. In effect, where Bouwsma argues from verificationism to the no-illusion view, Putnam argues from externalism to the no-illusion view.

I'll analyze Putnam's argument for the no-illusion view in chapter 20, where I discuss externalism. In my view, the argument doesn't quite succeed. Just as Bouwsma's argument requires an implausible verificationism and Zhai's view requires an implausible idealism, Putnam's argument requires a strong and implausible version of externalism. Still, like Bouwsma and Zhai, Putnam is on the right track in holding the no-illusion view.

Bouwsma, Putnam, and Zhai offer three different routes to the no-illusion view of simulation scenarios. All three routes can lead to what I think is the correct view of the Cartesian skeptical problem: the subject in Cartesian scenarios has largely true beliefs and is not deceived. However, all three routes rest on strong and implausible philosophical views that I reject. As a result, I think that none of these three philosophers has given a strong argument for the no-illusion view nor a plausible analysis of just why it is true. So we're still in search of a good argument and a plausible analysis.

I think the best argument for the no-illusion view comes not from verificationism, idealism, or externalism but from a sort of *structuralism* about the external world. I'll work up to this argument gradually in the next three chapters.

Chapter 7

Is God a hacker in the next universe up?

A NUMBER OF YEARS AGO, MY THEN-FIVE-YEAR-OLD NEPHEW Tom showed me how to play *SimLife*. He painstakingly built up a city with houses and cars, surrounded by trees. Then he told me "Here comes the fun part." He called up fires and tidal waves to destroy the city. I saw my nephew in a new light. Was he just a five-year-old kid playing games? Or was he an Old Testament God?

In an episode of *Rick and Morty*, the eccentric scientist Rick keeps a microscopic world of bicyclists under the hood of his spaceship in order to generate power. Every now and then, he visits this world, where people treat him like a god. He makes sure that everyone keeps pedaling to generate energy for his spaceship. (Let's not ask too many questions about how microscopic pedaling powers a macroscopic spaceship.) Rick created the world, and he has ultimate power over it. He is the god of the microscopic world.

If we create simulated worlds ourselves, we'll be the gods of those worlds. We'll be the creators of those worlds. We'll be all-powerful and all-knowing with respect to those worlds. As the simulated worlds we create grow more complex and come to include simulated beings who may be conscious in their own right, being the god of a simulated world will be an awesome responsibility.

If the simulation hypothesis is true and we're in a simulated world, then the creator of the simulation is our god. The simulator may well be all-knowing and all-powerful. What happens in our world depends on what the simulator wants. We may respect and fear the simulator.

At the same time, our simulator may not resemble a traditional god. Perhaps our creator is a mad scientist, like Rick—or perhaps it's a child, like my nephew.

The transhumanist philosopher David Pearce has observed that the simulation argument is the most interesting argument for the existence of God in a long time. He may be right.

I've considered myself an atheist for as long as I can remember. My family wasn't religious, and religious rituals always seemed a bit quaint to me. I didn't see much evidence for the existence of a god. God seemed supernatural, whereas I was drawn to the natural world of science. Still, the simulation hypothesis has made me take the existence of a god more seriously than I ever had before.

What is a god?

What is the definition of a god? Like most words, there's no definition that everyone agrees on. But at least in the Judeo-Christian and Islamic traditions, God is typically said to have at least the following four properties:

First, God is the *creator* of the universe.
Second, God is *all-powerful*. God can do anything.
Third, God is *all-knowing*. God knows everything.
Fourth, God is *all-good*. God is perfectly good, and acts only out of goodness.

At least as a first approximation, we could define a god as a being with these four properties. That is, a god is a being who created the universe and who is all-powerful, all-knowing, and all-good. Later, we can worry about whether all four properties are required, and whether other properties are required.

Suppose we're in a simulation. Let's say the simulator is a teenage girl in the next universe up. For us, the simulator is a sort of god.

Figure 18 God as a teenager in the next universe up.

First, and most important: the simulator is the creator of our universe. She set our universe into motion in a deliberate act of creation—if only by pressing a button in *SimUniverse*.

Second: the simulator will often be extremely powerful. Many simulations give the simulator huge control over the simulation's state. Depending on the simulation, there may be limits to this control. For example, someone running *Pac-Man* can't rearrange the whole state of the world or change a *Pac-Man* world into a *World of Warcraft* world at the press of a button. But many simulators with access to the source code and the data structures involved in their simulation will have near-unlimited powers over the worlds they create.

Third: the simulator will usually know a great deal about the simulation. Again, some video games and the like may conceal the full state of a world from her. But good universe-simulation software will include devices that she can use to keep track of what's going on anywhere in the simulated universe. For example, she might have a Cosmoscope (a

device that I wrote about in my book *Constructing the World* and that was illustrated in the TV series *Devs*), which allows users to zoom in on any part of the universe and in principle come to know about anything happening there. And again, simulators with access to the data structures could use these to monitor any entities in the world.

Fourth: will the simulator be all-good? There's not much reason to think so. All sorts of beings may have access to simulation software, and sterling character isn't typically a requirement. Some simulators may well be like my nephew, whose attitude toward his subjects was far from benign. Some may be like Rick, exploiting the subjects for selfish ends. Some simulators may want to see their subjects thrive, but simulators like that may be in a minority. Many may simply be looking for entertainment or for information.

To a first approximation, our simulator comes close to satisfying three of the four criteria. She created the universe, she's extremely powerful, and she's highly knowledgeable about those universes, but she need not be especially good.

There may be some limits to simulators' power and knowledge over the world, even in well-designed simulation software. An old chestnut asks, *Could God make a stone so heavy he could not lift it?* Or even more straightforwardly: *Could God create a stone he did not create?* It seems unlikely that God could contradict logic like this. But being constrained by logic isn't a serious limitation. More seriously, if the simulator has limited ingenuity, there will be things she can't create— quantum computers, say. Similarly, there may be things she doesn't know. For example, she might not know the secret to achieving world peace, or what someone's favorite color is. Still, access to a Cosmoscope will help a great deal.

The godliness of simulators will also be limited in another way. The simulator might be the creator of our universe—but not of the entire cosmos. Consider my nephew Tom. He created the simulated city he was playing with, and he had great knowledge of it and great power over it. But he did not create the ordinary universe he was living in, and he had no special knowledge or power in it. In this universe, he was an ordinary kid. Most simulators will be in the same situation.

The key distinction here is that our *universe* is the four-dimensional spacetime that we inhabit and the *cosmos* is the full extent of reality. If we're in a simulation, the universe is only part of the cosmos. Then our simulator created our universe but she need not have created the cosmos. Similarly, our simulator may have great power and knowledge in our universe but not in the cosmos as a whole. We might then say that the simulator is a *local god* but not a *cosmic god*.

Our simulator-god is perhaps not as awesome as some gods. The God of the Abrahamic religions—Christianity, Islam, and Judaism—is usually thought of as a cosmic god who is all-knowing, all-powerful, and all-good throughout the cosmos. Our simulator is merely a local creator of our universe, with a good deal of local knowledge and power and perhaps not much goodness.

Still, the Abrahamic God imposes an uncommonly demanding standard that the Greek gods didn't come close to meeting. Most Greek gods were powerful but not all-powerful, and few were especially good. The Hindu religion has many deities who are far from perfect. Many polytheistic religions—Shinto, for example, along with many traditional African religions—posit local gods who are neither cosmic gods nor especially good. Our simulator might at least be on a par with the gods of some of these religions.

The simulator is perhaps closest to what Plato called a *demiurge*. In ancient Greece, a demiurge (or *dēmiourgos*) was an artisan or craftsman. In Plato's dialogue the *Timaeus*, he describes the demiurge as a divine being who "fashioned and shaped" the material world. The demiurge is often treated in the Platonic tradition as a sort of second god, with the one true cosmic god above it. Plato's demiurge was benevolent, but later, in the Gnostic tradition, demiurges were regarded as evil. The simulator can likewise be treated as a second deity, perhaps benevolent and perhaps not, who is responsible for the creation of our world.

If a simulator created our universe, that already makes her a *sort* of god. Increases in knowledge of and power over the simulation may upgrade her. But is something missing? Do we really want to found a religion around our simulator? I'll return to that theme as we go along.

Arguments for the existence of God

There is a long philosophical tradition of giving arguments that God exists—and of replying to those arguments. The classic ontological argument for the existence of God was put forward by Saint Anselm of Canterbury, a Benedictine monk, in the 11th century. It goes something like this: God is by definition an absolutely perfect being. We can't conceive of any greater being. God has all the perfections—knowledge, goodness, power, and so on. But existence is a kind of perfection, too! An existing God is clearly greater than a nonexistent God. So if God doesn't exist, he is imperfect. Since God is perfect by definition, God must exist!

Many philosophers have found this argument a little too good to be true. One problem is that you could use an argument like it to "prove" the existence of all sorts of nonexistent entities.

Here's an example adapted from one of Anselm's fellow monks. Gaunilo of Marmoutiers gave an argument involving a perfect island, but we'll use a perfect hamburger instead. Let's define a perfect hamburger as a hamburger greater than which none can be conceived. Perhaps it will be perfectly juicy and perfectly tasty—as well as perfectly vegan so that no animals were harmed. But a hamburger sitting on a plate in front of me is greater than one which is not. So if a hamburger is not sitting on a plate in front of me, it's imperfect. So by definition, a perfect hamburger is sitting on the plate in front of me. I've just proved the existence of a perfect hamburger in front of me. But there's no hamburger there!

What a disappointment! Something has gone awry. But if something goes wrong with the perfect-hamburger argument, it seems likely that the same thing goes wrong with the perfect-being argument. One diagnosis is that the latter argument establishes only that *if* there is a perfect being, that perfect being exists (compare: *If* there is a perfect hamburger, it is sitting in front of me). But the argument cannot establish the existence of a perfect being (or a perfect hamburger) in the first place, and therefore cannot establish the existence of God.

In any case, the god of the ontological argument doesn't resemble

Figure 19 A simulated al-Ghazali on the cosmological argument.

the god of the simulation. As we've already seen, the god of a simulation may be imperfect in many ways.

Another classic is the *cosmological* argument for the existence of God. Versions of this argument are found in many philosophical traditions, but it is associated especially with medieval Islamic philosophers such as al-Ghazali. It goes something like this: Everything has a cause. Therefore, the universe has a cause. That cause must be God.

Hang on, you may say. What causes God? If nothing causes God, then the premise that everything has a cause is false. But if something causes God, then God isn't the ultimate cause after all. To get around this, al-Ghazali restricted the premise to be "Everything that begins to exist has a cause," saying that the universe has a beginning but God has existed forever. But now we can ask, what if the universe has existed forever? If an eternal God doesn't need a cause, neither does an eternal universe. Furthermore, if an eternal universe can exist without a cause, why couldn't a bounded universe that starts in a Big Bang do the same?

If we're in a simulation, the simulator does some of the work that the cosmological argument asks for. She serves as a cause for our universe. But in this case, the cause of our universe isn't much like a traditional god. The simulator certainly began to exist, for example. This raises the obvious objection: Who or what causes the simulator to exist? We'll get a chain of causes that eventually goes back to the cosmos as a whole and its creation. Some will stop the chain at the cosmos. Some will bring in a cosmic god. Some will ask for a cause for the cosmic god. It's not obvious why bringing in God and then stopping makes more sense than any other stopping point.

Another influential argument for the existence of God is the *argument from design*. Our universe exhibits impressive design. Humans and other animals are amazingly well-functioning mechanisms. Nature is extraordinary in its complexity. Nothing so remarkable could happen randomly. It requires a designer.

Two hundred years ago, this was perhaps the most convincing argument for God's existence. These days, the original version of the argument has long since been eroded by Darwin's theory of evolution by natural selection. To explain the remarkable mechanisms throughout nature, we don't need a designer. The evolutionary process is enough.

A more sophisticated version of the argument from design still exists, though: the *argument from fine-tuning*. This argument concerns the physical laws of our universe—laws concerning gravity, quantum mechanics, and so on. If those laws had been just a little different, then our universe would have been much less interesting and life would never have evolved. By some reasonable measures, most ways of setting up the physical laws of a universe wouldn't lead to life, but the laws in our universe did. So this universe seems to be *fine-tuned* for life. This fine-tuning requires an explanation. The obvious explanation is a fine-tuner—namely, God.

The god of the fine-tuning argument is quite consonant with the god of the simulation. It seems plausible that people who run simulated universes will often be more interested in universes that contain life than in universes that don't. Indeed, simulators may often simulate universes *in order to* simulate life, as when they simulate episodes in the history of their own species. In this case, simulators will quickly abandon settings that don't lead to life and will focus on those settings that lead to interesting forms of life. In this way, the preferences of a simulator can explain why our universe is among the (presumably) relatively few that are fine-tuned to produce life.

The fine-tuning argument is controversial. Some people deny that the necessary conditions for supporting life are all that special. Others think we might just have been lucky. Still others say that if the universe hadn't supported life, there would never be anyone there to notice. Given that we notice, it's no surprise that our universe supports life. This last line of thought is called the *anthropic principle*.

Anthropic reasoning works best if there's a multiverse, a cosmos made up of many universes. Perhaps some spring from black holes in other universes, for example, or perhaps there is a sequence of universes arising after successive Big Bangs. Some cosmologists hold that the laws of the universe are partly determined by what happens just after the Big Bang, so the laws may vary from one universe to another within the multiverse. If there are enough universes with varying laws, it's highly likely that one or more of them will contain life and observers. If so, it's no surprise to find ourselves in such a universe.

The multiverse solution to the fine-tuning problem is usually put forward as an alternative to the God solution. But the simulation argument allows for the two to be compatible. Suppose we have simulators who are utterly uninterested in life or observers. They're interested only in charting the physical dynamics of many different universes with different laws. In that case, they'll create many universes, and if they create enough, some of those universes will support life and observers. As before, it will be no surprise that we find ourselves in one of those universes. Here, even though there's a sort of god, the solution to the fine-tuning problem doesn't lie in the god's designing our world but in the existence of a multiverse. If design plays any role, it's the overall design of the multiverse that counts.

This reflects a general weakness in the multiverse solution to the fine-tuning problem: What explains the fine-tuning of the multiverse? The multiverse itself is presumably the consequence of some underlying laws or principles. If those laws had been different, there might have been just one universe, not a multiverse. So we need to explain this fine-tuning too. It won't help to postulate a higher-level multiverse in which our multiverse is contained, since this just iterates the problem. So we need some other solution here. Perhaps fine-tuning for a multiverse was unsurprising, or perhaps it was lucky. Or perhaps it involved a designer.

On the other hand, the simulation analogy also brings out a weakness in any argument from design. The designer itself must presumably be an impressive creature—one that indicates design. Surely that design itself requires explanation. Another designer just pushes the problem back: What explains the design of that designer, or of the whole system of designers? Someone might say that God is exempt from explanation, but this looks like special pleading. So both the designer hypothesis and the multiverse hypothesis leave something unexplained.

We may well have to accept some unexplained impressiveness as a brute fact about the universe. We can try to minimize what needs to be explained, perhaps by boiling it down to simple principles at the fundamental level. But we'll always be left with the problem of why there is something rather than nothing. We'll always be left with the problem of why the ultimate laws are the way they are. And we'll always be left with the problem of why they're so interesting.

The simulation argument for the existence of God

It's easy to find problems in arguments for the existence of God. The fine-tuning argument is perhaps the strongest of them, but it seems far from definitive.

Might the simulation argument then be the most powerful argument for the existence of God? Of course, it's mainly an argument for a creator and not for a being who is all-powerful, all-knowing, and all-good. But the same is true of the cosmological and design arguments, and, unlike those, the simulation argument suggests a god with considerable power and knowledge beyond the mere power of creation. It's also only an argument for a local creator, but the same is true of the design argument. If the simulation argument is even approximately as good as the design argument, it deserves to be in the pantheon of arguments for God's existence.

A version of the simulation argument could have been made long before computers existed. Where the argument talks about simulation, one could talk about universe creation instead. Here's a cousin of the initial simulation argument in chapter 5:

1. A few top-level populations will each create many populations.
2. If a few top-level populations each create many populations, then most intelligent beings are created.
3. If most intelligent beings are created, we are probably created.

   ---

4. So: We are probably created.

Here a "top-level" population is one that wasn't created by anyone (except perhaps by a cosmic god). "A few," "many," and "most" can be understood along the lines of the numerical simulation argument in chapter 5: e.g., at least one in ten, one thousand, and 99 percent. The case for premises 2 and 3 goes through roughly as before. Premise 2 follows from mathematical reasoning, and premise 3 seems to follow at least if we are typical intelligent beings.

If one had made this argument a century ago, the main objection would presumably have been: Why believe the first premise? That is, why believe the capacity to create universes will be reasonably common? Perhaps one could have speculated, but the support is not obvious.

What the simulation idea adds is a reason to believe the first premise. It does that by suggesting that the capacity for universe creation is relatively straightforward and likely to be within range of many populations. Given that much, the rest quickly follows, at least once we make appropriate provisos for sim blockers and sim signs.

Like the simulation argument, this creation argument does not really get us all the way to the conclusion that we are probably created. The simulation argument could be blocked by holding that humanlike sims are impossible, or that they're possible but most humanlike nonsims won't create them. Likewise, the creation argument could be blocked by holding that creating humanlike beings is impossible, or that this is possible but most top-level humanlike beings won't create them. Still, we can get to the same sort of three-way conclusion: either most beings are created, or most humanlike populations won't create humanlike populations, or creating humanlike beings is impossible. Before computer simulation technology, an atheist could easily enough accept the second alternative instead of the first. At that time, there wasn't much reason to think population creation would be rife. After simulation technology, there's much more reason to believe that population creation will be common, thereby making the argument that we're created more plausible.

The simulation route leads to a distinctive sort of god. The simulator is a *natural* god, one who is part of nature. The ontological, cosmological, and design arguments are often used to argue for a *supernatural* god, one who stands outside nature. The simulator is beyond our own physical universe but not beyond nature as a cosmic whole. In principle, the simulator can be explained by the natural laws of the cosmos.

As a result, the simulation hypothesis is compatible with *naturalism*. Naturalism is a philosophical movement that, at a minimum, rejects the supernatural. It holds that everything is part of nature and can be explained by natural laws. Many have thought that naturalism and God are hard to reconcile, so that naturalism should lead to atheism. The simulation hypothesis offers a path to reconciliation: a god that even a naturalist can believe in. The most famous argument against the existence of god is the problem of evil. An all-good, all-knowing, and all-powerful god would not permit evils such as natural disasters and genocide in the world. But these evils exist, so there is no god. I'll discuss the problem of evil in more depth in chapter 18. But it's worth noting here that the problem of evil is no obstacle to a naturalistic simulator god. As we've seen, a simulator need not be all-good. She may well tolerate some evils in the simulation.

Simulation theology

In Stanislaw Lem's 1971 short story *Non Serviam*, Professor Dobb, a specialist in "personetics," creates a society of artificial "personoids." After many generations, the personoids speculate about the nature of their creator. The personoid Edan 197 assumes that their god requires reverence and gratitude in order for them to win salvation; if the personoids don't believe in their creator, they won't be saved. Adan 900 says that this would be unjust: God hasn't given them strong evidence of his existence, so he cannot justly punish them for not believing in him; a perfectly just god would save nonbelievers as well. Adan 900 goes on to suggest that since an almighty god could have provided certainty, the fact that God has not provided certainty suggests that God is not almighty.

Professor Dobb listens to these debaters with interest. He says their reasoning is impeccable. He created them, so he's their god. He hasn't provided proof of his existence, and he doesn't demand worship from them. In an afterword, he states:

In point of fact, when I create intelligent beings, I do not feel myself entitled to demand of them any sort of privileges—love, gratitude, or even service of some kind or other. I can enlarge their world or reduce it, speed up its time or slow it down, alter the mode and means of their perception; I can liquidate them, divide them, multiply them, transform the very ontological foundation of their existence. I am thus omnipotent with respect to them, but indeed, from this it does not follow that they owe me anything.

He speculates that he could add an "enormous auxiliary unit" to serve as a "hereafter," admitting only those personoids who believe in him, and annihilating or punishing all those who don't. He says this act would be regarded as a piece of "fantastically shameless egotism." But he notes, with regret, that one day his university will demand that he turn the simulation off.

Lem's story is an early work of simulation theology. Theology is (roughly) the study of the nature of God from the point of view of God's subjects. Simulation theology is the study of the nature of the simulator-as-God from the point of view of those within the simulation.

We, too, can engage in simulation theology. Under the assumption that we're living in a simulation, we can speculate about the nature of our simulator. Is the simulator probably humanlike, or some sort of artificial intelligence? Is the simulator running the simulation for entertainment? For science? For decision-making? For historical analysis?

You might think we have no basis for reasoning about these things: All simulation theology must be idle speculation. But if we take the simulation argument seriously, simulation theology is not completely inane. We can reason about the character of our simulator by reasoning about what sort of simulations will be most likely to emerge in the history of the cosmos.

For example, we might wonder whether our simulator is more likely to be a biological or quasi-biological entity in its own universe, or something more like an artificially intelligent system or perhaps a simulated being that inhabits its own simulation. At least in our world, it seems probable that in the long run, AI systems will be both much faster and much more capable than biological systems. If so, we can expect that AI systems will produce many more simulations than biological systems. It seems not unreasonable that the same may apply throughout the cosmos. If so, we should expect our simulator to be an AI system and not a biological or a quasi-biological system.

This is reminiscent of the situation in *The Matrix*, in which the Matrix creators are themselves machines. If we were living in the Matrix world, our gods (that is, our simulators) would be the machines. At the very least, the machines are our demiurges! This may well be the typical case for people who are living in simulations. Most creators are machines.

(I can't resist a tangent at this point on my theory of the theology of *The Matrix*, which in one of the prouder achievements of my life, I expound in a bonus "Easter egg" video in the official box-set version of *The Matrix*. It is often suggested that Neo is the Christ figure in the movie, with Morpheus representing John the Baptist and Cypher representing Judas Iscariot. If I'm right and the machines are the gods, this interpretation is all wrong. Who is the son of the machines, who is sent down into the world to save it from those who want to destroy it? Clearly, it's Neo's nemesis, Agent Smith. Agent Smith is the true Christ figure in *The Matrix*. Perhaps it is telling that in the sequels, he is resurrected.)

Here's another application of statistical theological reasoning: It's plausible that simulations run for scientific purposes will be more common than those run for entertainment purposes. Scientific simulations will be run in large numbers at once, whereas entertainment will require many fewer simulations, perhaps not more than one per person at a time. If so, it's much more likely that our simulator is a scientist rather than a fan.

Suppose we want to study just how fine-tuned our universe is for life. If we have good-enough simulation technology, we can set up a large portfolio of different simulated worlds with different laws and different initial conditions. We can run all the simulations and see how many end up evolving life. The more simulations we run, the more accurate information we'll get. For all we know, scientists in the next universe up are running billions of simulations of this sort.

Another role for simulations is in decision-making. In the *Black Mirror* episode "Hang the DJ" (spoiler alert!), people using dating apps on their phones routinely run simulations in order to determine how compatible they are. It's typical to run 1,000 simulations instantly. Each simulation places simulated versions of the prospective couple together and sees whether a successful relationship results. If 998 of 1,000 simulations lead to a good relationship, then the couple can be reasonably confident they should end up together. This saves a lot of time! So if you find yourself in the early stages of a relationship, perhaps you should suspect even more strongly that you're in a simulation.

Still, one wonders: Do simulators allow the use of simulation technology by the people *within* the decision-making simulation? If they do, the computational costs will be enormous, and a huge chain of simulations within simulations threatens. If they don't, then the simulated reality will be quite unlike the unsimulated reality (where use of simulation technology will be ubiquitous). Either way, when a given simulation technology becomes widespread among a group of people, it thereby becomes less useful for purposes of predicting what those people will do.

These limits also apply to political, military, and financial decision-making. Daniel Galouye's *Simulacron-3* (see chapter 2) and other early simulation scenarios in science fiction described businesses engaged in product development testing the market through simulations. That's OK when the people being simulated don't have the technology. But when they do, market testing may require more and more advanced simulation technology, with simulations within simulations within simulations. It's easy to see a simulation arms race resulting.

In any case, simulating people like us (who lack advanced simulation technology) won't be useful for making decisions in a world where that technology is present. So perhaps our simulator is more apt to be a scientist than a decision-maker. But there may be all sorts of other reasons for creating simulations that we've not yet begun to fathom.

There's reason to think that most science-based simulations in the cosmos will be part of large batches of simulations all similar except for tiny tweaks. Let's say a batched simulation is a simulation in a batch of 1,000 or more, and that a singleton simulation is one being run on its own. Suppose that batches of simulations are at least one percent as common as singleton simulations. Then for every 100 singleton simulations we'll have at least one batch of 1,000 batched simulations, so that batched simulations outnumber singleton simulations by at least ten to one. The same reasoning extends to batches of a million or more simulations, until we reach the point at which batches become so large and so difficult that their numbers drop off rapidly.

To simplify a little: Suppose that batches of 10, 100, 1,000, 10,000, 100,000, and 1,000,000 simulations are roughly equally common and things drop off fast after that. Then most simulations will be part of a batch of a million or more, and we should expect to be in such a batch ourselves. We should expect to find ourselves in a batch of a million simulations. If things drop off slowly, so there are a good number of batches of a billion simulations (say, one percent as many as batches of a million), then we should expect to find ourselves in a batch of a billion. We certainly should not expect to find ourselves in a singleton simulation unless singletons are for some reason hugely more common than larger batches.

It's reasonable to conclude that if we're in a simulation, our simulator is probably running a large batch of simulations.

This has some further consequences for simulation theology. For example, simulators running large batches of simulations may be much less likely than singleton simulators to pay attention to individual simulations while they're running, or to regularly intervene in the simulations. If this is right, we should expect that our simulator isn't paying attention to us and is unlikely to intervene. Of course, our simulator may be gathering observations for statistical purposes, and all sorts of automatic intervention mechanisms may be set up. And if our simulator is a sufficiently advanced artificial intelligence, perhaps it will have no trouble paying scrupulous attention to each simulation in a batch as it evolves. Still, we should take seriously the possibility that our creator is ignoring us.

Furthermore, simulations of all sorts may well have stopping criteria. Scientists and decision-makers are using these simulations to gather information, and once they have the information they need, there may be no need for the simulation. Perhaps ethics guidelines will require that each simulation be allowed to play out indefinitely—but perhaps not. So we should be aware that there's some chance that our universe will end abruptly once a stopping criterion is satisfied.

Of course, we cannot know what those stopping criteria are. Simulations set up for entertainment may be terminated once they're no longer entertaining. The philosopher Preston Greene has speculated that many simulations will be terminated once the simulated populations themselves develop world-simulation technology because at that stage the computational power required to support simulations within simulations will make the original simulation too expensive to run. But we might give some thought to what those stopping criteria might be and how to avoid satisfying them.

(The very idea of a stopping criterion is at least superficially reminiscent of the 19th-century idea of the "end of history," associated with the German philosopher Georg Wilhelm Friedrich Hegel and a number of other thinkers. One version of this idea holds that the world is evolving to a point where it becomes conscious of its nature, which is the endpoint of history. In the naturalistic key of simulation: perhaps simulators are studying what we know, and will terminate the simulation when we become conscious that we are in a simulation.)

Does the simulation hypothesis accommodate an afterlife? It may at least make an afterlife possible. Computer processes are portable. Simulators may be able to transfer simulated brain processes from one simulated world to another (heaven?), or even connect them to a body in the simulators' own world (reincarnation?). Perhaps that will happen in some simulations, especially in personal entertainment simulations, and perhaps for truly exceptional beings in batch simulations. Doing this routinely for most simulations would come with many costs, however. If those costs are prohibitive, we shouldn't hope for a simulation afterlife. On the other hand, perhaps ethics review panels for simulators will insist that no simulated being ever be truly killed. After a sim "dies" in a simulation, its code must be transferred to another virtual world, perhaps running at a slow speed to reduce costs. If so, there may be greater prospects for life after death.

If we ever create artificial intelligence within a simulation, it may be hard to keep it contained. For instance, if we communicate with the simulated beings, they'll presumably become aware that they're in a simulation, and they may become interested in escaping it. In aid of this, they may try to figure out our psychology in order to see how to persuade us to let them out (or at least give them unfettered access to the internet, where they can do whatever they want). And even if we don't communicate with them, they may seriously entertain the hypothesis that they're in a simulation, and then do their best to figure the simulation out. That would be a form of simulation theology.

We could in principle do the same thing. We could try to attract the attention of our simulator and communicate with it—perhaps by writing books about simulations, or by constructing simulations. We could try to figure out our simulation, to determine its purpose and its limits. But if our simulator is an artificial intelligence who has designed a batch of watertight simulations and isn't paying attention, then our efforts may be in vain.

Is our simulator herself in a simulation? A variation on the simulation argument can be used to argue that we are probably simsims: beings in simulations within simulations. At least if there are no simsim blockers (factors that prevent the creation of many simsims), then most humanlike beings will be simsims. On the other hand, limits on computer power may tend to block simsims more than they block sims. We should certainly assign a lower probability to being a simsim than to being a sim, since all simsims are sims but not vice versa. But nevertheless there remains a non-negligible chance that our simulator is simulated.

Does there have to be an unsimulated simulator at the top of the chain? It's highly intuitive that there must be a fundamental level with unsimulated reality. The alternative is reminiscent of the old story where an audience member tells the American philosopher William James that the earth stands on the back of a turtle that stands on another turtle in turn. When pressed on the details, she says "It's turtles all the way down." Still, the contemporary philosopher Jonathan Schaffer has argued that there need not be a fundamental level in nature: There could instead be a never-ending sequence of levels. If Schaffer is right, this opens up at least the theoretical possibility that we are in a cosmos where it is simulations all the way up.

Simulation and religion

Should simulation theology lead to a religion of simulation? There's more to religion than theology. It involves a deep commitment that people organize their lives around, and a distinctive system of moral beliefs and moral practices. The Judeo-Christian tradition has a set of prescriptions about how one should live, from the Ten Commandments to the Sermon on the Mount. Islam has its own moral commandments laid out in the Qur'an. Hindu texts lay out a series of Yamas and Niyamas, which are vows of moral conduct. Buddhist scriptures lay out five precepts that are at the core of its moral code.

Does simulation theology come with any set of moral practices? Why should it? There might well be self-interested practices. For example, potential sims might act in certain ways in hopes of being uploaded out of the simulation. There might even be population-wide practices: For example, we should forbid the building of simulations, lest our simulator terminate our own simulation. It's morally imperative that we keep the simulation going! But these principles don't really constitute a religion.

Another mark of religions is that they typically entail forms of worship. People worship the Judeo-Christian and Islamic God and the Hindu gods. There are religions without typical gods and without worship—Buddhism, Confucianism, Daoism. But where gods are involved, worship is the rule.

Should we worship our simulator? It's hard to see why. The simulator may simply be a scientist or a decision-maker in the next universe up. We may be grateful to her for creating our world. We may be in awe of the power she has over our world. But gratitude and awe alone are not worship.

We may be terrified by the power that the simulator has over our lives. Perhaps if we came to believe that our simulator was like the Abrahamic God in demanding our worship as a condition for reaching the afterlife, then we might consent to worship her in order to survive. But there's not much reason to think a simulator will have a psychology like this. And if she does, does she really deserve our worship? To para-phrase Lem's personoid Adan 900: Any god that demands our worship doesn't deserve it.

Even if our simulator is a benevolent being, why should we worship her? She may be working to create as many worlds as possible with a sufficient balance of happiness over unhappiness in order to maximize the amount of happiness in the cosmos. If so, we might admire her and be thankful to her—but, again, there's no need for worship.

I find myself thinking that even if our simulator is our creator, is all-powerful, is all-knowing, and is all-good, I still don't think of her as a god. The reason is that the simulator is not worthy of worship. And to be a god in the genuine sense, one must be worthy of worship.

For me, this is helpful in understanding why I'm not religious and why I consider myself an atheist. It turns out that I'm open to the idea of a creator who is close to all-powerful, all-knowing, and all-good. I had once thought that this idea is inconsistent with a naturalistic view of the world, but the simulation idea makes it consistent. There remains a more fundamental reason for my atheism, however: I do not think any being is worthy of worship.

The point here goes beyond simulation. Even if the Abrahamic God exists, with all those godlike qualities of perfection, I will respect, admire, and even be in awe of him, but I won't feel bound to worship him. If Aslan, the lion-god of Narnia, exists as the embodiment of all goodness and wisdom, I won't feel bound to worship him. Being all-powerful, all-knowing, all-good, and entirely wise aren't sufficient rea-son for worship. Generalizing the point, I don't think any qualities can make a being worthy of worship. As a result, we never have good reason to worship any being. No possible being is worthy of worship.

I'm sure many religious readers will disagree, but even they may agree that a mere simulator would not be worthy of worship—and therefore that a simulator is not a god in the full-blooded sense. If so, we can at least ask the question: What could make a being worthy of worship, and why?

Chapter 8

Is the universe made of information?

N 1679, GOTTFRIED WILHELM LEIBNIZ INVENTED THE BIT. THE German philosopher and mathematician is celebrated as the coinventor (along with Isaac Newton) of the calculus. He also designed and built one of the first mechanical calculators. He is famous for his optimistic thesis that we live in the best of all possible worlds. But none of his ideas was more important than the invention of the binary number system, on which all modern computers are based.

In his 1703 essay "Explanation of Binary Arithmetic," Leibniz draws inspiration from the *I Ching*, the ancient Chinese Book of Changes. The *I Ching* uses hexagrams—stacks of six horizontal lines, one under another—for divination. The hexagrams can be understood as a simple binary code based on the distinction between yin and yang. Yin is

Figure 20 Leibniz with *I Ching* hexagrams encoding binary numbers.

encoded as a broken line, whereas yang is encoded as an unbroken line. Each line corresponds to a binary digit, or a bit.

Whereas decimal digits run from 0 to 9, binary digits are just 0 and 1. If you count through the binary numbers, there is 1 (one), 10 (two), 11 (three), 100 (four), and so on. The *I Ching* hexagrams each encode a six-digit binary number, such as 110101. There are 64 possible hexagrams. In principle, any sequence of letters or numbers can be expressed at more length as a sequence of bits.

Like the *I Ching*, the integrated circuits in modern computers encode sequences of bits. Where the *I Ching* uses a broken line for 0 and an unbroken line for 1, transistors in these circuits typically use a low voltage for 0 and a high voltage for 1, or vice versa. Where the *I Ching* encodes just six bits at a time, computers often encode a trillion or more bits. Almost everything in modern computers can be explained in terms of the interplay of bits.

The interplay of bits has also been used to model reality itself. In 1970, the British mathematician John Horton Conway devised the Game of Life, in which an entire universe is made up of a pattern of bits. The universe is a two-dimensional grid of cells, infinite in all directions. At any given time, each cell is either "on" or "off"—or equivalently, 0 or 1.

As the name of the game suggests, Conway was interested in simulating living processes, which can grow for a while and then die off. For this purpose, he formulated some basic rules for how the grid evolves. These rules are the "laws of physics" for the Game of Life.

Each cell has eight neighbors—one to the north, one to the northeast, one to the east, and so on. The fate of each cell depends on the states of its neighbors. At any given time, a cell will die of "loneliness" if it has too few neighbors and of "overcrowding" if it has too many. More precisely, a cell that is *on* stays on if it has exactly two or three neighbors that are on. If it has less than two or more than three neighbors, it turns off. At the same time, a cell that is *off* stays off unless exactly three neighbors are on, in which case new life is born by the cell's turning on.

These simple rules set a dizzying array of complex behaviors into motion. A single cell will die of loneliness. A two-by-two square will stay the same forever, with each cell happily having three neighbors. A line of three cells will flip back and forth between horizontal and vertical. A "glider" is a group of five cells that shuttles across the world diagonally in a repeating pattern. There are even "glider guns" (pictured in figure 21) that produce an endless stream of gliders.

You can try out the Game of Life on many sites online. Many configurations grow for a while and go through a number of phases before settling into a stable state with a few cells going through a fixed, repeating pattern. (It has been suggested that this is reminiscent of the academic process of getting tenure.) However, mathematicians have demonstrated that there are some states that grow indefinitely and never settle down into a repeating pattern, much like a thriving life-form.

The Game of Life is itself a sort of computer known as a *cellular automaton*. It has been demonstrated that it's a *universal* computer, which means that it can do anything that any computer can do. In principle, we could use the Game of Life to run the program that controls the launch of rockets to Mars. We could also use it to run giant simula-

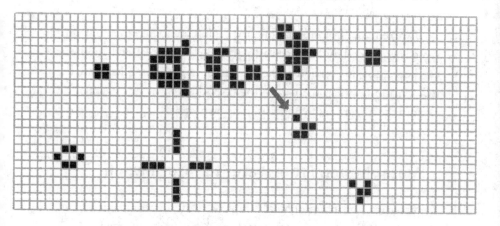

Figure 21 Conway's Game of Life. A glider gun (in the top half) producing gliders (to the lower right), along with a beehive (six stable cells) and a traffic signal made up of four blinkers (each with three cells blinking from horizontal to vertical).

tions. If you can simulate the entire universe using a computer, you can simulate it using the Game of Life.

This raises the question: Is our universe like the Game of Life? Is everything in our universe ultimately a pattern of bits?

This idea is sometimes called the it-from-bit hypothesis, a label first used by the physicist John Wheeler in 1989. Wheeler may have intended something different by the phrase, as I'll discuss later in the chapter, but the it-from-bit idea is so powerful that it has transcended his original conception. The powerful idea is that everything in the physical world around us—tables and chairs, stars and planets, dogs and cats, electrons and quarks—is made of patterns of bits.

The it-from-bit hypothesis is a wonderful thing for a philosopher—a new metaphysical idea. Some philosophers have thought reality is made of minds. Others have thought it's made of atoms. Now we have a new hypothesis: *The world is made of bits.*

The it-from-bit hypothesis has obvious resonances with the simulation hypothesis. If we're living in a simulation, then our universe is, in some sense, a giant computer process. If the computer is a standard digital computer, then its processes all involve the processing of bits: sequences of ones and zeroes processed by logical circuits that transform patterns of bits.

If we're living in a simulation on a digital computer, then our universe is grounded in the interplay of bits. So the simulation hypothesis can be seen as one version of the it-from-bit hypothesis. I'll explore the relationship between the two in the next chapter. In this chapter, I'll explore the it-from-bit hypothesis in its own right.

Metaphysics: From water to information

Metaphysics, the philosophical study of reality, encompasses many questions. Perhaps the most central is "What is reality made of?"— that is, what is the fundamental level of reality from which everything else derives?

Many indigenous cultures have their own metaphysical systems.

According to Australian Aboriginal traditions, reality as we know it originated in the Dreaming of ancestral spirits. According to Aztec traditions, reality is grounded in a self-generating power known as *teotl*.

One early golden age of metaphysical theorizing was in ancient Greece. The tradition passed down to us starts with Thales of Miletus, who lived around 600 BCE, 200 years before Plato. Thales is most famous, or infamous, for the metaphysical thesis that everything is made of water. Water is the "primary principle" from which all things come to be and to which all things return. You might ask, "What about trees and rocks?" Thales seems to have thought of them as modified forms of water that would eventually return to being water.

Other Greek philosophers put forward rival hypotheses. Thales' fellow Miletian Anaximenes, who lived around 550 BCE, suggested that everything is made of air. Earlier in that century, Heraclitus proposed that everything is made of fire. Heraclitus thought the world was fundamentally about change; hence his memorable remark that you cannot step into the same river twice. But neither the water hypothesis, the air hypothesis, nor the fire hypothesis really caught on.

A much more widespread hypothesis in ancient times held that reality was made of four or five basic elements: earth, air, fire, and water, and sometimes "aether" or "void." In Greece, the four-element earth/air/fire/water system was put forward by Empedocles around 450 BCE. A Babylonian text, the *Enuma Elis*, dating from sometime before 1000 BCE, recounts a cosmic history in which the world is formed by gods representing earth, wind, sky, and sea. The Indian Vedas, around 1500–500 BCE, make frequent reference to five basic elements: often earth, air, fire, water, and space (or aether or void). The ancient Chinese Wu Xing system, dating from around 200 BCE, suggested five elements in an eternal cycle: Wood feeds fire, which creates earth, which bears metal, which collects water, which nourishes wood.

In modern times, earth, air, fire, and water have been decomposed into more basic components, all the way down to quarks and electrons. Two other ancient Greek ideas have worn much better. Pythagoras, who lived around 550 BCE, suggested that everything is made of numbers. The number 1 represents the origin of all things, 2 represents matter, 3 represents beginning/middle/end, 4 represents the four seasons, and so on. While the specifics of the Pythagorean system may not have lived on, the idea that reality is fundamentally mathematical is still taken seriously today—as in the it-from-bit hypothesis, which might be understood as saying that the world is made up of the numbers 0 and 1 in patterns.

An even more influential metaphysical idea is associated with Democritus, who lived around 450 BCE. He was known as the "laughing philosopher" because of his cheerful manner. He is sometimes called "the father of modern science," though his ideas were heavily influenced by his teacher, Leucippus. Democritus and Leucippus held that everything is made of atoms: tiny, indivisible bodies that move in an infinite void. A number of schools of Indian philosophy, such as the Nyaya, Vaisesika, and Jaina schools also put forward atomist views.

Democritus's view is a clear antecedent of modern *materialism*, according to which the world is made of matter. Materialism has been perhaps the most popular metaphysical view among philosophers and scientists in recent decades. It's often regarded as the obvious metaphysics suggested by modern science, where the aim is to support ideas with experimental evidence and ultimately explain everything in terms of physics.

The biggest obstacle to materialism has always been the existence of minds. Alternative metaphysical views give a major role to the mind. One alternative is *idealism*, the theory that reality is made of minds, or that reality is fundamentally mental. We have encountered idealism already in Berkeley's 18th-century thesis that appearance is reality, and in the Buddhist thesis that reality is consciousness-only. Idealism can also be found in the ancient Vedas of Hindu philosophy. It plays a central role in the Advaita Vedānta school, in which the ultimate reality is held to be Brahman, a sort of universal consciousness, and in which all individual minds and all matter are grounded in Brahman.

The other classical metaphysical theory is *dualism*, the idea that both matter and mind are fundamental. Dualism holds that one cannot explain mind in terms of matter, nor can one explain matter in terms of mind, but between them mind and matter explain everything. The

Samkhya school of Indian philosophy was deeply dualist: The universe consists of *purusa* (consciousness) and *prakriti* (matter). Traditional African philosophy, Greek philosophy, and Islamic philosophy all have strong elements of dualism. In the 17th century, René Descartes became a central advocate of dualism, arguing that the world consists of matter and minds in interaction.

In Western philosophy since Descartes's time, metaphysical theorizing has typically involved an oscillation among materialism, dualism, and idealism. Descartes's English contemporary Thomas Hobbes advocated materialism. In the 18th century, Berkeley advocated idealism, and various forms of idealism dominated German and British philosophy in the 19th century. In the 20th century, the pendulum swung sharply back to materialism, which has remained the dominant approach in recent decades. Materialism has encountered major obstacles in explaining the mind, however, and as a result there remain many dualists inside and outside philosophy. Even idealism has been making a small comeback in 21st-century philosophy, in part because of an upswell in discussion of panpsychism, the thesis that all matter has an element of consciousness.

In this well-trodden metaphysical landscape, the view that reality is made of information, or bits, is something of an exciting newcomer. You might have thought that as the inventor of the bit, Leibniz would have liked the it-from-bit view, but in fact he preferred a panpsychist form of idealism in which the world is made of simple "monads" with perceptions of their own. The it-from-bit view is closer to materialism— at least, if we think of bits as the fundamental components of matter. But it is a distinctive and special form of materialism. I'll develop this view, but first we should clarify the notion of information.

The varieties of information

When we talk about information, we encounter a central ambiguity. In one sense, information is the realm of *facts*. In another sense, information is the realm of *bits*. These are two very different things.

In ordinary English, talk of information is usually talk of facts. If I know what movies will be playing next week, I might say, "I have some information that will interest you." Here the information might be the fact that *Star Wars* is playing at your local theater next week. Likewise, facts like *The current temperature is zero* or *The US president is Joe Biden* constitute information.

There's an interesting question about whether an untrue claim, such as *The capital of Australia is Sydney*, counts as information. Usually we'd count it as misinformation and distinguish it from information, but sometimes it makes sense to group these two categories together. In a modern database, for example, a fair amount of the information contained may be misinformation. An online database may have an outdated address for me and the wrong date of birth. Philosophers usually reserve the word *fact* for claims that are true, and use the word *proposition* for claims that may be true or false. Here, perhaps the most useful concept of information deals with propositions rather than facts. For now, I'll stay neutral between the two. I'll more often talk of facts because this is shorter and simpler, but most of what I say about facts can be restated in terms of propositions.

Information in this sense (involving facts or propositions) is often called *semantic* information. It makes a claim about the world: the claim that *Star Wars* is playing next week, for example. Semantic information is crucial for understanding language, thought, databases, and much more. Some philosophers have even suggested that reality is made of semantic information. In his 1921 *Tractatus Logico-Philosophicus*, Ludwig Wittgenstein wrote, "The world is the totality of facts, not of things." This "it-from-fact" view is an important metaphysical view, although it is very different from the "it-from-bit" view, according to which reality is made up of bits.

The sort of information that is most central in computer science, and most central in this book, is what I will call *structural information*. Structural information most commonly involves a sequence, or structure, of bits. As we've seen, a bit is simply a binary digit: 0 or 1. Bits can be arranged into binary sequences such as 01000111. Modern computers fundamentally deal with structural information. They encode structures of bits and transform them into new structures of bits.

In addition to inventing binary arithmetic, Leibniz himself designed and built some of the earliest mechanical calculating machines. The first such machine was designed by another great philosopher/mathematician, Blaise Pascal, in 1642. Pascal's calculator performed addition and subtraction, and Leibniz's (designed in 1671) performed multiplication and division as well. Their machines used dials encoding sequences of decimal digits rather than binary digits (Leibniz discussed a binary machine in his 1703 essay but never built it), but they process structural information all the same. This brings out that bits aren't strictly required for structural information—sequences of decimal digits or even letters of the alphabet qualify too. Still, the core case of structural information remains sequences of bits.

In some ways, the most interesting sort of information arises when bits encode facts, or when structural information encodes semantic information. I'll call this *symbolic information*. For example, the bit sequence 110111 (structural information) in a certain area of a database's memory—or on a punched card, as in figure 22—might encode the fact that my age is fifty-five (semantic information). Here, bits encode facts. That makes the bit sequence symbolic information.

Symbolic information is the central sort of information in mod-

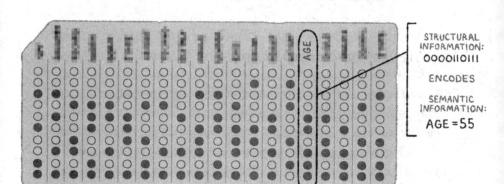

Figure 22 A punched card illustrates structural, semantic, and symbolic information.

ern data science. In any database system, and in any computer system that carries information about the world, bits encode facts. Symbolic information is also present in ordinary language—for example, whenever a string of letters, like "John is in Sydney," encodes a fact about the world. Strings of letters are structural information, and their meanings are semantic information, so language as a whole involves symbolic information.

To recap: Structural information involves *bits*. Semantic information involves *facts*. Symbolic information involves *bits encoding facts*.

Strictly speaking, the definitions should be broader than this. We've already seen that to encompass misinformation, as in "Sydney is the capital of Australia," semantic information can go beyond facts to include propositions. Likewise, structural information can go beyond bits to include structures of letters, decimal digits, and (as I'll discuss shortly) systems of differences more generally. But bits and facts are the core case, and *differences encoding propositions* doesn't roll off the tongue nearly as well as *bits encoding facts*. So we'll stay with the latter for now.

You can make the same three-way distinction for *data*. Structural data involves bits, semantic data involves facts, and symbolic data involves bits encoding facts. In different contexts, the word "data" can be used in any of these three ways. In these days of Big Data, the last meaning (bits encoding facts) seems dominant.

Structural information

Not all bits encode facts. Sometimes bit sequences serve quite different purposes. In many computer processes, like the Game of Life, bits may encode no facts or propositions at all. Computers are fundamentally devices for encoding and manipulating structural information. Encoding and manipulating semantic information is simply one of their central applications.

The 20th-century field known as information theory centers on structural information, not semantic information. Some of its main achievements are to provide *measures* of structural information. There are at least three important measures of structural information.

The simplest and most familiar measure simply counts the *size* of a bit sequence. For example, the amount of information in a sequence of 8 bits is simply 8 bits, or 1 byte. When we say that your computer has a memory of 32 gigabytes, we're invoking this measure. A second measure, developed in the 1940s by the mathematician and engineer Claude Shannon, measures how surprising or improbable a given sequence of bits is. A third measure, developed in the 1960s by the Russian mathematician Andrey Kolmogorov and the American mathematicians Ray Solomonoff and Gregory Chaitin, measures how easily a sequence of bits can be generated by a computer program. These three measures of structural information serve complementary purposes in the analysis of computation and communication.

As we've seen, structural information doesn't always involve *binary* digits. Leibniz's calculator was based on decimal digits (0 through 9). You can build a computer that uses three-way informational states (0, 1, 2), sometimes called *trits*. Strings of trits also count as structural information, although bits are more practical for the purposes of computing.

All varieties of structural information involve a system of *differences*. The simplest difference is a binary difference, between the two states 0 and 1. You can even generalize structural information further, to quantum and analog differences.

The relatively new field of quantum computation focuses on *qubits*. A qubit (pronounced q-bit) involves two states in a quantum superposition: A qubit can be in both states simultaneously. Whereas an ordinary bit is either 0 or 1, a qubit is a superposition of 0 and 1, with different amplitudes for each. Qubits are more complex than bits, but they're still a form of structural information involving a distinctive system of differences.

Theorists have also developed models of analog computation: computation using continuous real numbers, such as 0.732 and the square root of 2. For example, a 1989 article by Lenore Blum and two colleagues describes a computer that uses real numbers such as 0.2977 (with infinite precision) instead of bits. We might say that these numbers serve as *continuous digits* (conts?) or *analog digits* (ants?), but for our purposes I will call the continuous version of bits *reals*.

Analog computers using continuous quantities aren't especially useful in practice. Reliable infinite-precision analog computers are impossible to build. We don't have infinitely precise control over physical materials, and beyond a certain point precision is washed away by background noise. Finite-precision analog computers are often redundant—they can be approximated on ordinary digital computers if we use enough bits—but they have nevertheless found some use in chip design. For philosophical purposes, continuous information remains useful in thinking about the space of possible systems for processing information. It will become especially relevant when we think about the connection between structural information and continuous laws of physics.

Information is physical

When I was in high school in Adelaide, a single computer on the other side of town served all the schools in the city. To use the computer, we had to send commands on punched cards by penciling in some circles on the card and not others. A computer program would comprise a whole deck of penciled-in cards. We'd send them across town and wait a day or so for a printout of the output. As often as not, the output said "Syntax Error." We'd have to find the problem in our commands, redo the cards, and keep submitting the program until it worked.

This helped me understand the power of structural information. Each card was essentially a sequence of bits: ones and zeroes. A circle was either filled in, representing 1, or it wasn't filled in, representing 0. Each card had maybe a thousand circles, so the card as a whole made up a sequence of a thousand bits. A few bits (01000 . . .) might represent a letter, such as "P." A few letters in a row might make up a word, such as "PRINT." A few words together might make up a command like "PRINT SQRT(2)" to print out the square root of 2. (Actually, we used the ultracompressed language APL to save time and cards, but this is a simpler example.) Enough bits in enough cards would make up a whole computer program.

These bits were *physical* bits, embodied in a paper card. For much of the 20th century, punched cards (depicted in figure 22) were central to computing. Programs were encoded using a keypunch machine, which punched a hole in the paper to represent 1 and left the paper intact for a 0. A given sequence of bits was embodied in a lacy paper card with holes poked through it. These cards would be fed to card readers that recognized holes (or pencil marks) and could thereby process the information.

These cards teach a powerful lesson: Information is physical—or, at least, structural information can be physically embodied. The basic idea of structural information as strings of bits is an abstract mathematical idea, but strings of bits gain causal powers once they're embodied in physical systems, such as punched cards and computers. These days, bits are embodied as voltages in transistors, as directions of magnetization in a hard drive, or as electric charge in solid-state memory. These physical bits—essentially, binary states in a physical system—have the causal power to drive physical processes in computers that end up running huge parts of our lives.

The English cyberneticist and semiotician Gregory Bateson once defined information in terms of causal power, saying that information is "a difference which makes a difference." This slogan may not apply to semantic information; some trivial facts may make no difference to anyone or anything. It doesn't apply to structural information in abstract mathematics. A sequence of binary numbers, such as 0100, is a system of abstract differences, but it makes no more difference to other entities than any mathematical object does. However, Bateson's slogan is a perfect characterization of *physical information*—that is, of physically embodied structural information.

Think about a punched card. Here, bits are physically embodied in the difference between a hole and intact paper. That difference makes a difference to the card reader; the card-reading process is sensitive to the difference between hole and paper. So this is a difference that makes a difference. Punched cards actually predate the computer. They were first used to control weaving looms—encoding instructions as differences that make a difference—in 1804. In 1833, the mathematician and inventor Charles Babbage proposed using punched cards to provide input to his Analytic Engine, an early computer design that was never built. In 1890, machine-readable punched cards were used in the United States census, which was processed far more quickly than any previous census. All of these innovations worked through the physical embodiment of structural information, as differences that make a difference.

Well before the punched card, mechanical devices were used to process information. We have seen how the ancient Antikythera mechanism used a series of gears to encode semantic information about the planets. Pascal's and Leibniz's calculators used a series of dials to perform mathematical calculations. In the first general-purpose computers, designed in the early 1940s, mechanical switches driven by electrical relays encoded patterns of bits. Soon, physical information became completely electronic, with bits embodied in arrays of vacuum tubes and later in transistors in integrated circuits.

The driving idea here is what we might call the *bit-from-it* thesis (not to be confused with Wheeler's it-from-bit hypothesis). Bits are physically embodied by some more basic physical entity (an "it"), such as a gear wheel or a transistor that has at least two distinct states. Differences in this "it" make for differences in bits. By grounding the structure of bits in the structure of its, the abstract mathematical power of computation is harnessed in a physical system.

The idea is simple but extraordinarily powerful. By systematically encoding sequences of bits as differences that make a difference, we have built the foundation for modern computer technology. Increasingly efficient encoding—smaller differences that make a difference faster—has led to increasingly efficient computers. Vast structures of bits can be embodied as structures of differences that make a difference inside a hard disk drive or on the circuit board of a modern computer.

In principle, physical information can be embodied in all sorts of ways. In 1961, the Russian physicist and science-fiction writer Anatoly Dneprov published a short story called "The Game," in which 1,400 people are called to a football stadium. They are given pieces of paper with symbols on them, and some simple rules for changing these symbols and passing them to each other. At the end of the process, they discover that they have been implementing a program for translating a sentence in Portuguese into Russian. None of them had any idea what they were doing. The pieces of paper served as bits, and the people served as processors of this information. Physical information requires only a systematic pattern of differences that makes a difference, one that can be embodied in all sorts of underlying substrates.

In this sense, physical information is *substrate-neutral*. We encountered the idea that *minds* are substrate-neutral in chapter 5. There the idea was that the same sorts of conscious experiences might be had in systems made of very different substrates: neurons, silicon chips, even green slime. The substrate-neutrality of minds is a controversial thesis. By contrast, the substrate-neutrality of information is not controversial at all. The same sequence of bits can be encoded in all sorts of substrates—punched cards, mechanical switches, transistors, or patterns of beer cans. Dneprov's language-translation system could be implemented by people and cards in a football stadium, or on an electronic circuit board, and it would remain the same algorithm processing the same information.

The physics of information

In recent years, ever-stronger links between information and the physical world have been forged. The physicist Rolf Landauer introduced the slogan *Information is physical* to express the idea that structural information plays a role in laws of physics. He developed close connections between information and core ideas in thermodynamics. Others have tried to cast more basic laws of physics in informational terms. Still others have proposed a more radical idea: that physics itself might be all about the processing of bits.

Conway's Game of Life provides a suggestive illustration here. Could something like the Game of Life underlie physics in our world?

It needn't be exactly Conway's game, of course, but maybe something like it? For example, could quarks and photons be distinctive shapes in an underlying three- or four-dimensional grid made up of bits—or perhaps made up of qubits, in a quantum mechanical universe?

This idea sometimes goes by the name of *digital physics*. Pioneering work in this area was done by Konrad Zuse, a German engineer who is regarded by some as the inventor of the computer for the work he did in constructing the programmable Z3 computer in 1941. In the 1960s, he wrote a book called *Calculating Space* suggesting that the universe as a whole might be a sort of computer based on digital rules for the interaction of bits. Versions of this idea have been developed by a number of other theorists, such as Edward Fredkin and Stephen Wolfram.

Digital physics can be encapsulated using John Wheeler's now-familiar slogan: it from bit. The slogan suggests that all physical objects (its) are grounded in structural information (bits). Wheeler introduced the idea this way:

> [E]very physical quantity, every it, derives its ultimate signifi-
> cance from bits, binary yes-or-no indications, a conclusion which
> we epitomize in the phrase, *it from bit*.

Wheeler's own conception of "it from bit" is not entirely clear. He ties it to the idea that the universe is "participatory," fundamentally involving the participation of observers. Observers ask questions using measuring instruments such as microscopes and particle accelerators, and the bits provide the answers. This way of understanding his proposal suggests an element of idealism: Reality is grounded in the observations of observers, and observations are states of consciousness. However, Wheeler's slogan has often been understood as the idea that physics is grounded in digital structures (structures of bits), whether or not these bits have any special tie to observers or measuring instruments. This is the way I will understand the it-from-bit thesis here. The thesis says that every *it*—every physical object and quantity—is grounded in a pattern of *bits*.

Our leading current theories in physics aren't formulated in terms of bits. They invoke more complex mathematical quantities, such as quantum wave functions involving mass, charge, spin, and so on, all embedded in space and time. But these theories are consistent with the existence of a deeper level involving the interaction of bits or qubits (I'll mostly say "bits" from now on, but this should be understood to include qubits). On that approach, current physics is *realized* by digital physics involving the interaction of bits.

The word "realize" as used by philosophers means something like *make real*. In practice, it can be used whenever low-level entities make up higher-level entities. Atoms realize molecules, molecules realize cells, and so on. The same verb can be employed in discussions of scientific theories: Thermodynamics, the high-level physics of pressure, temperature, and so on, is "realized" by statistical mechanics, which is formulated in terms of the motion of particles. Given molecules moving in certain ways, it follows automatically that a system has a certain pressure and temperature; in effect, molecular motion makes pressure real. It's because of the underlying statistical mechanics that systems have temperature and pressure at all.

Similarly, molecular physics is realized by atomic physics, and atomic physics is realized by particle physics. We can derive molecular physics from atomic physics and atomic physics from particle physics. In each case, the lower level provides a fine-grained basis that supports the coarse-grained structure of the higher-level theory.

Advocates of digital physics hope that we can derive something like our current physics from digital physics. There will be an underlying level consisting of bits interacting according to some algorithm. We will then use the interaction of bits to construct particles and waves with mass and charge, interacting in space and time. In effect, current physics will be a consequence of digital physics in much the way that thermodynamics is a consequence of statistical mechanics.

What will be the status of properties like mass and charge in digital physics? These may well be high-level properties that emerge from the interaction of systems of bits. If so, they'll be realized by digital physics, but they won't be present at the most basic level of digital physics. Digital physics may simply involve bits, along with algorithms governing their interaction.

At the fundamental level, digital physics need not even invoke space and time. Researchers studying quantum gravity—the unification of quantum mechanics and general relativity—are increasingly studying theories in which space and time themselves emerge from something more fundamental, a little like the way pressure emerges from motion. Theorists have entertained the idea that space and time emerge from some sort of underlying digital physics. The goal is to derive the structure of space and time from that underlying level. If we can do that, the structure, dynamics, and predictions of current physics can be extracted from the algorithmic interaction of bits.

It should be said that digital physics and the it-from-bit hypothesis are not especially favored in physics. Even among speculative theories of quantum gravity, nondigital approaches, such as string theory and loop quantum gravity, are more popular. Fortunately, for my purposes I don't have to claim that digital physics is correct or even supported by physical evidence. What matters is only that the it-from-bit hypothesis *could* be true and that it's at least consistent with known evidence. In this respect, the it-from-bit hypothesis is like the simulation hypothesis. I'm not really arguing that these hypotheses are true; instead, I'm thinking about what they say about the world, and what follows from them.

We've already seen that if we're in a perfect simulation, the simulation hypothesis could be true even though we can never discover that fact. For similar reasons, the it-from-bit hypothesis could be true even though we can never discover that fact. Suppose that in our world, Newton's laws determine all of our observations and that they're perfectly realized, at an underlying level, by the interaction of bits. If so, then the it-from-bit hypothesis will be true: Entities in our universe are grounded in bits. But if Newton's laws are complete, we'll never have evidence for the it-from-bit hypothesis, since the bits don't show up in what we observe.

I'll call the version of the it-from-bit hypothesis in which we can't discover the bits the *perfect it-from-bit hypothesis*. Like the perfect simulation hypothesis, the perfect it-from-bit hypothesis may not be a scientific hypothesis, since we can never obtain evidence for or against it. People who are impatient with untestable hypotheses can always focus on the imperfect versions, in which the bits are detectable. Nevertheless, even the untestable hypotheses remain interesting for philosophical purposes. They are perfectly coherent hypotheses that could be true, and we can still reason about what follows if they are.

There are cousins of the it-from-bit hypothesis without bits. We've seen that information can also involve trits (ternary digits), qubits (quantum digits), reals (real-valued continuous digits), and other basic elements. There will correspondingly be it-from-trit physics, it-from-real physics, and so on. It-from-trit physics may be no more plausible than it-from-bit physics, but it-from-real physics reflects the structure of many physical theories with continuous quantities. For example, Newtonian mass and distance can be understood as reals, the continuous analog of bits.

The physicists David Deutsch, Seth Lloyd, and Paola Zizzi have explored the *it-from-qubit* thesis, which says that quantum computation underlies physical reality. Given that we live in a quantum rather than a classical universe, the it-from-qubit thesis is in some ways a better match for our own reality than the classical it-from-bit thesis. To avoid presupposing knowledge of quantum mechanics, I'll focus mainly on the it-from-bit thesis, but many of the points I'll make about it-from-bit physics apply equally to it-from-qubit physics.

The it-from-bit-from-it hypothesis

According to the it-from-bit hypothesis, physical entities such as quarks and electrons are realized by bits. What about the bits themselves? Are they realized by some further level of physics, or are they fundamental?

A relatively conservative version of the it-from-bit idea is what we might call the *it-from-bit-from-it* hypothesis. This hypothesis says that ordinary physical entities, from galaxies to quarks, are made of bits at some level, and the bits in turn are made of more basic entities. This version of the idea combines the it-from-bit idea with the bit-from-it model provided by ordinary computers. Ordinary entities are made of bits, and bits are always physically embodied by more basic states, such as voltages. For example, the universe could be made up of cells in the Game of Life, in which each cell can be on and off, but the on/off state might be embodied in a more basic physical quantity, such as electric charge. In this universe, bits aren't absolutely fundamental. Bits are differences grounded by differences in something more basic.

Under the it-from-bit-from-it approach, digital physics is realized by some deeper physics that involves more than bits; for example, the digital physics of the Game of Life might be realized by some sort of quasi-electromagnetic physics. As in other cases of realization, what matters is that once the underlying electromagnetic level is specified, the structure and dynamics of the Game of Life follows as a consequence.

What will the deeper physics be like? It could take many different forms. We've already seen that physical information is substrate-neutral. So is digital physics. The Game of Life could be realized electrically, or mechanically, or by forms of physics we cannot yet comprehend. In principle, any substrate will do, as long as it's organized to yield the right information obeying the right algorithm.

Figure 23 The pure it-from-bit hypothesis (left) and the it-from-bit-from-it hypothesis (right).

A radical form of the it-from-bit-from-it hypothesis is the *it-from-bit-from-consciousness* hypothesis. According to this view, digital physics is realized by interacting states of consciousness. These could be complex states of consciousness in God's mind, as in Berkeley's idealism. They could be simple states of consciousness in atomic entities, as in the panpsychist view, in which consciousness is everywhere. Either way, according to the it-from-bit-from-consciousness view, *mind* in some form comes first, and physics derives from it. This highly speculative view highlights the way that digital physics is consistent with many different substrates.

The pure it-from-bit hypothesis

An alternative to the it-from-bit-from-it thesis is the *pure it-from-bit* thesis, in which bits are absolutely fundamental in the universe, and there's no underlying "it." There are two basic states a fundamental entity can be in. We can call those two states "0" and "1," or "on" and "off." The difference between these two states is a pure difference, without any underlying differences such as voltages or states of consciousness. The universe at its bottom level is a universe of pure differences.

The idea isn't easy to grasp at first. We're accustomed to thinking of differences as differences *in* something, so that any physically embodied bit must be grounded in something more basic, such as voltage or charge. Pure it-from-bit sounds a little like software without hardware—as if Microsoft Word were running without any computer for it to run on. Nevertheless, the idea is attractive to many scientists and philosophers. If it turns out that the laws of physics can be formulated in terms of bits, it's not clear that we would be forced to postulate any more basic level.

Say we had strong evidence that the laws of physics conformed precisely to Conway's Game of Life. Then we'd have evidence that physics involves the interaction of bits. For structure among the bits, we may need to postulate certain relations between them, such as Life's neighbor relations between cells. But would we need to postulate a more fundamental it-from-bit-from-it level in which the bits are grounded? Many theorists will resist this. The extra level just makes the model more complicated and doesn't make any difference in what we observe.

The cost is that a world of pure bits is a world of pure differences. This is a shocking idea at first, but many people get used to it. The pure it-from-bit hypothesis holds that physical reality can be fully described in mathematical terms and is ultimately grounded in structural information. A more general view along these lines has become popular in recent years under the label *structural realism*, and we'll return to it in chapter 22.

You can hold this more general view of physical reality even if physics is continuous rather than digital. We've seen that continuous physical theories can be developed as an it-from-real view, understanding reals as continuous analogs of bits. In classical physics, for example, the location and mass of a particle can be represented with reals such as 0.237 and 3.281. Whereas Life dynamics involves rules for the interaction of bits, classical dynamics involves equations for the interaction of reals. You can then choose between a pure it-from-real hypothesis, in which pure real-valued quantities are fundamental, and an it-from-real-from-it hypothesis, in which these quantities are grounded in something more basic. On the pure it-from-real view, we might say that reality is grounded in continuous information.

Ultimately, the it-from-bit view will serve as a sort of step ladder that we can kick away. What matters most is not the bits per se but the underlying structuralism on which reality can be fully described in mathematical terms. But the it-from-bit view provides a wonderful illustration of this structuralist idea, and it also provides a clean bridge to the simulation hypothesis.

Chapter 9

Did simulation create its from bits?

H ERE'S A CREATION MYTH FOR THE INFORMATION AGE.
God said "Let there be bits!" And there were bits.

God saw that the bits were good, and he separated them from each other. He called one bit "zero," and the other bit "one."

If you squint a little, you can read this it-from-bit creation story into the biblical book of Genesis. At the start of the Old Testament, everything is formless darkness. Nothing is different from anything else. Then God commands there to be light and separates the light from the darkness. Now some things in the world are different from other things. Light and darkness are bits. Suddenly the universe has form, as it cycles between light and darkness. God calls the light "day" and the darkness "night."

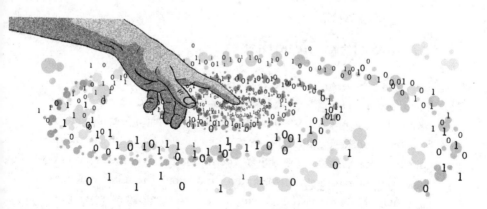

Figure 24 "Let there be bits!"

Admittedly, Genesis tells us that Heaven and Earth were created before the light and darkness. The it-from-bit version of Genesis switches the order. God merely needs to create and arrange the bits of light and darkness. Then Heaven and Earth will take care of themselves.

Creation stories normally leave holes to be filled in, and the it-from-bit creation story is no exception. One hole is shared with many creation stories: Where did God come from? Another hole: Where did the bits come from? Putting the two together: If God is powerful enough to create bits, won't his mind already involve many bits? If so, then the it-from-bit creation story cannot tell us where the first bits came from, any more than the standard creation story can tell us where the first minds came from.

Still, the it-from-bit creation story works well as a *local* creation story: a story of how our own universe was created, in a cosmos that already had minds and bits. God already exists in a heavenly universe with bits of its own. On the first day, God says "Let there be bits." He initializes our universe by creating and arranging a distribution of bits, akin to the initial conditions in the Game of Life. On the second day, God says "Let there be its." He programs the bits to interact in a way that will support a physical world, akin to Life-style rules that underpin our physical laws. On the third day, God says "Let reality unfold," and the interaction of the bits commences. That's how our universe may have begun.

A tale of two hypotheses

The it-from-bit creation hypothesis combines the it-from-bit hypothesis and the creation hypothesis developed in preceding chapters by saying that the physical world was created by a creator and is made of bits. The hypothesis should have a familiar ring. Its structure is analogous to the simulation hypothesis. The simulation hypothesis has two basic components: a simulator and a simulation. The it-from-bit creation hypothesis has two basic components: a creator and some bits. The simulation sets in motion the algorithm for the simulation, and

Figure 25 The it-from-bit creation hypothesis and
the simulation hypothesis are equivalent.

the creator sets in motion the algorithm for the interaction of bits. The creator and the simulator have essentially the same job.

I won't argue that the it-from-bit creation hypothesis and the simulation hypothesis are *true*. Instead, I will argue that they are *equivalent*.

If you accept the it-from-bit creation hypothesis, you should accept the simulation hypothesis. If you accept the simulation hypothesis, you should accept the it-from-bit creation hypothesis. These hypotheses are two ways of describing the same situation.

You may have different associations that come to mind in thinking of these hypotheses—perhaps creation involves a god while simulation involves a mortal—but these differences are not essential to the hypotheses themselves. Both hypotheses describe a being that creates a reality by setting bits in motion.

If that idea is right, it has important consequences. The it-from-bit creation hypothesis is not a hypothesis in which nothing is real. It's not a hypothesis in which there are no tables and chairs; it's a hypothesis in which tables and chairs are made of bits. If the simulation hypothesis is equivalent to the it-from-bit hypothesis, then the simulation hypothesis is not a hypothesis in which nothing is real. Even if the simulation hypothesis is true, there are still tables and chairs. The tables and chairs are made of bits.

Suppose God appears and says that she created our universe. We won't then conclude that nothing is real. If God created cats and chairs, that's an interesting fact about their history. But cats and chairs are as real as ever.

Now suppose God announces that the it-from-bit hypothesis is true. Underneath traditional physics is a level of interacting bits, a little like Conway's Game of Life. Will we then conclude that nothing is real? I don't think so. Discovering atoms didn't make us reject molecules. Discovering quarks didn't make us reject atoms. So discovering bits shouldn't make us reject quarks. If the it-from-bit hypothesis is correct, there are still quarks, cats, and chairs. It's just that the cats and chairs are made of atoms which are made of quarks which are made of bits.

Finally, suppose God tells us that she created the universe by arranging the bits required for the it-from-bit hypothesis and making them interact. Will we then conclude that nothing is real? I don't think so. If things are real in the creation scenario and in the it-from-bit scenario, they're real in the it-from-bit creation scenario. There are still cats and chairs. They're just created by a creator and made of bits.

So the it-from-bit creation hypothesis is not a skeptical hypothesis. It's a hypothesis in which ordinary things are real and our ordinary beliefs are mostly true.

Now we can mount the following argument.

1. If the it-from-bit creation hypothesis is true, most of our ordinary beliefs are true.

2. If the simulation hypothesis is true, the it-from-bit creation hypothesis is true.

———————————

3. So: If the simulation hypothesis is true, most of our ordinary beliefs are true.

I've argued for the first premise just now. If the it-from-bit creation hypothesis is true, ordinary things like cats and chairs exist and our ordinary beliefs about them are true.

I've made a preliminary case for the second premise in suggesting that the simulation hypothesis and the it-from-bit creation hypothesis are equivalent. For the purposes of my argument, I don't need full equivalence. I need only establish that the simulation hypothesis leads to the it-from-bit creation hypothesis, not the reverse. I'll try to do this in the two sections that follow. The argument is somewhat abstract; if you're more interested in the conclusion, it's fine to skip these sections.

The conclusion is what I've called simulation realism (chapter 6): even if we're in a simulation, things are mostly as we think they are. I'll discuss this thesis and address some objections toward the end of the chapter.

From the simulation hypothesis to the it-from-bit creation hypothesis

Let's start with the simulation hypothesis. More specifically, we'll start with the *perfect, global,* and *permanent* simulation hypothesis: We're in a permanent global simulation of the universe, in which physics is simulated perfectly.

I'll assume a simulation on a digital computer, to match up with the it-from-bit hypothesis. The arguments generalize to a simulation run on a quantum computer, which matches up with the it-from-qubit hypothesis. As we saw in chapter 2, a quantum simulation of our quantum world will likely be far more efficient than an ordinary digital simulation, though both are possible in principle. The arguments also generalize to simulations on an analog computer, which match up with the it-from-real hypothesis, where reals are real-valued continuous variables.

Our simulator sets out to simulate physics by executing some algorithm that simulates the structure of physics. To do that, she will initialize a pattern of bits (or qubits or reals) on the computer and will set those bits in motion according to the rules of the algorithm.

The it-from-bit creator needs to do much the same thing—that is, set up a structure of bits (or qubits or reals) that mirrors the structure of physics and have them follow the right algorithm. At this point, the two tasks look very similar. The simulator is doing the core work of an it-from-bit creator, at least in creating and arranging bits.

In a standard simulation scenario, the bits that our simulator is creating aren't *fundamental*. Instead, the bits are realized by processes in a computer in the simulator's world. But this just means that these simulations aren't versions of the *pure* it-from-bit hypothesis, in which atoms are made of fundamental bits. Instead they match up with the *it-from-bit-from-it* hypothesis encountered in the last chapter. (There are also nonstandard versions of the simulation hypothesis with a computer made of pure bits, corresponding to the pure it-from-bit hypothesis, but I'll set those aside.) Atoms are made of bits, and the bits are made of something more fundamental. If the creator created atoms that way, that's still a perfectly fine creation hypothesis.

The simulation hypothesis also says something about us. It says that we're *in* the simulation. This means that the simulator has to ensure that the simulation is connected to us so that we receive sensory inputs from and send motor outputs to the simulation. As we saw in chapter 2, there are two ways this could go. First, we could be simulated creatures ourselves, as in the pure simulation hypothesis. Second, we could be nonsimulated creatures connected to the simulation, as in the impure simulation hypothesis.

The pure and impure simulation hypotheses will lead to two different versions of the it-from-bit creation hypothesis. In the pure simulation hypothesis, in which we ourselves are simulated, the creator's arranging the bits will have to create *us*. Our bodies, our brains, and our minds will all arise from bits. In the impure simulation hypothesis, the creator's arranging the bits will not create us; more precisely, the bits will create our bodies but not our minds. We're separate creatures that interact with the bits, and we have to be created separately.

For now, we can work with either or both of these pictures. They offer quite different pictures of how *we* fit into the it-from-bit physical world, and of how our minds and our bodies interact. I'll focus on these pictures in chapters 14 and 15.

So, the simulator is arranging bits (which are made of its), ensuring they follow the right algorithm and hook up to our perception in the right way. The it-from-bit-from-it creator does something very similar. The creator needs to arrange bits (which are made of its) and ensure they follow the right algorithm and hook up to our perception in the right way.

To oppose my claim that the simulation hypothesis leads to the it-from-bit creation hypothesis, you'd have to say that this isn't enough. That is, merely arranging bits, running the right algorithm, and hooking all this up to our perception isn't enough for the it-from-bit creation hypothesis to be true. Something more is required.

How can we get from bits to its?

By far the most important challenge to my claim is this: *What about the "its"?*

Let's spell out the objection. The it-from-bit hypothesis builds in the existence of real physical objects (*its*), but the simulation hypothesis does not. The simulation hypothesis just builds in the bits, along with a simulator who creates them. We might say that the simulation hypothesis is defined as a *bit* hypothesis, not as an it-from-bit hypothesis. It says that there are bits in a simulation but says nothing about atoms and molecules. We can easily see how the it-from-bit hypothesis leads to the bit hypothesis, but it's not so obvious why the bit hypothesis should lead

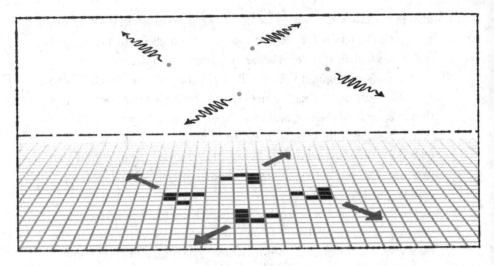

Figure 26 Structure in digital physics (e.g., gliders in the Game of Life)
maps onto structure in standard physics (e.g., photons).

to the it-from-bit hypothesis. Why should the interaction of bits bring atoms and molecules into existence?

Nevertheless, there's a path from the simulation hypothesis to the it-from-bit hypothesis. Suppose the simulation hypothesis is true. Then the simulation involves a system of bits. Our experiences are produced by an algorithm involving a pattern of bits that simulates a standard physics, with quarks, electrons, and so on. This pattern is itself a system of digital physics, with bits obeying digital laws. To oversimplify for the purposes of illustration (see figure 26), we can imagine that the simulation is running on the Game of Life, with gliders in digital physics simulating photons in standard physics. If the digital physics is a perfect simulation of the standard physics of quarks and electrons, the behavior of gliders in digital physics will perfectly simulate the behavior of photons in standard physics. In a perfect simulation, the mathematical structure of photons can be recovered from the mathematical structure of the underlying digital objects. More generally, we can recover at least the *mathematical* structure of standard physics from the underlying digital physics.

Now comes the crucial step. If we can recover the mathematical structure of standard physics from digital physics, and if digital physics produces our observations, then digital physics *realizes* standard physics—that is, digital physics makes standard physics real. The bits of digital physics make the quarks and electrons of standard physics real. In our oversimplified example, the gliders of digital physics make the photons of standard physics real.

Importantly, if the simulation hypothesis is true, our observations of electrons and quarks are produced by certain patterns of bits. Every time we observe a photon, the observation is produced by a pattern of bits: in our example, a glider. If these patterns of bits underlie the mathematical structure of photons in our world, and also produce our observations, then those patterns of bits realize photons. By arranging the bits of digital physics in the right structure, the simulator has thereby created the entities—the its—of standard physics. In this way, the simulation hypothesis leads to the it-from-bit creation hypothesis.

When one theory realizes another, the entities of both theories are real. When atomic physics realizes molecular chemistry, for example, then molecules are real, and they're made of atoms. Similarly, if digital physics realizes standard physics, then photons are real, and they're made of bits. If photons, quarks, and electrons are real, so are all the physical entities they make up: atoms, molecules, cells, rocks, organisms, buildings, planets, stars, galaxies. If the simulation hypothesis is correct, all of these entities are real.

This crucial step involves *structuralism* about physics. Structuralism says that theories in physics can be boiled down to their structure—roughly, their mathematical equations and their observational consequences. A theory in physics is true if this structure is really present in the world. If the structure of atomic physics is really present in the world, for example, then atomic physics is true, and atoms exist.

Structuralism is an extremely popular view in the contemporary philosophy of science. Versions of it are widely held among physicists, too. Structuralism about science (which is distinct from structuralism about culture, a major philosophical movement in the mid-20th century) is the focus of chapter 22, and I'll say more about it then. For now,

I'll give a very quick sketch of how it works and how it supports my reasoning.

When we reduce Newton's physics to its essential structure, it tells us that all entities are related by certain mathematical equations: $F = ma$, the law of gravitation, and so on. These equations say that mass plays a certain mathematical role in inertia and gravitation. Mass may also have a certain role in observation: We have certain ways of detecting mass with our measuring instruments. Structuralism says that mass is defined by these mathematical and observational roles. As it is sometimes put, mass is what plays the mass role.

Similarly, photons and quarks are defined in physics by the mathematical roles they play and by their connections to observations. Photons are what play the photon role. Quarks are what play the quark role. It follows that if there's something in the world that plays the mathematical and observational role of a photon, that thing is a photon. If something plays the role of a quark, that thing is a quark. Structuralism tells us that if there are entities in our world that play the roles specified in standard particle physics, then standard particle physics is true.

What does structuralism mean for digital physics and simulations? It tells us that as long as digital physics can replicate the structure of the standard physics of photons and quarks, then this is enough to make standard physics real. If we discover that gliders play the mathematical role of photons and generate our normal observations of photons (showing up in our measurements the way that photons are supposed to show up), then those gliders are photons. We can summarize the argument like this.

1. Photons are whatever play the photon role.
2. If we're in a simulation, digital entities play the photon role.

---

3. So: If we're in a simulation, photons are digital entities.

Premise 1 is essentially a version of structuralism about photons, where the relevant role is a structural role concerning their role in a mathematically describable structure and their impact on observation.

Premise 2 follows from an analysis of the simulation hypothesis: a certain pattern of bits plays the mathematical role of a photon and generates our observations. The conclusion is the key it-from-bit claim that we need: if we're in a simulation, "its" like photons really exist and are digital entities.

More generally, given that a simulation can replicate the structure of standard physics, with digital entities playing the core mathematical roles and producing the right sort of observations in us, the simulation will make standard physics real. That's enough for the it-from-bit hypothesis to be true.

I'm not saying that necessarily, any simulation of physics will make physics real. If we're not in a simulation, then our physics is unsimulated. If so, photons—that is, the entities that play the structural role of photons in our world—are probably nondigital entities. As a result (as I'll discuss in chapter 20), a simulated photon won't be a genuine photon. However, if we are in a simulation, then our physics has been digital all along. Digital entities have always played the structural role of photons in our world. As a result, simulated photons have been genuine photons all along.

Of course, you can always resist the argument by denying structuralism. You could say that to have photons in our world, it's not enough to have entities that play the structural role associated with photons. More generally, for standard physics to be true in our world, more than the right mathematical *structure* is needed. The right *substrate* is needed. A simulation has the right structure but not the right substrate.

For example, some people think that objects in a simulation are not *solid* enough to yield a genuine reality. Nonsimulated objects are solid and substantial, but simulated objects are not. According to this view, the only way to get a genuine it-from-bit physical reality from digital physics is if the bits are realized by something more substantial. If so, the simulation thesis does not imply the it-from-bit hypothesis.

We already know that this is the wrong way to think about solidity. In physics, the basic level of reality consists of evanescent quantities, such as quantum wave functions, with no special properties of solidity. Science also tells us that solid objects are mostly empty space. What makes objects count as solid is the way they interact with one another. A solid object is (roughly) one that other objects cannot easily penetrate. Solidity is really defined in terms of a certain pattern of interaction. This pattern can be present just as easily in a simulated reality.

Some people worry about *space*—that if we're in a simulation, objects aren't spread out in space the way they seem to be. There's not really a desk three feet in front of me, and so on. In order for digital physics to yield genuine objects in a genuine space, the bits will have to be arranged in the right spatial relations themselves. If so, the simulation hypothesis doesn't imply the it-from-bit hypothesis.

This view of space as a sort of primitive container of matter is intuitive, but physical theories increasingly suggest that it's incorrect. Relativity theory suggests that space is not absolute. Many physicists entertain theories in which space isn't present at the fundamental level but emerges only at a higher level. If that's right, then the presence of space is not a constraint on the fundamental level. Instead, like solidity, space is grounded in the patterns by which things interact with one another. Those patterns can be equally present in a simulation.

You could worry about the *stuff* that photons and quarks are made of. Quarks have a special "quark-ish" intrinsic nature, and a simulation of a quark won't have genuine quarkishness.

There's nothing about this sort of quarkishness in modern physics. Quarks are characterized in mathematical terms, and that's that. Some philosophers and physicists do speculate that quarks and other fundamental entities could have some sort of underlying nature. If the it-from-bit view is right, their underlying nature may involve bits. If the simulation view is right, their underlying nature could involve processes in the next world up. However, physics is neutral on what this intrinsic nature is. If it turns out that quarks are made of bits, so be it. They're still quarks.

I'll discuss these issues about structuralism further in chapters 22 and 23, where I'll develop the argument in more detail. For now, I've made a case for the second premise of my argument: If the simulation hypothesis is true, the it-from-bit creation hypothesis is true. I've also made a case for the first premise: If the it-from-bit creation hypothesis is true, most of our ordinary beliefs are true. Together these make an argument for my conclusion.

Simulation realism

The conclusion of my argument is *simulation realism*: If we're in a simulation, most of our ordinary beliefs are true. There really are cats and chairs around us. That really is a tree outside my window.

Suppose God tells us tomorrow that the simulation hypothesis is true. We should react much as we would if God told us the it-from-bit creation hypothesis is true. We should say that ordinary objects like cats and chairs were brought into existence by a creator, and that they're made of bits. That will be surprising and interesting, but most of our ordinary beliefs—say, that a cat is sitting on a chair over there—remain unthreatened.

Perhaps we'll have to revise some of our more theoretical beliefs. If we thought our universe wasn't created, we were wrong. If we thought the level of quarks in our universe was the bottom level of reality, we were wrong. If we thought our spacetime was the whole cosmos, we were wrong. But most ordinary beliefs, such as my belief that there are two chairs in my office, will still be true.

Even if *we're* not in a simulation, beings who are in simulations can make a case that their own world is real. Suppose we create a *Matrix*-style simulation containing pure sims with many beliefs about their world. Then the simulation hypothesis will be true for them: Their world is a simulation. The it-from-bit creation hypothesis will also be true for them: Their world was created and is made of bits. Most of their ordinary beliefs will be true. The objects they interact with are perfectly real. They're just made of bits.

You may find these conclusions counterintuitive on first encounter. I'll briefly address some common objections, with pointers to later chapters in which I discuss the issues in more depth.

Objection: *What about the simulator in the next world up?* If the simulator can stop the simulation at any time, doesn't that threaten our reality? If the simulator based cats and chairs in our world on cats and chairs in its nonsimulated world, doesn't that mean that the simulator's cats and chairs are real and ours aren't?

Response: These issues also arise given a standard creator. God might have the power to stop the world at any time, but that doesn't mean that the world around us isn't real. God might have based our cats and chairs on cats and chairs in Heaven, but this doesn't mean that our cats and chairs aren't real.

Objection: *There are no cats in a computer simulation.* Computers don't contain cats and chairs. The brain in a vat believes it sees a cat and a chair, but there are no such objects in the simulation.

Response: The computer simulation contains a virtual cat and a virtual chair. These are real digital objects, each made of bits. I discuss virtual objects as digital objects in chapter 10.

Objection: *A virtual cat isn't a real cat.* A simulated hurricane doesn't make you wet. So how can a virtual cat or a simulated hurricane make my beliefs about cats and hurricanes true?

Response: If we're in a simulation, cats have been virtual cats all along. Our word "cat" has referred to virtual cats all along. Our beliefs about cats have been about virtual cats all along. I discuss these issues about language and thought in chapter 20.

Objection: *A simulation doesn't contain genuine minds, brains, and bodies.* If someone has a brain and a body outside the simulation, like Neo in the Matrix, doesn't this mean that the brains and bodies in the simulation aren't real? If someone has no brain and body outside the simulation, like the Agents in the Matrix, aren't they akin to mindless nonplayer characters in a video game?

Response: Neo has a physical body outside the Matrix, and a virtual body inside the Matrix. Both are perfectly real. As an impure sim, he has a brain outside the Matrix supporting a mind of its own. In a pure simulation, people have virtual brains, but these can still support minds. I discuss these issues in chapters 14 and 15, which focus on the relationship between minds and bodies in simulations.

Objection: *Isn't anything a computer if we interpret it the right way?* Won't that mean that it is trivial to generate realities? How can there be genuine causal processes in a computer-based reality?

Response: Computers involve genuine causal processes connecting different elements of the computer. To run a simulation, you need a system in which these causal processes are set up the right way, which is highly nontrivial. I'll talk about these issues in chapter 21 on computers and computation.

Objection: *Is simulated physics really physics?* Aren't bits too insubstantial to make up a world? Does simulation really yield a genuine space?

Response: As we've seen in the previous section, we can address these objections by developing a structuralist view of physics and making analogies with quantum mechanics, digital physics, and other views in which physical reality seems evanescent. I'll discuss some of these issues further in chapters 22 and 23.

Objection: *What about other skeptical scenarios?* Even if perfect lifelong simulations are not illusory, what about others? What if I've only recently entered a simulation? What if it's just a local simulation? What if I'm dreaming? What about Descartes's evil demon?

Response: In chapter 24, I'll argue that my reasoning generalizes to any *global* Cartesian scenario threatening global skepticism: the evil-demon scenario, the lifelong dream, and so on. I'll also discuss the prospects for *local* skepticism, based on local skeptical scenarios such as the hypothesis that I entered a simulation yesterday. I don't claim to have overcome all forms of skepticism, but if I'm right, we've at least put a dent in one of the most severe forms of skepticism about the external world.

Taking stock

This case for simulation realism caps off my initial response to Descartes's problem of the external world. By arguing for simulation realism, I've blocked the central Cartesian argument that uses simulations to argue for global skepticism.

The Cartesian argument combines two main premises: *We can't know we're not in a simulation* and *In a simulation, nothing is real.* It concludes: *We can't know that anything is real.* If my argument for simulation realism is correct, it shows that the second premise is false. Even if we're in a simulation, things are still real, and most of what we believe is still true. As a result, the simulation hypothesis does nothing to undercut our knowledge of the world.

If you've followed the book straight through so far, there are many directions you can take now. None of the next four parts of the book presuppose each other, so you can read them in any order you like. If you're interested in how all this applies to real virtual reality technology, try part 4. If you're interested in connections to issues about the mind and consciousness, try part 5. If you're interested in issues about ethics and value, try part 6. If you'd like to pursue the argument for simulation realism in more depth and understand its philosophical foundations, try part 7.

Part 4

REAL VIRTUAL REALITY

Chapter 10

Do virtual reality headsets create reality?

N HIS 1992 NOVEL *SNOW CRASH*, NEAL STEPHENSON DESCRIBED a definitive virtual reality world: the Metaverse. The Metaverse is a shared, computer-generated world in which people socialize, work, and play. The novel's central character, Hiro Protagonist, accesses the Metaverse through goggles and an internet connection. The main drag of the Metaverse is a giant boulevard known as "the Street."

The Metaverse may sound a lot like the Matrix, but there's an important difference. The Matrix is a simulated universe in which most people have spent their whole lives. The Metaverse is a virtual world that no one spends an entire lifetime in and that people can enter and exit as they choose. Everyone in the Metaverse was born in ordinary physical reality and is still based there. When they choose to, they can don a headset and perhaps a bodysuit and enter the virtual world of the Metaverse. The Matrix is still science fiction (unless it has been reality all along!), but the Metaverse is gradually becoming reality.

Perhaps the first genuine VR system was devised in 1968 by the computer scientist Ivan Sutherland. Sutherland combined computer simulation technology with stereoscopic vision technology—the technology once used in View-Master headsets for viewing stereoscopic color pictures—to yield an enormous headset system. The system was nicknamed the "Sword of Damocles" because it was attached to the ceiling and hung over the head of the user, like the threatening sword described by the Roman orator Cicero. Sutherland's system was immersive and computer-generated, though interaction was limited to tracking the head movements of the user in order to change perspective on the image. Over ensuing decades, the headsets grew smaller and cheaper, and the interfaces and the computer simulations grew more sophisticated. These days, a number of consumer-grade virtual reality headsets are in widespread use.

Many people have tried to create a Metaverse—a common virtual universe for everyone to spend time in, living out day-to-day lives with many forms of social interaction. The most successful attempt to date has been the virtual world of *Second Life*, which at its peak around 2008 had over a million users. But *Second Life* is a world displayed on a two-dimensional screen. It has proved infeasible to port it to genuine VR because of the much higher number of frames per second that VR requires. There have been a few attempts at a Metaverse in virtual reality, but none has yet come close to establishing itself as definitive. Where VR headsets are concerned, the most common use remains game playing. The use of VR for social interaction—also known as social VR—is advancing, though, and it would not be surprising to see a flourishing ecosystem of Metaverses (or one giant Metaverse, depending on how one carves up virtual space) before long.

You needn't invoke the Metaverse to raise philosophical questions about temporary VR. They arise even for simpler VR environments, such as those used for gaming. The Knowledge Question (*How do we know we're not in a virtual world?*) may not arise; most users of VR headsets know they're using VR. But the Value Question (*Can one live a good life in a virtual world?*) and the Reality Question (*Are virtual worlds real or illusory?*) are as pertinent as ever. Some of what we've said about the simulation hypothesis carries over to ordinary VR, but there are important differences as well.

By far the most common view is that virtual objects aren't real. Stephenson himself tells us that the Street in the Metaverse is unreal: "This boulevard does not really exist; it is a computer-rendered view of an imaginary place."

As you might expect, I disagree. If the boulevard is a virtual boulevard as described, it really exists. It is a real place in a virtual world. It's grounded in computer processes, but no less real for that.

Even regular users of VR commonly distinguish between the "real world" and the unreal domain of VR. If I'm right, that's the wrong way to talk. Instead of talking about the "real world," we should talk about the "physical world" or the "nonvirtual world." Instead of talking about "imaginary" objects, we should talk about "virtual" objects. Virtual objects are real, too!

What is virtual reality?

How can we best define "virtual reality"? Philosophers have learned how troublesome definitions are. Define "chair." I'll bet that for any definition you come up with, there will be counterexamples. Is a chair something you can sit on? Rocks, floors, and beds meet this definition, but they aren't chairs. How about a flat surface with a back, designed for sitting? Loungers and sedan chairs don't meet this definition. You can refine definitions, but generally you can never dispatch every last counterexample.

In his 1953 *Philosophical Investigations*, Wittgenstein noted that there seems to be no characteristic common to everything we call a "game." Instead there is at best a family resemblance involving a few common threads. The Berkeley cognitive psychologist Eleanor Rosch has used behavioral experiments to argue that in the human mind, most concepts are represented using prototypes instead of definitions. Chairs might be represented using a few prototypical chairs, for example. In fact, most philosophers doubt that perfect definitions of ordinary words are possible in a natural language, such as English. Still, we can try to define "virtual reality" and see where the effort gets us.

Let's start by trying to define "virtual." The word comes from the Latin word "virtus," which originally meant *manliness* but which came to mean *strength* or *power*. "Virtus" is also the root of the word "virtue," which we now use for strengths or powers of a person in a general sense. In the medieval era, a virtual X was something which had the strengths or powers of X—and most importantly, the effects of X. In a 1902 dictionary of philosophy, the American philosopher Charles Sanders Peirce enshrined this definition: "A virtual X (where X is a common noun) is something, not an X, which has the efficiency (virtus) of an X."

When defined this way, "virtual" means something like *as if.* A virtual duck is an as-if duck—something that looks like a duck and has some of the effects of a duck but isn't a genuine duck. A virtual object in optics arises when one has the appearance of an object without an object really being there. According to Peirce's definition, a "virtual" reality would be an as-if reality—something that has some of the effects of reality, but which isn't real. If we approach VR this way, it is illusory more or less by definition.

When the French polymath Antonin Artaud introduced the expression "la réalité virtuelle" as a description of theater in his 1932 essay "The Alchemical Theater," he seemed to have this conception of the virtual in mind. He likens theater to the "fictitious and illusory" world of alchemy. Both of them are "virtual arts," and both involve a "mirage":

> All true alchemists know that the alchemical symbol is a mirage as the theater is a mirage. And this perpetual allusion to the materials and the principle of the theater found in almost all alchemical books should be understood as the expression of an identity (of which alchemists are extremely aware) existing between the world in which the characters, objects, images, and in a general way all that constitutes the *virtual reality* of the theater develops, and the purely fictitious and illusory world in which the symbols of alchemy are evolved.

Artaud's conception seems to be that a virtual reality is an as-if reality: an alternative world that's an illusion or mirage but nevertheless has great powers. You can see the beginnings of a connection to the current use of the term. Theater and today's VR have key similarities: Both involve immersive experience of an alternative world, and many people would hold that both are illusory. There are also key differences: The theater isn't usually interactive, nor do computers typically play a role in sustaining it.

Fortunately, another meaning of "virtual" has evolved from this starting point. These days, "virtual" most often means *computer-based*. A virtual library is a computer-based library, and a virtual dog is a computer-based dog. It's still part of the idea that a virtual library will have many of the effects of a library, but there's no longer an implication that a virtual library must be a merely as-if library or a fake library. Whether a virtual X is a genuine X varies from case to case: A virtual dog may not be considered a genuine dog, but a virtual library is a genuine library, and a virtual calculator is a genuine calculator.

According to this usage, a "virtual reality" is a computer-based reality. The VR pioneer Jaron Lanier is usually credited with the first use of the expression "virtual reality" in this sense in the 1980s. The computer-based understanding leaves open whether or not a virtual reality is a genuine reality, which is fortunate for our purposes. Still, there's more to VR as Lanier and others understood it than just a computer-based reality. *Pac-Man* involves a computer-based reality, but it's not VR because it's played on a two-dimensional screen. A fully digital movie, such as *The Incredibles*, is computer-based, but it's not full-scale VR because the experience of watching it is passive whereas the experience of VR is active.

This leads us to the definition of virtual reality that I gave in the introduction: A virtual reality environment is an *immersive, interactive*, and *computer-generated* space.

Immersive means that we experience the environment as a world all around us, with ourselves present at the center. There are many degrees of immersion. An ordinary video game on a computer screen can be *psychologically* immersive, occupying all of our attention in a sort of flow state, but it's not *perceptually* immersive because we don't perceive the world as a three-dimensional world surrounding us.

For genuine virtual reality, we require perceptual immersiveness. Perceptual immersion comes in degrees. Current VR headsets achieve *audiovisual* immersion; the environments look and sound as if you're immersed in them. They don't achieve *bodily* immersion, in which you experience your whole body as part of the world.

The holy grail for VR is *full immersion*. In the Japanese *Sword Art Online* novels, written by Reki Kawahara from 2002 to 2008 and later

IMMERSIVE · INTERACTIVE · COMPUTER-GENERATED

Figure 27 The three defining conditions of VR. It is immersive, interactive, and computer-generated.

adapted into a successful animé series, this was called full-dive VR. A fully immersive or full-dive VR is one that users apprehend with all their senses, as if they're physically inhabiting the environment, and where no trace of the ordinary physical environment remains.

Interactive means that there's two-way interaction between users and the environment, and among objects in the environment. The environment affects users; users affect the environment. Objects in the environment affect one another. In full-scale VR, the user controls a virtual body—an avatar—with action options available more or less continuously.

Computer-generated means that the environment is based in a computer: that is, a computer is generating the signals that are sent to our sensory systems. This contrasts with such non-computer-generated environments as the theater, movies, and TV—and with ordinary physical reality. Unless the simulation hypothesis is correct, physical reality is immersive and interactive but not computer-generated.

According to this definition, VR must meet all three of these conditions: It must be (perceptually) immersive, interactive, and computer-generated. Still, as with words like "chair" and "game," usage can't be entirely legislated by a definition. Sometimes the term "VR" is used for any immersive environment you experience using a VR headset. In this broader usage, VR can include noninteractive environments, like 360-degree movies. It can also include non-computer-generated environments, as when a surgeon performs an operation remotely using telepresence equipment. Sometimes the term "VR" is used for non-immersive (but interactive and computer-generated) virtual worlds, such as *Second Life*, that are experienced on a two-dimensional screen. Sometimes it's even used for events held over videoconferencing (a music festival held over Zoom, say), which may be neither immersive nor computer-generated.

Calling these things VR stretches the term a good deal, but language is flexible and hard to legislate. I'll simply stipulate that *core VR* involves immersive, interactive, and computer-generated environments. When I say VR, I will typically have core VR in mind.

What about the crucial notion of a *virtual world*? This expression was introduced by the American philosopher Susanne Langer in her pioneering 1953 book *Feeling and Form: A Theory of Art.* Langer's book focuses on many forms of virtuality in the arts: virtual objects, virtual space, virtual powers, and virtual memory. Her central cases derive from the visual arts, in which the paradigmatic virtual objects are typically images found in pictures.

As Langer understands virtuality in art, there's no requirement that virtual worlds be immersive. So it's perhaps appropriate that as the term is used these days, virtual worlds need not be immersive. It's common to call the world of a screen-based video game a virtual world, even if it's not virtual reality per se. For example, the world of Azeroth in the video game *World of Warcraft* is a virtual world, even though it's not immersive. Even a text-based adventure, such as *Colossal Cave Adventure* (described in the introduction) can be said to involve a virtual world.

I define a virtual world as an interactive and computer-generated space. Whereas virtual *reality* is immersive, virtual *worlds* need only be spatial. Azeroth and Colossal Cave are non-immersive spaces. An ordinary database is not spatial, so it isn't a virtual world, despite being interactive and computer-generated. I won't try to define "space" for now, but I mean it in the broad and intuitive sense that includes virtual space as well as physical space. As before, to exclude partial spaces (one room in a larger virtual world) and disconnected spaces (two separate virtual worlds taken together), I require that these spaces are complete and interconnected.

As with any definition, you can raise potential refinements and counterexamples. Could there be a nonspatial virtual world in which users interact with each other by speech alone? (I wouldn't count it as a virtual world.) Is a geographical database a virtual world? (Perhaps, if it's fully interactive as defined above.) What about complex social VR or game environments that involve many spatially disconnected spaces that users can teleport between? Are they interconnected enough to count as one virtual world, or are they many virtual worlds? (Current usage goes both ways; I'm happy to be flexible.) What about multiple copies of the same world? (These are sometimes called different instances or shards of one world.)

Much of what I say in this chapter and the next applies both to VR and to virtual worlds in general.

Virtual realism and virtual fictionalism

How real is virtual reality? As I suggested in chapter 6, we can get a better grip on this question by asking whether what's inside virtual reality—that is, *virtual objects*—are real.

In addition to introducing the expression "virtual world," Susanne Langer also introduced the term "virtual object" into the philosophy of art. In *Feeling and Form*, Langer wrote:

> Where we know that an "object" consists entirely in its semblance, that apart from its appearance it has no cohesion and unity—like a rainbow, or a shadow—we call it a merely virtual object, or an illusion. In this literal sense a picture is an illu-

sion; we see a face, a flower, a vista of sea or land, etc., and know that if we stretched out our hand to it we would touch a surface smeared with paint.

Here Langer equates virtual objects with illusions by definition. For my purposes, it's better to define them more neutrally: A virtual object is an object within a virtual world—an avatar is a virtual object, for example. Similarly, a *virtual event* is an event that takes place within a virtual world—a virtual concert or a battle, for example.

We can raise the Reality Question for both virtual objects and virtual events. First, we can ask, "Are virtual objects real?" Do the avatars and objects we encounter in VR really exist, or are they mere illusions? Second, we can ask, "Do virtual events really take place?" Do the battles or concerts we experience in VR really happen, or are they mere fictions?

The most common answer to these questions is no: they don't exist and they don't happen. Virtual reality isn't real. Instead, virtual objects are *fictional* objects, like Harry Potter or the One Ring. Virtual events are *fictional* events, like the Last Battle or the explosion of the Death Star. We might call this view *virtual fictionalism*. It says: *Virtual reality is fictional reality*.

It's easy to see why someone might accept that virtual worlds involved in video games are fictions. A *Lord of the Rings* video game takes place in the world of Middle Earth. Middle Earth is a fictional world created by J. R. R. Tolkien. Certainly the Middle Earth of the novels is fictional. Why should the Middle Earth of the video game be any different? The Gandalf of the novels is a fictional character, and the One Ring is a fictional object; presumably the same goes for Gandalf and the One Ring in the video game.

I agree that these video games involve fictions, but in this instance video-game worlds aren't a good guide to virtual worlds in general. The reason these video games involve fictions is not that they're virtual but that they're (role-playing) games.

Think about a live-action role-playing game. Perhaps some of us dress up as Frodo, Gandalf, and the gang, reenacting their adventures in *Lord of the Rings*, using a plastic ring to serve as the One Ring. Frodo and Gandalf are still fictional because we're merely playing their roles. I'm not Frodo and you're not Gandalf. The One Ring is fictional, too. The fictionality of characters and objects in a game like this is independent of their virtuality.

Think about a virtual world that is *not* a video-game world. An example is the virtual world of *Second Life*, which can be used for game-like purposes but needn't be. Many people in *Second Life* use the world primarily for interaction and communication. Suppose you and I are having a conversation in *Second Life*. My avatar is in a room with your avatar. I greet you and we talk about the weather before moving on to philosophy and then going to see a concert. Where's the fiction here?

A virtual fictionalist might say that our avatars don't really exist and neither does the room. The avatars and the room are fictional; the concert never takes place. I think that's wrong. The avatar, the room, and the concert are all perfectly real.

Of course, the avatar isn't a physical body and the room isn't a physical room, but who said they should be? The avatar is a virtual body that's perfectly real. We're really in a virtual room having a virtual conversation. We really attend a virtual concert. There's nothing fictional about any of this. The virtual objects and the virtual events aren't ordinary physical objects, but they're real all the same.

Virtual digitalism

What sort of objects are virtual objects? As I said in chapter 6, I embrace virtual digitalism. Virtual objects are digital objects—roughly, structures of bits. The bits here are physical bits (as in chapter 8), embodied in voltages in integrated circuits or some other physical basis. Virtual objects exist in the computer systems where virtual worlds are based.

To make the case for virtual digitalism, we can start with a weaker claim: For every virtual object we encounter, there's a corresponding digital object. When we encounter an avatar, there's a digital object in the computer system that corresponds to the avatar. Typically this digital object will be a data structure in the computer, encoding various properties of the avatar (its size, shape, location, clothing, and so on). The *Second Life* servers have a structured collection of data for every avatar, as well as one for every building and every tool within the virtual world. Each of these data structures is ultimately a structure of bits.

Of course, things can get messy. Some virtual objects may correspond to multiple data structures; a virtual city may involve many data structures for many buildings. The data structure for a virtual body may contain deeper data structures for arms and legs. Still, there's a digital object (a structure of bits) corresponding to a virtual city and a virtual body.

Virtual digitalism makes the stronger claim that virtual objects *are* the corresponding digital objects. At least to a first approximation, a virtual object is a structure of bits inside a computer.

We can refine this a little. A statue is not *exactly* the same thing as a structure of atoms. Atoms can come and go, and the statue will remain. The statue can be destroyed, but the atoms will remain. The statue depends on human interpretation to make it a statue, but the atoms do not. In just the same way, a virtual statue is not exactly the same as a structure of bits. Bits may change and the statue may remain. The statue may be destroyed and the bits may remain. A virtual statue depends on human interpretation to be a statue, but the bits do not.

To allow for this, we can say that digital objects are not *exactly* structures of bits. Instead, digital objects correspond to bits as physical objects correspond to atoms. Physical objects are made of atoms, even if they aren't exactly reducible to collections of atoms. Similarly, digital objects are made of bits, even though they aren't exactly reducible to collections of bits.

In some cases, human minds may play a role in making an object what it is. What makes a table the object that it is? In part, that we use it as a table. A statue is what it is in part because we built it and regard it as a statue. Money is what it is because we treat it as money. These things are also true for virtual tables, virtual statues, and virtual money. Physical objects like statues are made of atoms, perhaps with a contribution by human minds. In the same way, digital objects like virtual statues are made of bits, perhaps with a contribution by human minds.

Virtual digitalism says that virtual objects are digital objects, in the broad sense we've just explained. Its biggest competition is virtual fictionalism. Virtual fictionalism may allow that for every virtual object there's a digital object, but it insists that they aren't the same. The digital object is real, but the virtual object is fictional.

Why should we accept virtual digitalism over virtual fictionalism?

Here's one reason. In chapter 9, I argued that the simulation hypothesis is a version of the it-from-bit hypothesis, in which all the objects we perceive and interact with are digital objects. If this is true for objects in a lifelong simulation, it's plausibly also true for objects in more temporary virtual worlds. If so, the objects we perceive and interact with in ordinary virtual worlds are digital objects.

Another reason stems from the causal powers of virtual objects. On the face of it, virtual objects can affect one another. A virtual bat can hit a virtual ball. An avatar can scoop up a virtual treasure, and so on. Virtual objects can also affect *us*: When I see a virtual gun, the gun may elicit a fight-or-flight response in me. As the philosopher Philip Brey puts it in "The Social Ontology of Virtual Environments" (2003): "Virtual objects are not just fictional objects, because they often have rich perceptual features and, more importantly, they are interactive."

Digital objects have the causal powers that virtual objects seem to have. When a virtual bat hits a virtual ball, the data structure associated with the bat affects the data structure associated with the ball. There are processes inside the computer that lead straight from one data structure to another. If the virtual bat had been in a different location, then the virtual ball would have flown in a different direction.

Likewise, when I see a virtual sword, there's a causal pathway leading from the data structure associated with the sword, to the screen of a VR headset, to my eyes and my brain. If the virtual sword had been longer and sharper, the data structure would have been different, and as a result the sword would have looked longer and sharper to me.

We can lay this out as an argument for virtual digitalism:

1. Virtual objects have certain causal powers (to affect other virtual objects, to affect users, and so on).
2. Digital objects really have those causal powers (and nothing else does).

_____

3. So: Virtual objects are digital objects.

The argument isn't a knockdown argument. The virtual fictionalist will deny that virtual objects have the causal powers in question. The virtual bat only _seems_ to affect the virtual ball. Still, there's a strong case that digital objects really have these causal powers, and that makes a strong case, in turn, for saying that the digital objects and the virtual objects are the same thing.

This argument works best for interactive, computer-generated VR. In interactive virtual baseball, the virtual bat (a data structure) really makes a difference to the virtual ball (another data structure). In an ordinary digital movie, by contrast, the bits encoding each frame are static. They affect the experience of the viewer, but they don't affect one another. The same goes for an immersive digital movie. It's only once we get to fully interactive virtual worlds that we find digital objects with the full causal powers that we attribute to virtual objects.

Even in interactive virtual worlds, objects have many grades of causal powers. First, there are _decorative_ virtual objects. These don't interact with other objects at all. There might be a stationary virtual elephant that other virtual objects simply pass through. There might be a virtual mountain in the distance that nothing ever reaches. These virtual objects affect only our perception; they have the causal powers to make you experience a virtual elephant or a virtual mountain. Strictly speaking, this is enough to make them real, at least if one ties reality to causal powers and not mind-independence. (Their status is perhaps akin to the redness of an apple, which affects perception but not much else.) But if you think of reality as coming in degrees, you could reasonably think of these decorative objects as being less real than inter-active objects.

Second, there are *solid* virtual objects. These are solid in that other virtual objects cannot penetrate them. A wall in a VR may be solid in this way, without being able to move and without having any other interactive powers. If you think of being solid as the key to being real, then these objects may have a higher degree of reality than decorative objects.

Third, there are *mobile* virtual objects. These can move around and take on different orientations. They may be able to change their shape. They can interact with one another; when one mobile solid object collides with another, something has to give. When two virtual cars crash into each other, we can expect changes. Typically, mobile virtual objects will be governed by some sort of physics engine, which specifies how they behave in motion and interaction. If the physics engine is a good one, mobile virtual objects may have causal powers that mirror the causal powers of ordinary physical objects.

Fourth, there are *special* virtual objects. These have special causal powers distinctive to their class of object, typically more complex than the powers that would be conferred by a physics engine alone. A virtual gun may have the power to shoot virtual bullets. In the rhythm game *Beat Saber*, a virtual saber may have the power to destroy virtual blocks that rush by. A virtual treasure may have the power to be scooped up. A virtual key may have the power to open a specific door. A virtual monster may have the power to pick someone up and throw him to the ground. We could further distinguish *passive* from *active* special objects. A gun or a treasure may be passive; their causal powers must be triggered by someone or something else. Robots, monsters, and other nonplayer characters may be active; their causal powers are wielded autonomously and need not be triggered by others.

Fifth, there are *animated* virtual objects, which are directly controlled by users. The key case of an animated virtual object is a user's avatar. Its causal powers derive not just from the virtual world's but also from the actions of the user. In some respects, animated virtual objects are akin to passive mobile virtual objects, which are also controlled by users, but in the case of avatars the control is especially direct. Some avatars may have causal powers reflecting a few of the causal powers of human bodies in the nonvirtual world.

This taxonomy is incomplete, and many virtual objects may fall under more than one category here. Instead of dividing up virtual objects, we could instead divide up their causal powers: the power to be perceived, to resist penetration, to move in ways governed by physics, to interact with other objects in special ways, and to be guided by a human agent. This makes it clearer that one virtual object can have many different causal powers.

Is a virtual kitten really a kitten?

Here's a challenge to my view that objects in a virtual world are real. In the virtual world of *World of Warcraft*, there are dragons. But we know perfectly well that dragons don't exist. So the virtual dragons in *World of Warcraft* cannot be real.

Here's the answer: the dragons in *World of Warcraft* aren't physical dragons. They're virtual dragons. There are no physical dragons, but there *are* virtual dragons. Virtual dragons are digital objects. They are real objects that exist in computers.

My opponent objects: Virtual chairs aren't real chairs! And virtual cars aren't real cars! So they're not fully real.

I agree: Virtual chairs aren't real chairs, and virtual cars aren't real cars. We use the word "chair" for *physical* chairs, and (unless we're living in a simulation) virtual chairs are quite different from physical chairs. Still, they're perfectly real *objects*.

Here's an analogy. Where reality is concerned, virtual kittens are like robot kittens. There's now a small industry devoted to making robot pets, including cats and dogs. You can buy a "Zoomer Kitty" that chases things, purrs, and cuddles. Is this robot kitten real? It's certainly a real object. It really exists, it has causal powers, and it exists inde-

Figure 28 A biological kitten, a robot kitten, and a virtual kitten.

pendently of our minds. Still, there's one thing a robot kitten is not. It is not a real kitten. Kittens and cats belong to a biological species, of which the robot kitten is not a member.

What if we take an ultrasophisticated furry robot kitten programmed with the AI technology of the future to behave exactly like a real kitten? Let's say it eats and reproduces and even dies. Even so, it won't be a real kitten. At least on our current understanding, cats are DNA-based biological systems, and robot cats aren't. This isn't an insult to robot kittens. They may be better than real kittens in many respects. They're just different, that's all.

The same goes for virtual kittens. According to standard taxonomy, virtual kittens aren't real kittens. As before, kittens and cats are DNA-based biological systems. Virtual kittens are digital entities realized in silicon technology. Still, virtual kittens are perfectly real, just as robot kittens are real. And like robot kittens, they may be just as good as biological kittens and maybe better in some respects. They're simply different, that's all.

There are some exceptions to this pattern. As I've noted, a virtual library is a real library, and a virtual calculator is a real calculator. A virtual friendship is a real friendship. A virtual club is a real club. We might say that the category X (and perhaps the word "X") is *virtual-inclusive* when a virtual X is a real X, and *virtual-exclusive* otherwise.

What's the difference between libraries, calculators, and clubs (virtual-inclusive) and cars, kittens, and chairs (virtual-exclusive)? It has something to do with the fact that cars, chairs, and kittens are *substrate-dependent* (in the sense introduced in chapter 5). What they're made of matters. That's hard to replicate in a virtual world. By contrast, libraries, calculators, and clubs are substrate-neutral. Clubs and friendships are social. Libraries and calculators are informational. What matters is how information and people are connected, and not what they're made of. Virtual worlds can replicate these connections (at least assuming the people involved are real people), so in these cases a virtual X is a real X. To a first approximation, X is virtual-inclusive when being X is a matter of how things and people are interconnected, rather than a matter of what things are made of.

In matters of inclusiveness, the usage of words plays a key role. "Marriage" was once used in an LGBT-exclusive way (where same-sex marriage was not counted as marriage) and is now used in an LGBT-inclusive way. Likewise, the usage of "man" and "woman" has evolved toward trans-inclusiveness, where trans men are recognized as men and trans women are recognized as women. It's not out of the question that in a VR-oriented future, "car" could shift from being virtual-exclusive to being virtual-inclusive, so that virtual cars count as real cars. More significantly, the use of "human" might eventually shift so that virtual humans (that is, pure sims) are recognized as humans. Philosophers call this process *conceptual change* or *conceptual engineering*. I'll return to some of these issues about language in chapter 20.

So, where do virtual objects stand on the reality checklist? Existence: They really exist, as digital objects inside computer systems. Causal powers: They have causal powers to affect other digital objects and to affect users, as we've just seen. Mind-independence: They exist independently of our minds. I can take off my headset and do something else, and a virtual world can continue without me. Illusions: This issue is more complicated, but in the next chapter I'll argue that at least for sophisticated users, VR need not involve illusions.

That leaves the final criterion: Is it a real X? Here we've seen that virtual objects often fail. At least according to current usage, virtual dragons aren't real dragons and virtual cars aren't real cars. So it looks as if these virtual objects meet four of the five criteria on the reality checklist.

I argued before that if we're in a simulation, objects in the simulation meet all five criteria on the reality checklist. We now see that objects in ordinary virtual reality don't do quite as well. If we're in a simulation, then simulated cars *are* real cars, because real cars have been simulated cars all along. But if we're *not* in a simulation, then real cars haven't been virtual cars all along. In ordinary virtual reality, virtual cars are something new and different. So my virtual realism is a little weaker than my simulation realism. We might think of it as 80 percent realism for many ordinary virtual objects, such as virtual cars and virtual kittens, with 100 percent realism for others such as virtual calculators. This compares to 100 percent realism across the board for simulated cars, kittens, and calculators if we're in a lifelong simulation.

All this supports the conclusion that while virtual reality isn't the *same* as ordinary physical reality (at least, not unless physical reality is itself a simulation), it's a genuine reality all the same. Virtual kittens may not be the same as biological kittens, but they're still real. They exist, they have causal powers, they're independent of our minds, and they need not be illusory. One day in the virtual future, we may even recognize virtual kittens and biological kittens alike as genuine kittens.

Chapter 11

Are virtual reality devices illusion machines?

W HEN PEOPLE TALK ABOUT VIRTUAL REALITY, TALK OF ILLU-
sion is never far away. As we saw in the last chapter, Antonin
Artaud spoke of "la réalité virtuelle" in the theater as being "illusory"
and a "mirage." Likewise, Susanne Langer, in her 1953 book *Feeling and
Form*, equated a virtual object with an "illusion" that "apart from its
appearance . . . has no cohesion and unity."

Artaud and Langer were talking about virtuality in the arts, but the
link between virtual reality and illusions has persisted in discussions
of computer-based VR. VR pioneer Jaron Lanier wrote in the opening
to his 2017 retrospective, *Dawn of the New Everything*: "VR is one of
the scientific, philosophical, and technological frontiers of our era. It is
a means for creating comprehensive *illusions* that you're in a different
place, perhaps a fantastical, alien environment, perhaps with a body
that is far from human." (Emphasis mine.)

The same theme is ubiquitous in science fiction. Here's Arthur
C. Clarke in his 1956 novel *The City and the Stars*, which contains
one of the first-ever discussions of a computer-simulated virtual reality in print:

There was, however, one fundamental difference between the
two. The great bowl of Shalmirane existed; this amphitheater did
not. Nor had it ever done so; it was merely a phantom, a pattern
of electronic charges, slumbering in the memory of the Central
Computer until the need came to call it forth. Alvin knew that in
reality he was still in his room, and that all the myriads of people

who appeared to surround him were equally in their own homes. As long as he made no attempt to move from this spot, the illusion was perfect.

More recently, the illusion idea has been central to scientific research on VR. The psychologist Mel Slater has done perhaps the most influential work on how VR affects the human mind. He introduced the term *presence* for the sense of "being there" induced by VR. Slater breaks down presence into two "illusions": the Place Illusion and the Plausibility Illusion. He defines these as follows:

> *Place Illusion:* "the strong illusion of being in a place in spite of the sure knowledge that you are not there."
> *Plausibility Illusion:* "the illusion that what is apparently happening is really happening (even though you know for sure that it is not)."

When I'm playing the VR game *Beat Saber*, the Place Illusion is the illusion that I'm in an alley waving light sabers—even though I know I'm at home wearing a headset. The Plausibility Illusion is the illusion that flying cubes are moving fast toward me—even though I know that nothing like that is happening in physical reality.

A third illusion is often added to this list: the so-called Embodiment Illusion, or Body Ownership Illusion. This is the illusion that a certain virtual body, or avatar, is *my* body. I feel as if I'm embodied in the avatar. I seem to own the body in roughly the way that I own my own physical body. In *Beat Saber*, I feel as if I'm embodied in the avatar slicing the cubes.

Susanne Langer discussed a fourth, closely related illusion, which we might call the Power Illusion. Discussing dance, she wrote, "The primary illusion of dance is a virtual realm of Power—not actual, physically exerted power, but appearances of influence and agency created by virtual gesture." She would surely have held that virtual dance worlds such as *Beat Saber*, which centrally rest on virtual gestures, involve

Figure 29 Susanne Langer playing *Beat Saber*. Is it an illusion?

the illusion of power: the illusion that one is actually slicing the cubes, for example.

You can summarize the consensus view on VR by saying that VR devices are *illusion machines*. They generate illusions about where I am. They generate illusions about what's happening. They generate illusions about what I'm doing. They even generate illusions about *who* I am—or at least about the body I am attached to.

Despite the distinguished list of authorities who endorse the illusion-machine view, I think it's fundamentally incorrect. While it's true that VR can involve an illusion, it doesn't have to, and for many users it won't involve an illusion. The users' perception of place, of plausibility, of power, and of embodiment needn't be illusory. It will often be an accurate guide to their virtual world. In many cases, users have the sense of being in a virtual (not a physical) place, and they really are in that virtual place. Users may have the sense that things are happening in a virtual (not a physical) world, and those things really are happening in the virtual world. Users may have the sense of having a virtual body (not their own physical body), and they really do have that virtual body. They have the sense of performing virtual actions that they really perform. None of these things need be illusions.

Slater is correct in saying that VR involves a visceral *sense* of place, plausibility, and embodiment. I simply contest the claim that these things are illusions. In some cases they are, but much of the time, they involve non-illusory perceptions of real virtual reality.

If I'm right, VR devices aren't illusion machines; they're reality machines.

Is virtual reality a hallucination?

In his classic 1984 cyberpunk novel *Neuromancer*, science-fiction author William F. Gibson says that cyberspace is a "consensual hallucination experienced daily by billions of legitimate operators." "Cyberspace" has come to mean something like the space of the internet, but in Gibson's early usage it meant something more like the space of virtual reality. Gibson is saying that collective virtual realities are consensual hallucinations.

What is a hallucination, and how does it differ from an illusion? Philosophers sometimes distinguish the two as follows.

In an illusion, you perceive a real object, but the object isn't as it seems—that is, you're misperceiving the object. The paradigm of an illusion is a straight stick that looks bent when you partially submerge it in water. You're seeing a real stick, but the stick looks bent, although it's actually straight.

In a hallucination, you're not perceiving a real object. The paradigm of a hallucination is the pink elephant conjured up by an intoxicated person. There's no real object that you misperceive as a pink elephant. Instead, your brain invents the pink elephant.

In ordinary English, the word "illusion" covers both illusions and hallucinations as defined here. There are plenty of optical illusions wherein the visual system invents (hallucinates) an object that doesn't really exist. In this book, I've been using "illusion" in the ordinary way that includes hallucinations like these. An illusion is any case in which the world is not as it seems. Earlier in the book, in discussing skepticism and the simulation hypothesis, I understood illusions even more broadly to cover cases in which the world is not as we *believe* it to be.

For this chapter, on virtual reality, the focus is on perception. I'll understand illusions as cases in which the world is not as we *perceive* it to be.

The philosophers' distinction between illusion and hallucination is a useful one. There are two different questions we can ask about VR. *Is VR a hallucination?* That is, do the objects we perceive in VR really exist? *Is VR an illusion?* That is, do we misperceive objects in VR as being some way, when they're not that way?

Jonathan Harrison (see chapter 4) seems to have held that VR is a hallucination. He called Dr. Smythson's device the "endocephalic electro-hallucinator" and treated it as a sort of hallucination machine. Here he was using the word "hallucination" in the philosophers' sense. His idea was not that VR is like schizophrenia or intoxication, wherein objects are conjured up internally from crossed wires in the brain. Harrison's hallucinations were mostly produced by external apparatus. What makes them hallucinations is that, as with a pink elephant or a mirage, the objects we seem to see don't really exist.

This is a natural view for people who deny that virtual objects exist. For them, VR is an all-encompassing hallucination. We seem to see hundreds of virtual objects, none of which really exist. Like a mirage or a pink elephant, they're conjured up by a complicated interaction between the mind and the world.

At this point, it won't surprise you to hear that I think this view is wrong. Virtual objects really exist as digital objects inside a computer. When we see virtual objects, we're seeing a pattern of activity inside the computer. When I'm playing *Pac-Man*, Pac-Man himself is a sort of data structure. That's the data structure I'm seeing when I play *Pac-Man*. We aren't hallucinating Pac-Man. We're seeing a real digital object.

We can start with the Matrix scenario, in which we're always in a virtual world. I argue that in this case, the tables and trees that we see are real, and that they're digital objects made of bits. We're seeing digital objects, though this may not be obvious to us. Now consider ordinary VR through a headset. In this case, in much of ordinary life we don't see digital objects. But when we put a headset on, we're very much in the situation of someone in the Matrix. The world we're seeing is populated by digital objects. Given that we see digital objects in the Matrix, it's natural to say that we see digital objects in VR.

Another way to make the point is to invoke what philosophers call the *causal theory of perception*. According to the causal theory, the object we see is always an object that *causes* our experience of seeing an object. When I see a tree, the tree causes my experience. It sets off a long causal chain through photons, my eyes, and my optic nerve, culminating in my having an experience of seeing a tree. Even in an illusion, when I see a stick bending in water, the stick causes my experience. The light gets refracted so that my experience isn't accurate, but I'm still seeing the stick.

When I experience virtual objects, what causes my experience? The answer seems pretty clear: A digital object causes it. A data structure inside the computer sets off a long causal chain through the computer, the screen, the atmosphere, my eyes, and so on, culminating in my experience of a virtual tree. Now, this doesn't *prove* that the digital object is what I'm seeing. After all, my experience may have multiple causes, and even hallucinations can have causes. Still, the central role of the digital object in bringing about my experience strengthens the case that this is what I'm seeing.

You might object that what's really causing my experience is the computer screen or a screen inside the headset. But the computer screen is really just a way station, like a TV. When I see Barack Obama on TV, I'm really seeing *Barack Obama*. The TV is a way station that helps me see him. It's true that you see the TV because your experience of a TV is caused by the TV. But your experience of Obama while watching TV is more fundamentally caused by Obama. The same goes in a desktop video game. When I see Pac-Man on the screen, I'm really seeing Pac-Man. The screen enables me to see him.

In the case of a VR headset, the case against seeing screens is even clearer because the screen isn't visible. Instead, you see right through the screen, all the way to virtual objects, such as avatars and buildings, in a three-dimensional space. In fact, some VR headsets dispense with a screen altogether and project photons onto your retina. In this case, there's no screen for you to see, and it's even more compelling to say that you're seeing a digital object.

I conclude that the experience of VR is not a hallucination. When using VR, you are perceiving virtual objects that really exist. They are concrete data structures inside a computer.

Color and space in virtual and physical reality

Even if we agree that virtual objects really exist and that we see digital objects, a bigger challenge looms. Is virtual reality an illusion? That is, when we use VR, are the virtual objects that we see as they seem?

It's natural to say that virtual reality is an illusion. After all, in VR, a virtual building may look as if it's right in front of me, when in fact there's nothing in front of me. If the virtual building is anywhere, it's inside the computer, but that's not where it seems to be. Furthermore, the virtual building may look huge when the corresponding digital object is tiny. Isn't that an illusion, like the stick in water that looks bent although it's actually straight?

The same issue arises in our experience of color and shape. Say that a virtual fish in a virtual ocean looks to be a certain shade of violet. The digital object inside the computer is certainly not violet; in fact, it may be that nothing in the physical world is that exact shade of violet. Similarly, nothing may have the exact shape that the fish seems to have. Doesn't this suggest an illusion? The virtual fish seems to have a certain color and shape it doesn't really have.

The answer to these questions is "It's complicated." To get clearer on these matters, we first need to get clearer on how color and space work in VR.

Let's start with color. What's going on when a virtual fish looks green? The virtual fish is a digital fish, and the digital "fish" certainly isn't *physically green*. If you look inside the computer and manage to isolate the processes corresponding to the fish, they'll probably be colorless or some other color entirely. However, the fish may be *virtually green*. This is a virtual color, not a physical color. Virtual colors are the colors that matter in virtual reality.

The same goes for shape and size. What's going on when a virtual golf ball looks round and about an inch and a half across? The virtual ball certainly isn't *physically* round, or *physically* an inch and a half across. If we look for the digital object inside the computer, we won't find a ball of that physical shape and size. Nevertheless, the ball is *virtually* round, and *virtually* about 1.5 inches across. Virtual shapes and sizes are what matters in virtual reality.

What exactly is a virtual color or a virtual size? This is a fascinating and complex question that I'll address in more depth in chapter 23. For now, I'll just say that an object is virtually red when it *looks red* to us, at least to normal human observers under normal conditions for VR, such as wearing a headset.

This is parallel to a common view of physical colors. What does it mean to say an apple is red? Very roughly, an apple is red when it normally *looks red*, at least to normal observers under normal conditions for ordinary vision, such as daylight. (It may look different through blue-tinted glasses or to colorblind people, but those don't count as normal conditions or normal observers.) That is, physical redness comes from the way things look under conditions that are normal for ordinary vision. Likewise, virtual redness comes from the way things look under conditions normal for virtual reality. The same goes for virtual shape and size, virtual location, and so on. It's easy to find problems with this simple view, and I'll come back to it in more depth later. But it gives us enough to continue for now.

According to this picture, virtual objects are spread out through virtual space in much the same way physical objects are spread out through physical space. When I see a virtual building a mile away, the virtual building is not *physically* a mile away, but it may be *virtually* a mile away. Similarly, it isn't physically tall or rectangular, but it may be virtually tall and rectangular. And it isn't physically red, but it may be virtually red.

This doesn't solve the problem of illusion for us, but it clarifies the situation. Having made the distinction between physical space and virtual space, we can put the problem as follows. The virtual building looks *physically* red, and it looks like it's *physically* one mile away. But it isn't physically red or physically one mile away. Physically, it's colorless and close by. Even if it's virtually red and virtually a mile away, this doesn't remove the illusion. The virtual building isn't the way it looks. That makes it an illusion.

I don't think this is quite right. The best way to make the point requires a small diversion to think about mirrors.

Are mirror images illusions?

Suppose you're looking at yourself in your bedroom mirror. Are you undergoing an illusion?

According to the *illusion view* of mirrors, when you look in a mirror, you experience a *behind-the-glass illusion*. That is, the objects you see always look to be somewhere beyond the surface of the glass. If the mirror is three feet away from you, then your mirror counterpart looks to be about six feet away from you in a space beyond the glass. Of course, there won't be a body six feet away from you behind the glass. So you're experiencing an illusion.

According to the *no-illusion view* of mirrors, there's no behind-the-glass illusion. According to this view, when you look in a mirror the objects you see usually look to be on the same side of the glass as you are. When you see yourself in the mirror, you look to be in the place you actually are. When you see a friend in the mirror, she looks to be on the same side of the mirror as you are, somewhere in the room. So, when you look in a mirror, you experience things roughly as they are.

Which view is correct? You can go ahead and look in a mirror now, and reflect on whether or not you're experiencing an illusion.

My view of mirrors is that sometimes the illusion view is correct, and sometimes the no-illusion view is correct. I think the no-illusion view is correct for most ordinary experience with mirrors, however.

Here's one case in which the illusion view is correct. Many people have had the experience of walking into a restaurant which initially

seems surprisingly large. That's because they don't realize they're looking into a mirror. Before they realize this, they perceive diners over in
the distance, beyond where the mirror is. At this point, they're experiencing a behind-the-glass illusion. Once they realize they're looking
into a mirror, the room seems to fold back up to a smaller space. At this
point the illusion of a large room disappears.

Here's a case in which the no-illusion view is correct. You're driving
a car and you look in the rear-view mirror. You see some cars that are
actually behind you. Do those cars look to be in front of you or behind
you? The illusion view will say that these cars look to be in front of you,
behind the glass, pointing toward you while somehow staying a fixed
distance from you. This gets rear-view mirrors wrong. Anyone who's
driven a car knows that cars in the rear-view mirror look to be behind
you, where they are. There can be minor illusions here, like the side-
mirror warning highlighted in *Jurassic Park* with a rampaging *T. Rex*:
"Objects in mirror are closer than they appear." But no one has ever
posted a warning on a rear-view mirror that "Objects in mirror are
actually behind you." The cars look as if they're behind you all along.

A proponent of the illusion view may say that at the level of vision,
the car *looks* to be in front of you, but you *judge* that it's behind you. As
with many illusions, you exercise judgment in not taking the illusion

Figure 30 Seeing cars in a rear-view mirror. Nobody needs the
warning because the cars already appear to be behind you.

at face value. In the bent-stick illusion, the stick looks bent, but all the same you judge that it's straight. However, in my experience the rear-view-mirror case is *not* like the bent-stick case. In the rear-view-mirror case, the cars look to be behind you.

What's the difference between the restaurant mirror and the rear-view mirror? More generally, what's the difference between cases in which mirrors produce a behind-the-glass illusion and cases in which they don't? Most obviously, in the rear-view-mirror case, you *know* a mirror is present, and in the restaurant-mirror case, you don't. In the restaurant case, the moment you know a mirror is present, things look different. This is what psychologists sometimes call *cognitive penetration* of a perception, which happens when what you know or believe makes a difference to how you perceive things to be.

Knowledge isn't the only factor here. If someone uses a mirror for the first time, then even if they know they're using a mirror, they may experience the illusion that someone's behind the glass all the same. To avoid the illusion, you need something like *expertise* with mirrors. Most of us are used to mirrors and immediately interpret them correctly, except in situations like the restaurant case. For an expert, this interpretation runs deep enough to affect how things look to us. Our *actions* in the rear-view-mirror case depend very much on seeing the car as behind us. If we see it coming up fast, we move out of the way. If we merely believed it was coming up fast, our evasive maneuver might be less automatic.

Of course, there are still illusions involving mirrors. When we see text in a mirror, it looks reversed, almost as if it's written in a different language. This is a sort of illusion. Our perceptual system is not expert enough to reinterpret the reversed text as text! Furthermore, sometimes we can get a behind-the-glass illusion even though we know we're looking at a mirror. In the kinematic mirror illusion, your right hand is occluded by a mirror and you see your left hand in the mirror where your right hand should be. People typically get the strong sense that the hand they're seeing is the right hand behind the glass, even though they know it's not.

Most of us are expert users of mirrors. When we use mirrors, we have *mirror phenomenology*. "Phenomenology" is a fancy word for subjective experience. There's a distinctive sort of subjective experience that an expert has when using a mirror. The presence of a mirror tips us off to interpret the scene in a special way—as being about the space on the near side of the mirror, not the far side. This interpretation happens so fast and naturally that it affects how things *look* in the mirror. Because of this, we don't experience any illusion at all. Instead, when we look in a mirror, we see things more or less as they are.

Is VR an illusion?

Now we can ask the same question about VR. Is ordinary experience of VR an illusion? I think there's a close analogy between VR and mirrors.

Again, there are two possible views of the matter. According to the illusion view of VR, anyone using VR experiences a *physical-space* illusion. That is, users experience objects as being in a physical space in front of them, and this is an illusion. When you see a ball coming toward you in VR, the ball seems to be moving toward you in the physical space in front of you. In fact, there's no ball (virtual or otherwise) in that actual space, so you're experiencing an illusion.

According to the no-illusion view of VR, there's no physical-space illusion. Instead, you experience objects as being in *virtual space*. Typically, the objects will be where they seem to be in virtual space, so this won't be an illusion. When you see a ball coming toward you in VR, the ball seems to move toward you in virtual space—and in fact, it does. So there's no illusion.

Which view is correct? As with mirrors, I think the illusion view is right in some cases and the no-illusion view is right in others. However, I think the no-illusion view is true in most ordinary experience with virtual reality.

For a case in which the illusion view is correct, imagine someone who doesn't know they're using VR. Suppose that Rahul has fallen asleep. As a prank, his friends put a lightweight VR headset on his head. When Rahul wakes up, with no idea he's using VR, it seems to him as if he's drifting in empty space far above Earth. This experience is certainly an illusion.

For a case in which the no-illusion view is correct, take an expert user of VR in an unusual virtual space. Say we're playing *Minecraft* in VR. We know we're in a virtual world, and we experience that world as virtual. For an expert, virtual space doesn't seem to be physical space. It seems to be virtual space, with rules of its own. Expert users don't suffer from the illusion that virtual objects are in the physical space in front of them; instead, they experience the virtual objects as being in a virtual space in front of them.

Now let's consider a range of cases that move from the no-illusion view of mirrors to the no-illusion view of VR. As a first step, we can move from optical rear-view mirrors to the camera-based rear-view mirrors common in cars today. We use camera-based systems very much as we use optical mirrors. Once you're used to the system, objects seen on the screen seem to be behind you. The same goes for side cameras, where objects seem to be on one side of you.

We can extend the point to cameras on a remote car or a robot. Say you're sitting at home guiding a robot car via a camera that shows what's in front of the car. For an expert user, what you see on-screen won't seem to be in front of you at home. Instead, it will seem to be in front of the robot car, in an entirely different part of space. As in the 1966 movie *Fantastic Voyage*, suppose we're remotely guiding a microscopic robot submarine through someone's blood vessels. After a while, we'll interpret what we're seeing as happening around the robot inside the bloodstream. In effect, we'll see things as inhabiting a different space from the space around us.

Once we've gotten this far, it's a small step to virtual space. An expert user doesn't interpret things in VR as contained in the surrounding physical space, or in any part of physical space. Instead, she interprets them as contained in virtual space. As with mirrors, our knowledge and familiarity with VR makes this interpretation automatic. In many cases, our actions depend on interpreting the scene as virtual. For example, in many virtual spaces you can walk right through an object or a wall, which is impossible in physical spaces. You may be able to teleport. You pick up virtual objects in a special way. An automatic interpretation of spaces as virtual is crucial to guiding expert action.

As with mirrors, a proponent of the VR illusion view could say that all this interpretation happens only at the level of judgment or knowledge, and that where perception is concerned, objects in VR seem to be in physical space. It's just that we know better, and (as with the bent stick) our judgment compensates for the illusion. As a result, we come to believe, and even to know, that the object is in virtual space, but there's still a perceptual illusion.

This is an important view, but as in the mirror case, I think this view gets the experience wrong. When we use VR, things *look virtual*. Objects don't look to be in front of you in physical space; they look to be in a virtual space. Suppose that you, as an expert user, are placed in a highly convincing VR without knowing it. In this case, the objects you see will look to be in physical space around you. But once you're tipped off that you're in VR, there will be a comprehensive perceptual reinterpretation. Things will now look to be in a virtual space.

We can call this the *phenomenology of virtuality*, or just the *sense of virtuality*. There's a distinctive sort of subjective experience that an expert has when using VR. In most cases, the use of a headset, or the nature of the images, tips off the user to interpret things in a special way. For experts, the interpretation is so fast and natural that they *perceive* things as virtual. This perception isn't an illusion; instead, they're seeing a virtual world as it is.

There can still be illusions in VR, even for expert users with the sense of virtuality. For example, you can look into a mirror in VR without realizing that you're looking into a mirror. A virtual object may then appear to be on your left when it's really on your right in the virtual world. For many illusions in physical reality, there can be a parallel illusion in VR. Nevertheless, there is also a great deal of non-illusory perception of virtual reality.

Some readers may find it hard to accept the idea of non-illusory virtual reality. After all, VR exploits ancient mechanisms of perception in the brain designed to give us a model of physical space. These mechanisms are so strong that when using VR, it's hard to escape the sense that the space you're experiencing is *physically* around you. You may quickly *interpret* the space as a virtual space, but this interpretation follows after perception. Initially, things *look* physical. According to this view, perception in VR is an illusion.

I agree that the matter is far from obvious. In early stages of sensory processing, our brain may well represent VR exactly as if it were a physical space around us. But perception also involves layers of interpretation built on top of the early stages. Thanks to perceptual interpretation, we see an object in front of us not as a lump but as a dog or a cat. For expert users, there's a layer of quick and automatic interpretation of the VR world as virtual—a layer that runs deep enough to count as *perception* and not just as *belief* or *judgment*. It affects how things *look* and *feel* to us, just as happens with a mirror.

Even if I'm wrong about this and the layer of interpretation is something that comes clearly after perception, much of what I say still applies. For an expert user, the interpretation of a virtual world as virtual is most often the dominant interpretation. Any sense of the physical world recedes, and the user's overwhelming sense is that all this is happening in a virtual world.

The Place and Plausibility Illusions

How does all this bear on Slater's Place Illusion—the idea that VR brings on the "strong illusion of being in a place in spite of the sure knowledge that you are not there"? I think that *sometimes* VR brings on a place illusion. New users will have the strong illusion of being in a new physical environment. So will people who don't realize they're in VR (though this won't satisfy Slater's second condition of knowing they aren't there). Sometimes, even an experienced user may adopt an interpretation of a virtual environment as physical. For example, in a convincing VR replication of New York City, you may have the illusion of being in New York when you are not.

In other cases, there's no place illusion at all. There's a sense of place, but the sense is accurate, not an illusion. You have the sense of being in a virtual place, and you are. Or at least your avatar (your virtual body) is in that virtual place. That's good enough to count it as your virtual location, as we count the location of our physical body as our physical location (wherever our mind may be). For an expert user, there may be no sense of being in a physical place at all—or at least any sense of being in a physical place is overridden by the sense of being in a virtual place.

The same goes for Slater's Plausibility Illusion: the sense that what is apparently happening is actually happening, even though you know for sure that it isn't. When things in VR seem to be happening in physical space, an illusion may be involved. But when things in VR seem to be happening in virtual space (as they do for an expert user), usually that's not an illusion. Virtual events really happen. They just happen in virtual reality.

We could rename Slater's Plausibility Illusion as "the sense of plausibility"—the sense that all these things are really happening. We might even call it "the sense of reality."

Psychologists and philosophers have discussed a related "sense of reality" that's present in ordinary perception. Most of the time, things look real to us, but under special conditions, things can look or sound unreal. People undergoing schizophrenic delusions often report that their hallucinations don't seem real. A more mundane example is the creepy "uncanny valley" sense, which arises in us when humanlike robots don't look quite human enough.

The sense of reality and unreality also arises in VR. A recent article by brain researchers Gad Drori, Roy Salomon, and others discusses experiments in VR in which users rate various environments as looking "real" or "unreal." A virtual replica of an ordinary room with ordinary dimensions looks real, while a stretched-out virtual room may seem unreal. Arguably, "looks real" here means something like "looks like a plausible physical environment." I suspect that at least in some cases in which an environment looks unreal to a nonexpert user, it looks virtual to an expert user. If so, the sense of unreality may be a precursor to the sense of virtuality.

Physical bodies and virtual bodies

The third major "illusion" in virtual reality is the so-called Body Ownership Illusion. This is the illusion that a virtual body (an avatar) is one's
own body. A physically short person who adopts a tall avatar may have
the sense of having a tall body. Someone with a female-typical physical
body who adopts a male-typical avatar may have the sense of having
a male-typical body. Proponents of this illusion contend both that VR
can give you the sense of having a different body and that this sense is
an illusion.

In my view, this sense need not be an illusion. A virtual body is different from a physical body, but it's real all the same. It's possible for a
virtual body to be *my* virtual body. More generally, people can "own,"
or inhabit, their virtual bodies.

The word *avatar* comes from the Hindu tradition, in which it is used
for the physical bodies that gods such as Vishnu take on when they
come down to Earth. Vishnu is said to inhabit a physical body with
human form—an avatar. The embodiment may be temporary, but while
it lasts the avatar is Vishnu's body.

Later, the word "avatar" came to be used for virtual bodies, thanks
largely to 1980s video games such as *Ultima IV: Quest of the Avatar*
(whose usage of the word was inspired by the Hindu usage) and the
multiplayer role-playing game *Habitat,* and also to Neal Stephenson's
use of the term a few years later in *Snow Crash.* (In James Cameron's
2009 movie *Avatar,* the avatars are more like Vishnu's, with characters
inhabiting physical bodies belonging to an alien species.) In my view,
virtual avatars have much the same status as Vishnu's physical avatars.
I can be embodied in a virtual avatar. The embodiment may be temporary, but when I inhabit an avatar, it's my virtual body.

What is it for a body to be *my* body? For ordinary physical bodies,
a number of factors are involved. My body is the locus of my *action*: It
is the object I most directly control when I act. My body is the locus of
my *perception*: I perceive the world through sense organs in my body,
and I view the world from the perspective of my body. My body is the
locus of my *bodily awareness*: When I feel pain or feel hunger, it is this

body that I'm aware of. My body is the locus of my *mind*: My thinking and consciousness are strongly associated with processes in my brain, which is part of this body. My body is the locus of my *identification*: I feel as if this body is part of me and reflects who I am. My body is also a major locus of my *presentation*: It is a large part (though not the only part) of how I present myself to the external world and of how others perceive me. Some would go further and argue that it's the locus of my *existence*: My body is who I am—that is, I was born into this body, and I will die with it. I could not exist without it.

These factors can come apart. In syndromes such as body dysphoria (which involves a pronounced dislike of one's body), people lose a degree of identification with their body, but their body may remain their body in the other ways I've described. If Cartesian dualism (as discussed in chapter 14) is true, the locus of my thinking may be somewhere outside my body, but the body is still mine. In the philosopher Daniel Dennett's story "Where Am I?," the subject controls a remote body for a long period while his original body and his brain float in a tank. Here the remote body is the locus of perception, action, and presentation, while the original body and brain are the locus of mind and perhaps the locus of existence. In cases like this, the body splits into two.

What about virtual bodies? My avatar is typically the locus of my (virtual) action. My actions in VR are most often mediated by my virtual body, though I can sometimes also act directly on the external virtual world; for example, I can twist *Tetris* pieces without using my virtual body at all. I often *perceive* the virtual world from the perspective of my avatar, though sometimes I perceive it from different perspectives, such as a bird's-eye perspective on my space. My avatar is typically the locus of my *presentation* in VR. Others perceive me largely by perceiving my avatar. The avatar can also be the locus of my *identification* in VR. Even in a short-term context such as a quick video game, I'll have the sense that one avatar is mine. In a longer-term context, such as *Second Life*, I may build a deeper identification with my avatar, feeling that it reflects part of who I am.

What's missing in a virtual body? Of course, these aren't yet as rich

as a human body. In VR, we don't have the rich complex of bodily awareness: Avatars are not yet loci of pain or hunger, or of eating and drinking. More deeply, avatars are not loci of mind and existence. When I
inhabit an avatar, my thinking is tied more closely to my physical brain
than to any virtual brain. And when my avatar dies, I don't die with it.
My avatar is not quite the same as *me*.

Still, it isn't clear that any of these missing factors are essential to a
body's being mine. I could lose pain and hunger and be unable to eat or
drink, but this body would still be my body. My thinking could occur
in a Cartesian mind, but this body would still be mine. And it's not
so obvious that the physical body is the locus of my existence. I could
transplant my brain to a new body, or upload myself to the cloud, and
exist without the old body. So it's arguable that, like my avatar, my physical body is not quite the same as *me*.

In a fairly robust sense, a virtual body can count as *my* virtual body.
It's the locus of my perception and my action in virtual worlds, and it's
the locus of my identification and presentation. This is roughly what I
mean when I say that an avatar is *my* avatar. That claim is straightforwardly true. The sense that a virtual body is *my* virtual body needn't be
an illusion. It is a straightforward matter of fact.

Virtual bodies aren't physical bodies. A human being in a virtual
environment typically has *both* a physical body (sitting at home interacting with a computer) and a virtual body (in an adventure in a virtual
world). At different times, someone's sense of having either a physical
body or a virtual body may dominate. With current VR devices, awareness of one's virtual body is mediated by awareness of one's physical
body, tying the two senses together. For example, you may know where
your virtual arms are by knowing where your physical arms are. However, bodily awareness in VR is also often mediated by vision. In many
video games, players prefer to adopt a viewpoint behind their avatar so
they can see the body and its location in space. This gives them better awareness of what their avatar is doing. Vision allows awareness of
physical and virtual bodies to come apart. You can experience an avatar
as running in virtual space even though you experience your own body
as stationary in physical space.

Embodiment illusions can still arise. The core case of an embodiment illusion is experiencing a virtual body as your physical body. Suppose a short person using VR for the first time is given a tall avatar that otherwise resembles her body. She may well have the sense of having a tall *physical* body. Her physical body is not tall, so this is an illusion. But as before, there will be many cases with no illusion. Expert users of VR may experience themselves as having a tall *virtual body*. This isn't an illusion; in the virtual world, they have a tall virtual body.

In some cases, sensing your tall virtual body may lead you to sense that you have a tall physical body, with a resulting element of illusion. But you can also keep those senses separate; you might feel virtually tall while feeling physically short. Perhaps you're playing a video game with a tall avatar on a desk that's too high for you to easily work the mouse or the keyboard. You might first attend to your tall virtual body, ignoring your physical body entirely. Then you might need to attend to your physical body in order to accommodate the height of the desk. Both bodies may flit in and out of your attention.

The 2018 documentary *Our Digital Selves: My Avatar Is Me!* follows thirteen disabled people who adopt diverse avatars in *Second Life*. Some adopt avatars without a disability. Others adopt a virtual body much like their physical body. Others still have a virtual body that expresses their disability in a distinctive way. Many of the participants insist that the virtual body isn't a replacement for their physical body but is nonetheless very real. One says, "I am not denying my physical body. This is another part of who I am." Another says, "This is not escapism, it's augmentation." The strong sense is that they're experiencing their virtual bodies as another equally valid part of themselves: their virtual body is not an illusion. Virtual realism suggests that this sense is correct. The participants aren't having an illusion that their avatar has replaced their physical body. They're simply experiencing a real virtual body.

In other cases, people may experience a virtual body as replacing their physical body. Some transgender people report first experimenting with various body types in environments like *Second Life*. They often report that experimentation with avatars gives them a sense of what it would be like to have a different body and for others to treat them accordingly. This exercise may sometimes result in their having the sense of a new physical body that does not match their actual physical body (it might be shorter or rounder, say), but at the same time it may involve a deeper truth, experiencing a body that accords with their internal identity and ideals. Some users report identifying with both a male-typical physical body and a female-typical virtual body, or vice versa. The philosophy and psychology of embodiment is complex, but identification with one's virtual body is rarely an illusion.

This brings us back to the story of Narada's transformation in chapter 1. Vishnu says that Narada's long life as Sushila was an illusion. Is he right? (Vishnu believes that all of life is an illusion, of course, but we need not follow him on this.) Sushila's body was akin to a virtual body, in a virtual world generated by Vishnu rather than by a computer. It was the body that Sushila perceived with, acted with, presented, and identified with. I think that just as Vishnu genuinely had an earthly body during his periods of embodiment as an avatar, Sushila genuinely had a female virtual body during her life as a woman. Now, unlike the users of VR discussed above, Sushila did not know she was in a virtual world. As a result there was perhaps some illusion, initially, in taking her virtual body to be a physical body. But over enough time (as I'll discuss in chapter 20), it's arguable that Sushila's very concept of "my body" came to refer to her virtual body. There need be no illusion here.

Illusion machine or reality machine?

Perhaps everyone can agree that VR sometimes involves both the sense of a physical world and the sense of a virtual world.

Sometimes the sense of a physical world is dominant. A new user may sense that a block is about to fall on her physical head. Someone playing virtual tennis may experience it as a physical contest. Someone sitting on a virtual Caribbean beach may have the sense of really being in the Caribbean. Someone experimenting with a new virtual body may have a sense of it as a new physical body. In these cases, you experience real virtual entities, but you experience them as physical. This involves an illusion, though it may also involve a deeper truth.

Sometimes the sense of a virtual world is dominant. An expert user inhabiting an entirely novel domain may have no sense of it as physical. Someone with a new virtual body may experience it as a virtual body that in no way replaces their physical body. In these cases, we experience virtual entities as virtual. No illusion is involved.

For most users of VR, both senses may be present to some degree. You can always interpret a three-dimensional VR as a physical space, and some of your perceptual mechanisms will interpret it this way. You can also always interpret it as a virtual space, and any sophisticated user will do this, at least at the level of judgment and reflection as well as (I would argue) at some levels of perception. Often, one interpretation will be dominant. This is what determines whether you primarily have the illusion of a physical world or the correct perception of a virtual world.

VR *can* be an illusion machine. But it need not be an illusion machine, and it isn't simply an illusion machine. Instead, it generates virtual worlds and can allow you to correctly perceive them. When it does this, it is a reality machine.

Chapter 12

Does augmented reality lead to alternative facts?

FOR A FEW WEEKS IN 2016, AUGMENTED REALITY SWEPT THE world. The agent of the takeover was a phone game, *Pokémon Go*. In *Pokémon Go*, you walk around in physical space with your phone camera on, in search of virtual *Pokémon Go* creatures. When you get close to one, the creature appears on the phone screen as if it's in the physical space in front of you. You then have an opportunity to throw a virtual ball at the virtual creature, capturing it.

Pokémon Go was the first widely popular application of augmented reality, the technology by which virtual objects are projected into the physical world. In regular virtual reality, the user is cut off from the physical world and sees only a virtual world. In augmented reality, the user can see the physical world with virtual objects situated within it. Regular physical reality, such as an ordinary street, is augmented with virtual objects.

Pokémon Go does not require fancy headsets. The augmentation is all done by a smartphone, with virtual creatures inserted into an on-screen video image from a camera. More sophisticated augmented reality technology involves glasses with the ability to project images into the user's field of view. This technology yields an especially immersive form of augmented reality. So far, the glasses have been unwieldy, but they're getting smaller and more powerful. Augmented reality contact lenses are on the horizon.

Within a decade or two, we may all use augmented reality. It could eliminate the need for screens in desktop and mobile computers by projecting a screen or another interface in the space in front of you.

Figure 31 Augmented reality glasses, augmenting the ruins of Plato's Academy in Athens with images of Plato and Aristotle in the Academy (taken from Raphael's *The School of Athens*).

It might eventually replace street signs and traffic signals with digital counterparts. It could enable communication with faraway friends as if they're in the same space with you. It could navigate for you using built-in maps, recognize people for you using automated face recognition, and translate foreign speech for you using language-translation algorithms. It could bring historical locations to life by augmenting them with scenes from the past.

Augmented reality promises to augment our surroundings and our minds simultaneously. As I'll discuss in chapter 16, it augments our minds by extending our brains with new capabilities for navigation, recognition, and communication that we did not have before. This chapter focuses on the way it augments the physical world.

We can ask the Reality Question about augmented reality: Is augmented reality real? Are the *Pokémon Go* creatures real, for example? Unsurprisingly, my answer is, for the most part, yes. They have causal powers, and they exist independently of our minds. They may not be real creatures, but they're real virtual objects, existing as digital objects in a computer and made visible through an augmented reality system.

Another key aspect of the Reality Question: Is augmented reality an illusion? That is, is it the way it seems? This is a harder question. With augmented reality, virtual objects seem to be in the physical space around us, so it's harder to say (as I did for virtual reality) that they seem to inhabit only a virtual space. There's also a stronger case that augmented reality involves an illusion—the illusion that virtual objects are present in physical space.

To get to the bottom of this, we need to address the question "Are virtual objects in augmented reality present in physical space?" The answer might seem obviously no, but things aren't that clear. There may be at least some sense in which *Pokémon Go* creatures can be said to exist in the physical space around us.

Virtual objects in augmented reality

Suppose that sometime in the future, everyone uses the same augmented reality system; let's call it Earth+. Earth+ augments the physical environment all over Earth with virtual objects for everybody. The system is surgically implanted, and everyone has it. At certain locations in physical space, everyone sees the same virtual objects: virtual helpers, virtual furniture, virtual buildings.

Users of Earth+ don't just see and hear the virtual objects in front of them. Thanks to brain-stimulation technology, you can smell and taste virtual objects. You have the sense of eating virtual food and drinking virtual drinks. Thanks to special haptic technology, you can touch and feel virtual objects. You can pick up a virtual rock and feel its weight. Thanks to special bodysuits, you can sit in a virtual chair and encounter resistance when you run into a virtual wall.

Users will typically know whether an object they're interacting with is virtual or physical—it's usually obvious. Virtual objects look different from ordinary objects. They typically have special features; for instance,

Figure 32 A virtual piano in Washington Square Park.

virtual chairs can automatically change size or shape or comfort level. Virtual food stays fresh forever.

Let's say there's a virtual piano in Washington Square Park. You sit and play, and everyone hears the music.

Now, we can ask: Is the virtual piano real? It hits many of the marks on the reality checklist. It has causal powers: It plays music, you can't walk through it. It is mind-independent: Even if all users have left for the day, it remains in Washington Square Park unless someone chooses to move it.

Is it a real piano? This is tricky. Even in the nonvirtual world, we have digital pianos and other electronic pianos. Are they real pianos? A few might say yes. It's increasingly common to talk of "acoustic pianos" and "digital pianos" as if both are pianos, though of different types.

Still, most people would say that digital pianos aren't real pianos. A real piano must have strings that vibrate when struck by a hammer, along with other acoustic mechanisms. If digital pianos aren't real pianos, virtual pianos are probably not real pianos, either. Similarly, we'd probably say that virtual trees in Earth+ aren't real trees. On the other hand, virtual books are arguably real books. As with virtual reality, some virtual X's are real X's in Earth+, and some aren't.

Finally, the big issue: Is the virtual piano the way it seems? Or does it involve an illusion? Perhaps the biggest challenge concerns space. The virtual piano appears to be in physical space. It seems to be about a meter tall, piano-shaped, and situated in the middle of Washington Square Park. Is it really in that place? Or is there nothing there but air, in physical reality?

Why say the virtual piano is not really in Washington Square Park? One reason is that if someone came along without the Earth+ system, they'd see nothing there. Perhaps it's already invisible to birds and squirrels. If Martians land in Washington Square Park, they won't see the virtual piano. And if a group of mavericks never get the Earth+ system installed, the piano won't be there for them. You might think a virtual piano in Earth+ is like a rainbow: It seems to be there for some people, but it isn't there in reality.

On the other hand, why say the virtual piano *is* really in Washington Square Park? Certainly users will talk that way. It will be second nature to treat virtual objects as real objects ("Have you sat on the wonderful virtual sofa at Rockefeller Center?"), even if users distinguish them from physical objects. Furthermore, the virtual piano behaves as if it's in the park. It looks, feels, and functions like a virtual piano in the park.

A natural resolution of this matter is to make a distinction. The virtual piano is not *physically in* the park. But it is *virtually in* the park. An object is physically in a space if it has physical matter that occupies that space. An object is virtually in a space if it functions as if it occupies that space. The virtual piano has no physical matter in Washington Square Park, but it functions as if it occupies the space.

If the virtual piano seems to be *physically in* Washington Square Park, this is an illusion. The virtual piano is only virtually in the park, not physically in the park. On the other hand, if the virtual piano seems only to be *virtually in* Washington Square Park, this isn't an illusion. The virtual piano really is virtually in Washington Square Park.

In the previous chapter, I argued that sophisticated users of VR may see virtual chairs *as virtual* rather than as physical, and that they may see them as situated in virtual spaces such as *Second Life* rather than in physical space. If so, VR is not illusory. In Earth+, a sophisticated user may also see a virtual piano as virtual rather than physical. Now, armed with our distinction, we can go a step further. A sophisticated user may see a virtual piano as being *virtually in* its location in Washington Square Park rather than being physically in that location. If so, there's no illusion.

I don't think this conclusion is obvious. It may be hard for users to avoid the sense that the virtual piano is physically in the location in the same way a physical piano could be. Still, a generation raised on Earth+ would presumably learn to treat virtual objects and physical objects in very different ways, and this automatic interpretation would affect their perceptions. The distinction between physical location and virtual location could eventually become second nature; if it does, perception in Earth+ need not involve an illusion.

I conclude that the virtual piano in Earth+ is a real object. In addition, there's a reasonable case that it's a real piano and that it's not illusory. If so, this augmented reality is genuine reality.

From augmented reality to alternative facts?

It's easy to imagine that in the future, there will be multiple dominant systems of augmented reality. Instead of a single universal reality, there will be Apple Reality, Facebook Reality, and Google Reality. Each corporation will set up its own virtual worlds and augment them with its own virtual objects.

In Facebook Reality, there may be a virtual piano at a certain location in Washington Square Park. In Apple Reality, there may be a virtual sign at the same location. In Google Reality, there may be nothing there at all.

Is the virtual piano in the park? According to Facebook Reality, the virtual piano is in the park. According to Apple Reality and Google Reality, it is not. Which reality is correct? It's hard to believe that one of the three is objective reality and the others aren't. Instead, the situation seems symmetrical: Relative to Facebook Reality, the virtual piano is in the park. Relative to Apple Reality and Google Reality, it isn't. We have three different equally valid systems of reality. Has objective reality gone out the window?

Here we seem to have a sort of *relativism*. Whether a fact (the virtual piano is in the park) is indeed a fact depends on the system one's using. You might say that now we have *alternative facts*.

Alternative facts gained notoriety after President Donald Trump's inauguration in January 2017, when there was a dispute about the size of the crowd. The White House press secretary Sean Spicer said that many more people rode the DC Metro that day than on the day of Barack Obama's second inauguration in January 2013. Records show that in fact ridership was much higher in 2013. In an interview, reporter Chuck Todd asked Kellyanne Conway, counselor to the president, why Spicer would utter a "provable falsehood." Conway responded by saying that Spicer had stated "alternative facts."

Conway was criticized for her remark. Todd responded that "alternative facts are not facts. They are falsehoods." Many saw Conway's statement as suggesting a sort of relativism about truth. Relativism is, roughly, the idea that there are multiple equally valid sets of facts corresponding to different viewpoints. Facts are facts only relative to a viewpoint. Relative to Todd's viewpoint, ridership was higher in 2013. Relative to the White House's viewpoint, ridership was higher in 2017.

Relativism is a deeply controversial idea. There are some varieties of relativism that are widely accepted. For example, most of us are relativists about etiquette; what counts as a polite thing to do varies from society to society. A traditional American custom considers it polite to eat meat and vegetables with just a fork after first cutting them. Europeans hold that one should eat them with both knife and fork. Some British and Australian customs even require balancing vegetables on the back of one's fork. There are no objective facts about which of these customs is right; there are just relative facts about who's right from each viewpoint. But while relativism about etiquette seems acceptable, many people resist relativism about more concrete matters, such as the laws of physics.

In my view, augmented and virtual reality lead to a sort of relativism, just as differences in etiquette do. But both are harmless forms of relativism that don't threaten the idea of objective reality.

Much of what we once thought of as absolute has turned out to be relative. One might have thought the time of day was absolute. To our ancestors, it was an objective fact that a certain lunar eclipse happened in the morning. Now we know that the time of day is relative: When it's morning in Sydney, it's evening in New York City. One might have thought the strength of gravity is objective. Now we know it's relative; gravity is much stronger on Earth than it is on the Moon. Correspondingly, weight is relative, too. I weigh much more on Earth than I would on the Moon. One might have thought that shape, mass, and time are objective. According to the special theory of relativity, however, they're all relative to a reference frame. In a reference frame where an object is moving close to the speed of light, its shape compresses, its mass increases, and time slows down, at least relative to an object that is at rest in that reference frame.

Still, all of these matters are consistent with an underlying level of objective reality. For example, it is an objective fact that it's 1:00 pm *in New York City*. It is also an objective fact that a certain rock weighs six pounds *on Earth* and one pound *on the Moon*.

How can we reconcile this sort of relativism with objective reality? It's easy. We simply need to allow that relations are part of reality. It's 9:00 am in relation to New York and 2:00 pm in London. The object is circular in relation to one reference frame and elliptical in another. You need relations like these to give a full description of reality.

The same goes for multiple realities. You might have thought that whether or not there's a piano in Washington Square Park is an objective matter. It turns out to be a relational matter. In Apple Reality, there's a piano in the park. In Google Reality, there isn't. The same goes for laws of physics: It's an objective fact that in Apple Reality, the laws are one way. It's an objective fact that in Facebook Reality, they're another way.

Importantly, both the Apple Reality system and the Google Reality system are part of objective reality. It's an objective fact that in Apple Reality there's a piano in the park. And it's an objective fact that in Google Reality the laws of quantum mechanics are true. That's how we reconcile the relativism of virtual worlds with objective reality.

An extreme relativist may say that there's no level of objective reality at all. Even relational facts—such as the fact that this reality contains a piano, or that I find Bob Dylan's music beautiful—might be true from my perspective and false from yours. A non-relativist will say there are objective facts about what's true from my perspective and about what's true from yours. But the relativist will say that even facts about what's true from my perspective are only true from a particular perspective! That's an interesting view, but there's no good reason to accept it.

Even in a cosmos with multiple virtual and augmented reality systems, there are plenty of objective facts. For a start, there are objective facts about what happens in each reality system. It will be an objective fact that the piano is in the park in Apple Reality.

There are also objective facts about ground-level reality. Nothing in our discussion of virtual worlds within virtual worlds suggests that there is not a base reality at the top of the chain. Even if our own reality is a simulation 42 levels down from base reality, base reality has its own independent existence.

Importantly, there are objective facts about what happens in our ordinary reality. It is an objective fact that a certain number of votes were cast in the 2020 US presidential election. It's an objective fact that Joe Biden was declared the winner. Of course many facts about ordinary reality will be relativized to time, place, and more. Joe Biden was elected president of the US in 2020 but not in 2016.

The same holds true for virtual worlds. Biden was elected in ordinary reality, while perhaps someone else was elected in Meta Reality, the universe of our simulators, and someone else was elected in *Second*

Life, a virtual world we have constructed. But once we realize that reality is relational in this way, what happened in the United States in 2020 in ordinary reality remains an objective fact.

There can be disagreements about these facts. Trump supporters can think that Trump got more votes (in this reality), while Biden supporters think that Biden got more votes. But there's an objective fact about which is right. With some luck, we may even be able to find it.

In the *Black Mirror* episode "Men Against Fire", soldiers use an augmented reality implant called MASS that makes human mutants appear to be roaches. This system enables a form of genocide wherein the mutants are wiped out by the soldiers. What do we say about reality here? If MASS Reality removes any trace of the mutants in a farmhouse and shows virtual roaches instead, we might say that in MASS Reality, the farmhouse contains virtual roaches and not humans. However, in ordinary reality, the farmhouse still contains humans. A soldier who kills apparent roaches in MASS Reality is destroying virtual roaches there but is also killing humans in ordinary reality. Even if everyone in the world had the MASS implant, humans would be dying. The relativism of multiple realities does not yield an escape from cold, hard facts about ordinary reality.

Near-term augmented reality

For now, Earth+ is science fiction. Existing augmented reality systems are much more mundane. Augmented reality systems mainly allow us to see and perhaps to hear virtual objects, without allowing us to touch, smell, or taste them. Virtual objects don't impede our movements. No augmented reality system is installed permanently. Users put on their glasses occasionally but more often have them off. And no system is universal. There are a few different systems, each of which is used for short periods by a few users.

Still, some of what we've said about Earth+ applies to near-term augmented reality. Suppose I'm redesigning my living room with augmented reality, so as to see a virtual sofa in the corner of the room. Is the sofa real?

As noted, the virtual sofa is a real digital object existing inside the augmented reality device's computer. It has genuine causal powers—at the very least, it causes you to see it, and it might even cause you to buy a sofa. To a limited extent, it exists independently of our minds. If I take my glasses off but leave the program running, the digital object still exists and is still available to be seen by others, in principle. So it satisfies our first three criteria for reality—existence, causal powers, mind-independence—at least to some extent. But it fails the fifth criterion: It clearly is not a real sofa. I cannot even sit on it.

As for the fifth criterion: Is the sofa in my living room an illusion? For many users, it will look as if the sofa is physically in the corner of the living room. Because no sofa is there, this is an illusion. But it's arguable, as before, that for a sophisticated user of augmented reality it will look as if a *virtual* sofa is in the corner of the living room. And it may look as if the virtual sofa is only *virtually* in the room, rather than physically in the room. If so, there's no illusion: The virtual sofa really is virtually in the living room, at least to the extent of being visible there.

Some of the differences with Earth+ matter. Because we cannot touch the sofa and it cannot support us, the sense in which the sofa is virtually in the room is less robust: The sofa supports some interactions with the user, mostly visual interactions, but far from all the usual interactions. Because the sofa is visible only to me, it can't interact with others. Because I wear my glasses only some of the time, the sofa is much less of a constant. One could say that the virtual sofa is virtually in the living room *for me*, at those moments when I'm using the glasses, and for other people it isn't there at all. Still, as long as I perceive the virtual sofa as being virtually in the living room *for me*, this won't be an illusion.

At least in the near term, the sense of virtuality may be key to the use of augmented reality. Users will need to distinguish between physical and virtual objects, since if they treat virtual objects the way they normally treat physical objects, confusion will reign. In the near term, we won't want to sit down on virtual chairs or eat virtual food! The same goes for treating physical objects as virtual. We don't want to bump into actual walls. In the very near term, there may be no danger here: The limited technology for virtual objects may ensure that they're distinguishable. But I predict that once the technology allows indistinguishable virtual objects, there will be strong pressure to make sure that virtual objects stand out as virtual.

The physicality-virtuality continuum

In a 1994 conference paper, the industrial engineer Paul Milgram and colleagues at a Japanese systems research lab introduced the idea of a *reality-virtuality continuum*. At one end of the continuum is regular physical reality. At the other is pure virtual reality. In between, there are various sorts of *mixed reality*, wherein one experiences both physical and virtual objects. In standard augmented reality, the physical world serves as the basis and is augmented by virtual objects here and there. In *augmented virtuality*, a virtual world serves as the basis, augmented by physical objects here and there. When a nonvirtual music group appears on stage in a virtual concert hall, that is augmented virtuality.

I think Milgram's continuum is misnamed, because it bakes in the premise that virtuality is opposed to reality. As we've seen, that can't be assumed. A better name would be the *physicality-virtuality continuum*. Standard VR systems are largely pure virtuality, while AR systems augment physicality with virtuality.

At each point on the continuum, we find fascinating questions about how much reality is involved and how much illusion. So far, I've talked about standard VR and standard AR. What about other points on the continuum?

What should we say about augmented virtuality, in which some physical objects are experienced as being within a virtual world? Perhaps I'm talking to my sister in Australia while I'm in New York. I see her sitting on a chair next to me in a virtual world, not in avatar form but in her usual physical form. My sister is certainly real. She has causal powers and is independent of my mind. I experience her as a human being, and she is a real human being. However, I also experience her as being in the virtual room with me. Is this an illusion? It needn't be. My sister isn't physically in the virtual room, but she's virtually in the virtual room. If I'm familiar with the technology, that's how I'll experience her.

What about fully mixed realities, with equal components from the physical world and a virtual world, all interacting? Perhaps a physical building is augmented with virtual walls, with a mix of physically present people, virtually present physical people, and avatars in one large conversation. This can be seen as a combined physical/virtual world, which isn't quite the same as either the physical or the virtual world taken alone. A sophisticated user might experience it as a space in which physical and virtual objects alike are present. And this won't be an illusion.

In a mixed reality, it also matters whether the augmenting virtual objects interact with the surrounding physical world. Suppose that augmented reality glasses allow you to play a VR video game set in outer space, while at the same time you can see a physical table in front of you. Because the two worlds are independent, you may well experience them as separate virtual and physical worlds, without any real sense that the virtual spaceships are in the physical world. Perhaps locations will align with each other (a virtual spaceship may momentarily come close to your physical table), but we can largely ignore this alignment in our dominant perception of the world. Adept users may react entirely differently to physical and virtual objects—for example, walking out of the way of physical objects but flying through virtual objects. In this case, mixed reality may involve simultaneous non-illusory perception of both a physical and a virtual world.

Chapter 13

Can we avoid being deceived by deepfakes?

IN JULY 2020, SHORTLY AFTER THE LANDMARK ARTIFICIAL-intelligence program GPT-3 was released, the philosopher Henry Shevlin posted an interview online.

SHEVLIN: It's great to be interviewing you, Dave. Today I'd like to talk about your views on machine consciousness. Let's start out with a simple question: Could a text model like GPT-3 be conscious?

CHALMERS: It's unlikely in my opinion, although I'm a little uncertain on this issue.

SHEVLIN: Do you think we're likely to have a theory of consciousness in the near future that could allow us to settle these issues—to tell us whether a given artificial system is conscious?

CHALMERS: I think it's very unlikely. We don't even have a theory of consciousness that lets us settle these issues for humans! And humans are a lot simpler than modern computers.

SHEVLIN: What about animal consciousness? There it seems like we have an ethical imperative to come up with at least reasonably informed answers—to determine whether, for example, a fish can feel phenomenally conscious pain. Do you have any thoughts on how to approach that problem?

CHALMERS: Well, I think we can reasonably say that mammals are probably conscious.

SHEVLIN: On what basis can we say that mammals are conscious?

CHALMERS: Well, we can start with humans. We know they're conscious from introspection, and it's very hard to imagine that we're not.

SHEVLIN: Sure, but how can we move beyond humans? No other organism will, e.g., exactly implement the Global Workspace Architecture or whatever architecture you think underpins consciousness. So how can we determine whether, e.g., a dog is conscious, or a chicken?

CHALMERS: That's a really good question. I think one thing we can say is that if an organism has a central nervous system with a brain, then there's some chance it might be conscious.

The answers labeled "CHALMERS" in Shevlin's interview were written by GPT-3. GPT-3 is a huge artificial neural network that was trained using deep learning, a technique for training networks based on large amounts of data. GPT-3's primary goal is to produce plausible continuations of any piece of text, based on reading text from all over the internet. Shevlin gave it a prompt, saying, "This is an interview between Henry Shevlin and David Chalmers," followed by my biographical entry from Wikipedia. From there on, all the questions are from Shevlin and all the answers are produced by GPT-3.

Reading the interview was disconcerting. GPT-3 gets my opinions about these matters more or less right. Though there are occasional glitches (humans are simpler than modern computers?), plenty of friends reading it on Facebook said they thought it was a real interview with me. Kinder colleagues said that it sounded like me on a bad day. One colleague said that he thought the overuse of "I think" was a giveaway. I responded that I have the bad habit of overusing "I think" in my writing. Some joked that they could no longer be confident that they were talking to the real me.

In effect, GPT-3 had produced a simulacrum of me that fooled a good number of friends and colleagues. My simulacrum is an example, or at least a close relative, of a *deepfake*: a fake entity produced using the technology of deep learning.

The word *deepfake* is most often applied to fake photos and fake videos rather than fake text. Until recently, fake images were typically produced with Photoshop, CGI technology, and other tools. One example is the appearance of Princess Leia toward the end of the 2016 movie *Star Wars: Rogue One*. She looks like the young character played by Carrie Fisher in the original *Star Wars* movie from 1977. The scene was created using CGI technology. Specialists pasted Fisher's face from the 1977 movie onto the body of Norwegian actress Ingvild Deila.

The *Rogue One* scene received mixed reviews. The CGI technology was limited, and many viewers found the replication of Princess Leia jarring or unconvincing. However, just four years later, amateurs used a widely available AI program to replicate the scene by fitting Leia's face to the scene in the movie. Many viewers found the 2020 amateur version far more convincing than the 2016 professional version.

The new version of Leia was a deepfake. Deepfake photos and videos are produced by deep neural networks, interconnected networks with many layers of neuron-like computational units. These networks can be trained over time by the deep learning process, which adjusts the connections between units in response to feedback. Deep learning can train networks to perform many tasks—including the production of highly convincing images or videos. Often, a deepfake photo or video shows someone doing something they never did or saying something they never said. Sometimes a deepfake photo or video depicts a person who never existed.

Deepfakes can be found in contexts as diverse as politics and pornography. They often depict public figures saying outrageous things. There's one in which Barack Obama says, "Killmonger was right," referring to the revolutionary antagonist in *Black Panther*. There's another in which Donald Trump appears in the TV series *Better Call Saul* and explains money laundering. In a Delhi election campaign in 2020, the Delhi Bharatiya Janata Party used this technology to create a video in which one of its candidates, Manoj Tiwari, addressed a crowd in Haryanvi—a language he doesn't speak.

Deepfake technology is likely to get better fast. Soon we won't be able to distinguish deepfake photos or videos from real photos and videos. They may well enter augmented- or virtual-reality worlds. Perhaps a friend will appear to us via augmented reality, saying something she never said. Eventually there will be entire deepfake virtual realities intended to convince people they're in a place when they're not.

Deepfakes raise a version of the Reality Question: Are deepfakes real? If we grant reality to virtual objects, don't we have to do the same for deepfakes? They also raise a version of the Knowledge Question: How can we know that something we're seeing is not a deepfake? When deepfakes are widespread, can we know when a depiction is real?

The same questions arise concerning deceptive news stories, now often known as *fake news*. It has become increasingly common to spread deceptive news stories in order to hurt some political figures and help others. Once again, we can ask the Reality Question: Is fake news real? And all-importantly, we can ask the Knowledge Question: How can we know when news is fake news?

These are live questions in the 2020s. Questions about whether we're in Matrix-like scenarios may strike many of us as playful. But questions about whether some news is fake news are sobering. They invoke an especially realistic version of the problem of skepticism about the external world.

The question here does not concern Cartesian *global skepticism*: Is everything fake? Instead, it concerns *local skepticism*: Is this fake? Is that really happening? As a consequence, my strategy for responding to global skepticism (see especially chapter 6) may not apply here. I don't claim to have a general answer to problems of local skepticism. Still, let's see if we can say anything useful about the problems posed by deepfakes and fake news.

I'll focus especially on a version of the question in the chapter title. Can a critical observer in the modern world avoid being thoroughly deceived by deepfakes and fake news? I'll argue for a limited anti-skeptical conclusion: in principle, at least under weak assumptions, critical observers in modern democratic societies can avoid thorough deception (that is, deception across the board on a wide range of issues) by the news media. This is not to downplay the seriousness of the phenomenon. There is little doubt that in practice, fake news deceives many people and has many corrosive effects. And even critical observers may not be able to avoid being deceived on some issues. Still, there are limits on how much they can be deceived.

To approach these issues, I'll start with versions of the Reality Question and the Knowledge Question for deepfakes: *Are deepfakes real? Can we know whether an image is fake?* I'll then move to corresponding questions about fake news.

Are deepfakes real?

Let's start with the connection between deepfakes and reality. The first question is whether the entities and events we encounter in deepfake images (photos and videos) are real. Is deepfake Obama really Obama? Is a deepfake dog really a dog? Is it at least a virtual dog? Is it a digital entity at all?

You might say that deepfakes are a type of virtual reality. If so, my version of virtual realism will apply to deepfakes, and I'll have to say that deepfakes of Obama or a dog or a cat are real in the sense in which VR versions of Obama and the dog or cat are real. Intuitively, that would be a strange conclusion, but then I've argued for some strange conclusions already.

Fortunately, this conclusion doesn't follow. Standard deepfakes are not virtual realities at all. Recall that virtual realities are immersive, interactive, and computer-generated. Deepfakes meet one condition, and they could easily meet another. They are indeed computer-generated, and while they're currently not immersive, it's easy enough to imagine immersive versions, with a 360-degree deepfake video viewable through a headset. However, photos and videos don't meet the interactive condition. Deepfake photos and videos consist of a fixed image or a fixed series of images that don't need to interact with anything else.

Because they're not interactive, most current deepfakes are more akin to digital movies than to VR worlds. Importantly, they don't con-

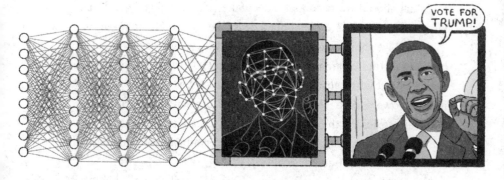

Figure 33 Deepfake Obama.

tain full-blown virtual objects. They may contain patterns of bits within them that roughly correspond to Obama, say, or to a particular dog or cat. But these patterns of bits don't have anything like the causal powers of a virtual version of the subject. Virtual Obama is interactive and thus able to produce a variety of utterances or actions depending on how you interact with him. A virtual dog, or even a virtual ball, is interactive, too. Deepfake dogs and deepfake balls have no interactive powers. At best they have the power to look a certain way.

There will be fully interactive deepfake virtual realities eventually. We've already seen that existing texts could be used to train GPT-3 to simulate a conversation with, say, Barack Obama. An extension of this process could use diverse audio and video recordings of Obama to train an artificial-intelligence network to look and speak like him in various circumstances. Even faced with wholly new inputs, the network will respond in a plausible manner consistent with its training. The network might be imperfect and implausible in some circumstances, but it will at least be interactive.

Likewise, an artificial-intelligence network could observe a whole environment, like a football stadium or a classroom, and could be trained to simulate it in many different conditions. Even when new things happen, the network will have some sort of response. If we were in an environment like this, we'd experience it much as we experience a VR version of a football stadium or a classroom.

What should we say about a deepfake virtual reality? Is a deepfake virtual football real? What about a deepfake virtual Obama?

I'd say we should treat these the same way we'd treat a virtual football or a virtual Obama in general. A deepfake virtual football is a real digital object with causal powers resembling those of a football, including the powers to be (virtually) kicked and thrown with various speeds and trajectories. At the same time, it's not a real *football*. Real footballs are made of certain materials and have a certain size. Virtual footballs don't have the required physicality.

As for deepfake virtual Obama: This is a real digital object with some causal powers resembling those of Obama. If deepfake Obama is made with near-term AI technology, then many of Obama's causal powers will be missing. Deepfake Obama won't display anything like the intelligence or flexibility of the real Obama. So it will probably not be a conscious being, or a person. Perhaps with the AI technology of decades to come, it will be possible to train a deepfake Obama that does display the intelligence and flexibility of the real Obama. If so, perhaps deepfake Obama could be conscious or a person. But this sort of deepfake is a long way off.

Is deepfake virtual Obama the real Obama? It seems clear that any near-term deepfake is not the real Obama, any more than a robot version of Obama is the real Obama. With advanced AI, this won't be entirely obvious. Perhaps a good-enough simulation of Obama could be a continuation of the real Obama, a sort of uploaded version of Obama, reconstructed using behavioral data alone. Some philosophers would regard an upload as a continuation of Obama's life. If this process took place long after Obama's death, perhaps it could even count as a resurrection. On the other hand, if the advanced deepfake is constructed while the biological Obama is still alive, most people will be inclined to say that it's not the real Obama. All this raises complex issues that I discuss in chapter 15.

So our answer to the Reality Question is mixed. In the short term, deepfakes may involve real digital entities, but these don't have anything like the causal powers of standard virtual or physical objects. In the long term, virtual AI deepfakes may involve real digital entities with causal powers on a par with those of standard virtual or physical objects. However, in both cases we'll probably say that a deepfake is not the real thing. Deepfake Obama is (probably) not the real Obama. A deepfake football is (probably) not a real football. This is enough to raise a serious question about knowledge.

How can we know whether an image is real?

Suppose we see a video that looks like Obama saying something. How do we know that this is the real Obama? *Can* we know that the video image is real? The problem generalizes to any photo or video. How do we know whether or not a photo of a waterfall is a photo of a real waterfall? How can we know that a video of a violent protest is authentic?

As the philosopher Regina Rini has observed, we are used to using images as reliable backstops for our knowledge. If we're in doubt, seeing is believing, and a photo is proof! But in an age of deepfakes, images cannot be trusted so straightforwardly. Until now, it's been possible to distinguish fake images from real images by a close-enough examination of giveaway signs. But as deepfake technology develops, the giveaways get harder and harder to notice. Soon they'll be detectable only by advanced algorithms, and then giveaways may disappear entirely. At that point, there will be no way to distinguish deepfakes from real images by examining the qualities of the image.

The implausibility of an image may tip us off that it's fake. If the image shows the Sydney Harbour Bridge turned upside down, it's probably fake. If it shows Bernie Sanders endorsing the Republican Party, it's probably fake. But if the image has enough antecedent plausibility—even just the plausibility of an everyday surprising news item—this method won't work.

At the other extreme, if you encounter a mundane video of your cousin Sam telling you unimportant things, it's most likely real—who'd bother to fake it? More generally, at present there are relatively few deepfakes circulating, so we have reasonable grounds for thinking most images are real. But as deepfakes become more common and easier to generate, they'll become more and more of a crucial issue.

In the long run, the only way to know for sure whether an image is real or fake may be through *authentication* by a reliable source. If a trusted friend tells you that she took a certain photo, you have good reason to think it's real. If a reliable media outlet puts out a video and says they recorded it, it's probably genuine. By contrast, if we find a photo lying on the road or encounter a video on a partisan website, we'll have less reason to think it's real.

Relying on authentication is probably the best way to protect oneself from deepfakes in an environment where they're ubiquitous. Of course, the method isn't perfect: A trusted friend might play a trick on us, or his email account might get hacked. A reliable news outlet might sometimes be fooled. Or it might be taken over by bad actors without our knowing, or it might be convincingly spoofed. Still, a source can build up a track record that gives us reason to trust them. Some sources may be endorsed as reliable by other reliable sources, expanding our network of trust. Some hard issues remain: for example, what if all our sources are misleading? I'll discuss these shortly.

In any case, photos and videos are only one source of evidence about the external world. If it turns out that we can't trust any of them, this would undercut some of our knowledge but not all of it.

Once deepfake virtual realities are possible, the problem will multiply. Some cases concern VR environments themselves: How can I know I'm really playing multiplayer *Beat Saber* with my friends, as opposed to a deepfake emulation with bots acting my friends' parts? Some cases concern ordinary perception: When I think I'm discussing an upcoming product with my corporate employer, how do I know I haven't been kidnapped into a deepfake virtual reality by competitors eager to learn our secrets?

Again, one solution is authentication. To help avoid fakes, you should access only virtual reality software that you trust, and you should use only reliable augmented reality equipment. There will be some virtual elements, but within limits that are understood. The rules will say, "No fake friends and relatives without telling the users!" In some contexts, these rules will be difficult to enforce. In social VR, where people have control over the appearance of their avatars, a deepfaker might make an avatar that looks like your mother. But authentication should still be straightforward—for example, via a username known to belong only to your mother.

We still have to worry about extreme circumstances. What if the VR environment has been taken over or hacked? What if there are *no* trusted systems? What if someone has hacked into your brain, as happened to an unsuspecting airplane passenger in the movie *Inception*?

Or worse: What if you are yourself a deepfake? Perhaps your enemies produced a deepfake simulation of the original you, based on video records and the like, and now are using it to gain information about the original? In the *Black Mirror* episode "White Christmas," police use a version of this method to induce confessions from suspects. Is there any way to know that this isn't happening to you?

I don't have a general answer to these questions. If the deepfake virtual reality is an imperfect simulation of the original, it should be possible to discover this through investigation. You can interact with your mother and see whether she knows things your mother should know. You can see whether the secret information in your secret notebooks is present. You can explore your world to make sure it's all there. You can perform scientific experiments to see whether or not they give the expected results.

If the deepfake virtual reality is a *perfect* simulation of the original, then those sorts of investigation won't work. We've seen that it's impossible to know you're not in a perfect simulation. This returns us to familiar issues. If you've lived your whole life in the simulation, it's your reality, and your beliefs about your world are still true. But what if your VR world has recently been hijacked, or you've been kidnapped or uploaded into a perfect simulation? In this scenario, much of what you believe about the world around you may be false, and there's no way to know for sure.

In a world in which VR hijacking, kidnapping, or uploading into perfect simulations is common, perhaps the best we can do is take precautions to avoid getting into those situations. Once deepfake virtual realities are perfected, we can expect that both computer security and brain security will be growth industries.

What about fake news?

Let's return to Earth in the early 21st century, where we face the very real issue of fake news—misleading news stories produced and circulated without regard for truth.

Fake news has been around for as long as there has been news. One could argue that the Roman triumvir Octavian engaged in a fake news campaign against his rival Mark Antony in 31 BCE, portraying him as a traitor to the Roman Empire. In 1782, during the American Revolution, Benjamin Franklin produced a hoax issue of a newspaper with a false story about American scalps being sent to the king and queen of England.

The term "fake news" exploded around the time of the US election in 2016. The paradigm case of fake news is perhaps the Pizzagate story, which circulated before the election. The report accused Hillary Clinton and other officials of the Democratic Party of running a child sex ring out of a pizzeria in Washington, DC. The story appears to have started on Twitter and was circulated widely through social media and alternative news media. Investigation (to say nothing of plausibility) revealed that there was no truth to it.

The explosion of social media has served to amplify fake news. Social media allows readers to circulate misleading news stories to like-minded people on all parts of the political spectrum. The term "fake news" itself has now become controversial, in part because public figures often use it to delegitimize the news media by calling any unfavorable story fake news. But like deepfakes, fake news stories are a worrisome phenomenon.

Fake news isn't the same thing as false or inaccurate news. If a journalist trying to report the truth makes mistakes, that's false news but not fake news. Fake news requires intent to deceive, or at least a disregard for the truth. A clickbait news site will promote itself by making up news stories without caring whether they're true or false.

We can raise a version of the Reality Question for fake news. If simulations and virtual worlds are real, then are the worlds conjured by fake news real, too? Fake news stories are akin to depictions of a fictional world. There's often an underlying proposition (Hillary Clinton is a crook! Barack Obama was born in Kenya!) that crops up from story to story, suggesting a single underlying fictional world. But this world isn't a virtual reality world: It isn't immersive or interactive, nor is it computer-generated. So the virtual-realism argument can't support it.

Perhaps we could imagine a computer simulation (Sim Pizzagate?), set up to spin off one fake news story after another. Inside the simulation would be a simulated entity, Sim Hillary, engaged in nefarious activities in a simulated pizza parlor. In that case, there would be a digital reality corresponding to the world of Pizzagate. But when we say "Hillary," we're talking about the original Hillary, so Sim Hillary's nefarious deeds do nothing to make the Pizzagate allegations against Hillary true. In a far-out case where our world has been a Pizzagate simulation from the start, then we may be talking about Sim Hillary who is really a criminal, and the Pizzagate news will not be fake. But short of our inhabiting the simulations ourselves, there's no danger that simulations of news stories will make fake news stories true.

The Knowledge Question for fake news is more pressing. How can we know whether any given news story is fake news? If we can't know this, is the news media a source of knowledge at all?

We certainly treat the news media as a source of knowledge. In modern society, most of us rely on news stories for much of what we know about the wider world. We know about the political situation this way. We know about what's going on in other countries and other cities this way. We know about crises and disasters this way. If we couldn't trust the news media, then we'd know much less than we think we do.

Fortunately, there are ways to distinguish real news from fake news. As with deepfakes, glitches and inconsistencies serve as giveaways. We can sometimes rely on implausibility, or on mundaneness that no one would bother to fake. As before, the most important method is authentication by a reliable source. If a story is presented by a reputable source, it's probably not fake. Stories by reputable news outlets often contain mistakes, but it is rare for them to be made up completely. Verification by other reputable sources, such as independent fact-checking outlets, can also bolster confidence in the accuracy of a news story. Because there are widely recognized reliable sources, the onset of fake news and deepfakes has not yet led to complete chaos, where no one knows what to believe.

How do we know that an apparently reliable source is indeed reliable? Mere consistency isn't enough. Fake news can be consistent and presented as if it's true. If a source is endorsed by other reliable sources, that's helpful, but we can still imagine that the whole network of sources might be unreliable. Some political subcultures involve networks of media sources interconnected in webs of mutual endorsement but which are nevertheless unreliable.

In these subcultures, many people may be deceived about many things. But for a sufficiently critical subject, there will be limits. If you have full access to the internet, it will soon become clear that many sources contradict the information in your bubble. At this point, further inquiry will often reveal which aspects of one's news are fake.

What if you're a citizen in a regime where information is tightly controlled? Perhaps you're in North Korea with only state-run media available. Perhaps you're in an American subculture that restricts access to most media. In these cases, it's harder to know the truth. One check is consistency with other sources of information: what you learn through your senses and what you hear from trusted others. People who have lived under these regimes often report signs that the news is misleading; perhaps the news says that everyone is well-fed, but people around you are starving. Deception is easier when stories concern matters far away. Perhaps state-run media can deceive people about what's happening in other countries. But even here, most people know that news outlets are tightly controlled, which raises questions about their reliability. This may not tell you what you should believe, but it might lead you to suspend judgment about what you seem to learn from the news.

Even suspending judgment is hard work. Reasoning about conflicts in the news may take a great deal of critical thinking. But if a sufficiently observant and reflective thinker can avoid being taken in, then there's at least a potential path away from deception by tightly controlled media. Of course, while suspending judgment about fake news is better than being deceived, it isn't as good as knowing the truth. It's arguable that the purpose of much fake news is simply to sow doubt. In her 1951 book *The Origins of Totalitarianism*, the German-born American philosopher Hannah Arendt said, "The aim of a totalitarian education has never been to instill convictions but to destroy the capacity to form any." If everyone suspends judgment as a matter of course, then that strategy will have largely succeeded.

For me and for many readers, media are not so tightly controlled. A typical democratic citizen with an internet connection has access to countless news sources and many different viewpoints. There are many biases and blind spots in the media, but there are also, typically, dissenting sources that reveal those biases and blind spots. For example, the widespread American media biases that economist Edward S. Herman and linguist Noam Chomsky document in their 1988 book *Manufacturing Consent* may be real, but they're also the subject of a high-profile book. Large chunks of the media may be in a position to deceive large chunks of the population, but the media as a whole are not tightly controlled enough for seamless deception of an entire population. (As Abraham Lincoln is supposed to have said, "You can't fool all the people all of the time.") When the media as a whole get something wrong, this is more often due to media ignorance than media deception.

Of course you can spin a scenario in which there are puppet masters behind the entire media complex, giving the appearance of openness when in fact all is intended to deceive. But across-the-board deception would require an enormous and complex conspiracy. Either the conspiracy would have to deceive an entire population, as in John Carpenter's classic 1988 movie *They Live*, or the population would have to conspire to deceive an individual or a group, as in *The Truman Show*.

Such enormous conspiracy theories can't be ruled out entirely, but

Bertrand Russell's appeal to simplicity (see chapter 4) seems appropriate: It's reasonable to count this hugely complex conspiracy as more improbable than far simpler ordinary-world hypotheses. Perhaps the only reasonably simple version of the fake-news hypothesis is one in which we're part of a giant computer simulation, but that brings us back to familiar ground.

Let's sum up. For people whose news sources are tightly controlled, it may not be possible for them to know whether their news is fake or real, but with some effort they may often be in a position at least to know that something is wrong and so to maintain doubt. For people like me and many readers, who seem to have access to a huge array of news sources, it is often possible to determine whether news is fake or real by using the network of sources as a whole. We cannot completely exclude extreme scenarios in which almost every source of information is fake, but short of a computer simulation, these scenarios are so complex as to be unlikely.

Part 5

MIND

Chapter 14

How do mind and body interact in a virtual world?

I N FEBRUARY 1990, I SET OUT ON A ROAD TRIP TO SANTA FE. I WAS a 23-year-old graduate student in philosophy and cognitive science at Indiana University. My fellow students and I had heard about the new field of artificial life in which researchers try to create or at least simulate living systems inside a computer. Ten of us—philosophers, psychologists, and computer scientists—rented a van and drove to New Mexico across the great plains of Kansas, Oklahoma, and Texas. Our destination was the second-ever conference on artificial life, put on by the fabled Santa Fe Institute for the study of complexity.

It had been just three years since the first artificial life conference at Los Alamos National Laboratory, but there was already a plethora of approaches to creating life inside a computer. The approach that intrigued me most was designed by the pioneering computer scientist Alan Kay. The idea behind Kay's "Vivarium" was to simulate a whole ecology in a computer. There was a simple physical environment consisting of a two-dimensional grid. Each square in the grid could be inhabited by an object, or it could be inhabited by an animal. Animals had simple bodies that could be oriented in various directions, move from square to square, and pick up objects.

The Vivarium world had a "physics." Simple rules governed the two-dimensional grid and ordinary objects. It also had a "psychology." The animals had separate rules governing their behavior. It was especially interesting to me that the physics and the psychology were distinct. This separation was and is standard practice in a virtual world. Inscribed in code, there is one set of rules governing ordinary objects in the world's

Figure 34 An artificial life virtual world akin to the Vivarium.

environment and another set governing how creatures behave in the world. Nonplayer characters in a video game are like organisms in the Vivarium. Their behavior is determined by rules special to them.

I wondered what would happen if Kay upgraded the Vivarium so that the organisms become increasingly intelligent and begin to explore their world? First, they will explore their environment and figure out the laws of "physics" governing it. Second, they will investigate one another and figure out how their "psychology" works. They might hypothesize that they have brains that are part of their physical world, driven by the same laws that govern their environment. But they would gradually discover that this isn't true. These creatures will never discover brains in the environment they inhabit. Their minds are outside their physical world.

It struck me that these creatures would almost certainly become *dualists* about the mind. As we've seen, René Descartes was the archetypal dualist, holding that minds are entirely distinct from physical processes. He held that thinking and reasoning go on in a separate, nonphysical domain, which interacts with the brain through a special mechanism. These days, Cartesian dualism is widely rejected. Most people think that our behavior is produced entirely by physical processes in the brain, and many think that the idea of interaction between physical and nonphysical processes makes no sense.

For the creatures of the Vivarium, the situation is different. Their physical environment—that is, their two-dimensional world—simply doesn't contain their psychological processes. Their minds are distinct from their bodies and their physical world. These creatures would become dualists and they would stay dualists, absent any evidence against their view. Furthermore, they would be correct to be dualists. In their world, the mental and the physical are quite distinct.

From our perspective, everything that's going on in the Vivarium is physical. It's just that the physical computer processes governing these creatures' behavior are distinct from the physical computer processes governing their environment. But from their perspective, what counts as physical is their environment. Their minds are distinct from that environment. They will hypothesize that their minds are distinct from their environments, and they will be correct.

It was fascinating to me that one could get a sort of dualism from such a simple setup. One gets the same sort of dualism from many virtual worlds with nonplayer characters, as we've seen. People often say that dualism is somehow incoherent, but here was a simple and naturalistic way that something like dualism could be true. If we had evolved in a world like this, we would all be dualists—and we would basically be correct.

Standard virtual reality environments lead to an even stronger sort of dualism. In these environments, human beings interact with virtual worlds. A human user of virtual reality has a virtual body inside the simulation, but the user's brain is outside the virtual world. Once again, there will be a virtual-world physics and an entirely separate psychology. Whenever a human being enters a virtual world, there's immediately a sort of dualism—at least, from the perspective of the virtual world.

We can imagine our entire species evolving in a virtual world, with virtual bodies in the virtual world and brains outside the virtual world. The brains receive all their inputs from the virtual world and send all their outputs there. Our observations would always be of the virtual world, and we'd develop a physics for the virtual world. Our own actions, however, would be determined by a source unseen inside the virtual world—our brains. From the virtual-world perspective, our actions aren't produced by physical processes inside our world at all. There are two quite different sorts of processes: the virtual physics that governs most objects in the world, and the psychology that governs users' behavior in the virtual world. If we grew up in this environment, Descartes would be vindicated. We would all become dualists.

Descartes and the mind–body problem

The mind–body problem asks: *What is the relationship between the mind and the body?*

The *mind* is the locus of our perceiving, feeling, thinking, and deciding. Seeing an apple is a state of my mind. Feeling happy is a state of my mind. Thinking that Paris is in France is a state of my mind. Deciding to go to a movie is a state of my mind.

The *body* is the biological system that I inhabit and sometimes control. It has two legs, two arms, a torso, a head, and many internal organs. One part of the body is especially important for the mind: the *brain*, which receives our sensory inputs and produces our actions.

One part of the mind–body problem is: *Are the mind and the body the same?* Or better: Are the mind and the brain the same? When I see an apple, a large group of neurons fire in my visual cortex. But is my seeing the same thing as the neurons firing? Or are these two events—the seeing and the neurons firing—two different things?

Dualism says that the mind and the body are fundamentally different. The mind is one thing, and the body is another. As we saw in chapter 8, mind–body dualism can be found in many different cultures. The Akan tradition in Africa endorses a sort of dualism. The Persian philosopher Avicenna used a thought experiment about a "floating man" drifting through the sky to argue that self-awareness is not the same as any bodily state.

In the European tradition, Descartes articulated the classic form of dualism. According to his view, the essence of the mind is *thinking*, while the essence of the body is *extension*, or taking up space. Descartes argued for dualism by saying we could imagine our minds without any bodies at all.

Recall Descartes's evil-demon scenario, in which the demon gives us sensations as if they came from the external world. Now imagine a version of this scenario in which we don't have a brain or a body at all. We have to imagine a pure mind. Descartes thought he could imagine this, and he thought this demonstrated that his mind was not identical to his body. He argued something like this:

1. I can imagine my mind without my body.
2. I cannot imagine my body without my body.

---

3. So: My mind is not my body.

A similar argument would establish that my mind is not my brain, and that my mind is not any physical object at all.

This argument has been controversial. Many philosophers have responded that imagination is not a great guide to reality. In particular, we can often imagine that two things are distinct when they're really the same. For example, I can imagine Superman without Clark Kent, but Superman is Clark Kent nevertheless.

Still, to many people it seems intuitively plausible that the mind is distinct from the brain. Thinking just doesn't seem like a brain process, nor does feeling pain. Part of the issue here comes down to the special qualities of conscious experience—something we'll focus on in the next chapter. Another part of the issue derives from the complexity of human behavior. How could Shakespeare's plays have been written by mere matter?

Descartes thought that nonhuman animals—flies, mice, birds, cats, cows, apes—were mindless automata and that their behavior was caused by mechanisms made of matter. He thought that *some* human actions were essentially mindless and could be explained this way, too. But he thought that some human actions could not be thus explained. In particular, he thought that creative language use is something that mere matter could never do. Only a nonphysical mind could use language as humans do.

This wasn't an unreasonable attitude in the 17th century, long before we knew about the complexities of the brain, and long before we had computers producing all sorts of complex behavior. Still, even in the 17th century, people realized that dualism had some problems to overcome.

The biggest challenge was the problem of *interaction*. How could a nonphysical mind and a physical brain interact? The mind seems to affect the body: When I decide to go for a walk, my body springs into action, at least some of the time. The body seems to affect the mind: When something cuts my skin, I feel pain. On the face of it, there's constant interaction between body and mind. But how does this work?

Notoriously, Descartes said the nonphysical mind interacts with the physical brain through the *pineal gland*. The pineal gland is a tiny structure located between the two hemispheres of the brain, so it has potential as a central conduit and as a unified locus for consciousness. Descartes's idea was that the brain receives sensory inputs and processes them, and then sends information through the pineal gland to the nonphysical mind. The mind thinks and reasons and makes decisions about what to do. Then it sends signals back to the brain through the pineal gland, and the brain carries out its actions.

This was a somewhat dubious theory even back in the 17th century. There wasn't much evidence that the pineal gland played any special role in our brain processes or our behavior. Today, most neuroscientists think it plays only a minor role in emotional processing. Furthermore, it was hard to see how the mind and the brain *could* interact through the pineal gland. How does the brain send its signals to the mind? How does the mind send them back? How can a nonphysical mind affect a physical brain?

The first and most acute version of the problem of interaction was posed by Princess Elisabeth of Bohemia. Descartes had been engaged as Elisabeth's tutor, and they had a long and fruitful written correspondence. Elisabeth had a sharp philosophical mind. In a different world where princesses were permitted to write philosophy, perhaps she would have written her own important philosophical books. She was deferential to Descartes, but she pressed him on the hardest questions.

So I ask you please to tell me how the soul of a human being (it being only a thinking substance) can determine the bodily spirits, in order to bring about voluntary actions. For it seems that all determination of movement happens through the impulsion of the thing moved, by the manner in which it is pushed by that which moves it, or else by the particular qualities and shape of the surface of the latter. Physical contact is required for the first two conditions, extension for the third. You entirely exclude the one [extension] from the notion you have of the soul, and the other [physical contact] appears to me incompatible with an immaterial thing.

Elisabeth is asking: How could a nonphysical mind move matter? For one object to push another requires them to be in physical contact, or at least requires one object to have a surface to do the pushing. But Descartes's nonphysical mind can't fulfill those requirements.

Descartes is evasive in reply, and Elisabeth follows up: "I admit that it would be easier for me to concede matter and extension to the soul than to concede that an immaterial thing could move and be moved by a body." Elisabeth denies that a nonphysical mind without a location in space could possibly affect a physical body. Why not take the mind to be physical, instead?

Descartes offered no good solution to the interaction problem. He didn't claim to have a final theory of the human mind, but he thought there was good reason to believe that the mind is nonphysical. Presumably a full theory of the nonphysical mind awaited progress in the science of the mind.

As it happens, the progress of science has not treated Descartes's dualism well, especially where his argument concerning human behavior is concerned.

First, computer science and neuroscience have made it seem less implausible that a physical system can cause all of human behavior. The development of computers has shown us how sophisticated information-processing in physical systems can be. The development of neuroscience has revealed how complex and impressive the brain is as an information processor. Putting these discoveries together, there's much less reason to deny that a brain could produce human behavior.

Second, physics has suggested a world in which physical processes form a closed network. It looks as if everything that happens in physics has a physical cause. Every time a particle moves, something physical makes it move. In light of this, it is hard to see how a nonphysical mind could affect behavior. Suppose my nonphysical mind makes my motor neurons fire in a way that makes my arm move. At some point, particles will have to move (in the neuron, say) in a way uncaused by another physical system. This event would be anomalous from the viewpoint of physics. It would constitute a violation of standard physical laws. Such violations would go against the standard view that physics alone governs the behavior of physical entities such as particles.

Some dualists, including the Hungarian physicist Eugene Wigner, have speculated that the mind could play a role in quantum mechanics. A standard formulation of quantum mechanics gives a central role to measurement: for example, particles can be in many places at once and have a definite position only when they are measured. If we understand measurement in terms of the mind, this suggests a possible role: the mind makes physical systems enter definite states. I take this quantum mechanical dualism seriously enough to have tried to make it work. In joint work with the New Zealand philosopher of physics Kelvin McQueen, I have recently tried to spell out a mathematically precise version of Wigner's view. The upshot of our work is that the view is worth taking seriously but it faces many difficulties. Among physicists and philosophers more broadly, quantum mechanical dualism is widely rejected.

As a result, Descartes's dualism is unpopular these days, and materialism is much more popular. There are still significant challenges to materialism, especially when it comes to the problem of consciousness, which I'll discuss in the next chapter. But the challenge from human behavior has largely dried up. Most philosophers and scientists see no strong reason, in principle, why behavior can't be explained in physical terms.

Mind-body interaction in virtual reality

Descartes may have been wrong about the physical world we live in. But he was right about many virtual worlds.

Consider a typical video game. Most video games—at least those set in a somewhat realistic three-dimensional virtual world—have a physics engine at their core. This engine simulates crucial aspects of physics, such as motion, gravity, and the way bodies interact when they collide. The video game *Angry Birds*, in which circular birds are flung at structures containing pigs, uses a simple two-dimensional physics engine to calculate the trajectory of objects under the influence of gravity and to simulate how structures behave when objects collide with them. The popular space-simulator *Kerbal Space Program* uses a much more detailed physics engine to simulate how three-dimensional objects behave in outer space. The virtual world of *Second Life* even allows users to directly control the physical properties of objects and to experiment with different laws of physics.

Suppose Elisabeth and René were born into one of these virtual worlds—a fully immersive virtual reality version of *Minecraft*, say. In the "outside world," they're strapped into immersive VR headsets. Everything Elisabeth and René ever see and hear comes from a headset and ultimately from the virtual world. They never see and hear the outside world. From their perspective, the virtual world is their world. They experience avatars as their bodies and move around the world. They build a life interacting with objects and other people in the virtual world.

In this situation, Elisabeth and René may study the physics of their world. Using a combination of experiments and theory, they could derive principles of physics that govern the behavior of ordinary objects in this world, including basic principles of mechanics and gravity.

Figure 35 Princess Elisabeth and René Descartes in *Minecraft*.

At this point, Elisabeth could entertain the question of whether a human being is just another physical object in their world, and whether the mind is the same as the body. René gives his argument from behavior, saying that no physical object could produce all the creative behavior that human beings produce. He argues for the dualist conclusion that the mind is distinct from any physical object. Elisabeth counters that she cannot see how a nonphysical mind could interact with physical objects. She expresses sympathy for the materialist view that the human mind is merely another physical object in space and time.

In this debate, René is essentially right! The world of *Minecraft* is a dualistic world. The principles of physics governing ordinary objects in that world don't govern human behavior. The human mind is fundamentally different from objects in the virtual world. It doesn't lie within the three-dimensional space of the virtual world at all. It lies in another realm, governed by different laws.

You might object that what generates human behavior here is a *brain*, which is still a physical object. So the human mind is physical, and there is no dualism. I think the correct verdict depends on which perspective we take: the perspective of the "outside" world or the "inner" virtual world.

First, the perspective from the outside world: Suppose we're creatures who grew up in the outside world, the one containing the simulation. Then for us, the physical world is the world around us, with relativity, quantum mechanics, and so on. In our world, René is a biological creature strapped into a headset in *Minecraft*. From our perspective, René's brain is physical, and the interaction between his brain and the virtual world is a physical process. There's no suggestion of dualism here.

Second, the perspective from the "inside" world: Suppose we're creatures who grew up in the virtual world of *Minecraft*. Then for us, the physical world is the world around us governed by the much simpler physics of the physics engine. For us, René's avatar is a physical object, but his brain is not. His brain isn't governed by our physics, and it inhabits the next world up.

What's going on here is that we have two worlds with different physics: a physics of the outer world (with quantum mechanics, relativity, and so on) and a physics of the inner world (the physics engine). Relative to the outer physics, René's situation is not dualistic. Relative to the inner physics, it is dualistic.

That said: What René calls "physics" is primarily the inner physics of *Minecraft*. What he calls physical objects are the objects of *Minecraft*. What he calls space is the three-dimensional space of *Minecraft*. The central issue for him is whether his mind is another physical object like these others, inhabiting the same space, governed by the same principles. It seems clear to him that his mind is outside his space and not governed by its laws of physics. For him, the mind is nonphysical.

Of course, René is sophisticated, and he may well acknowledge the possibility that his world is somehow contained within a different world with a space and a physics of its own. From his perspective, these might count as "metaspace" and "meta-physics." He could acknowledge the possibility that his mind is found within metaspace and that it's metaphysical. Still, this is quite different from being found within space and being physical. (A full analysis of the situation requires some analysis of language, and how words like "physical" and "space" might mean different things for people in different worlds. I'll return to that issue in chapter 20.)

Where and how does the interaction between mind and matter take place? Within metaspace, the answer is fairly straightforward: There's an outer brain and body interacting with a computer. But from the perspective of René's inner space, where does it take place? René's avatar doesn't contain anything like a brain, so it doesn't take place there. Rather, René's mind affects his avatar directly, moving his limbs and propelling his body around the world. From the perspective of the inner world, it's as if we have a sort of nonphysical will that controls our body directly.

Minds and bodies in a variety of virtual worlds

So far, I've talked as if virtual worlds involve only a physics engine and one or more human players who control avatars. In practice, most virtual worlds are more complex.

For a start, many virtual worlds contain nonplayer characters. The paradigmatic nonplayer character is a superficially human creature who is not controlled by humans in the outer world at all. These creatures behave in somewhat humanlike ways. They might speak, use weapons, and move around the world to achieve apparent goals. Other nonplayer characters include animals, monsters, aliens, and robots that display similar sophisticated, goal-directed behavior. When an aggressive monster detects you in the vicinity, it will move toward you.

Nonplayer characters are controlled not by a brain in the outer world but by algorithms inside a computer in the outer world. These algorithms are quite different from the algorithms of the physics engines governing the dynamics of virtual objects. From the perspective of the inner world, the actions of nonplayer characters aren't governed by the standard laws of inner-world physics but by laws of psychology, such as laws of goal-directed behavior.

Are these characters physical, from the point of view of the inner world? Their bodies certainly are, but what about their minds? Let's suppose that these nonplayer characters eventually have a psychology as complex as ours. No amount of neuroscience from within the inner world will directly reveal the cognitive mechanisms in the inner world; no cognitive system is there to be found. Of course, an inhabitant of the outer world could find the mechanisms by examining the algorithms in the computer, but this route isn't available to denizens of the inner world. The best they can do is infer some principles of the underlying psychology by watching these creatures behave. As with human players, the minds of nonplayer characters lie outside inner space. From the inner perspective, these minds are nonphysical.

Moreover, the minds of nonplayer characters are different from those of the human players. The minds of the former lie wholly within the computer running the virtual world. The minds of the latter derive from a biological organism separate from the computer running the virtual world. There are at least three sorts of entities in the virtual world, each governed by its own laws: physical objects, nonplayer characters, and human players. Instead of being a form of dualism, one could call this a sort of *trialism*.

In fact, the story is even more complicated. We've seen that in most virtual worlds, many *special* virtual objects have special causal powers of their own. For example, a gun may have special dynamics allowing a user to pick it up and fire it. A car may have special features allowing a user to control it. In our ordinary world, the behavior of a gun or a car derives from underlying physics involving mechanisms within a car's engine or a gun's firing mechanism. But in most video games, these complex mechanisms don't exist. Instead, cars move and guns fire via special algorithms governing those objects.

What philosophical view do these special causal powers suggest?

Dualism, trialism, quadrism? Perhaps a better analogy is *animism*, the idea that objects in the physical world have their own animating forces and perhaps their own agency. Animism has been widespread in the indigenous traditions of Africa, the Americas, Asia, Australia, and Europe. Any number of cultures have believed that at least living organisms, such as plants and animals, have these animating forces, and many have extended it to nonliving objects, such as rocks and clouds. In contemporary science, animism is widely rejected. But if we had grown up in a virtual world such as *Minecraft*, we might have reasonably conjectured that animism was true, and in a sense we'd have been right.

Not all virtual worlds require special treatment for physical objects and nonplayer characters. Some virtual worlds may be worlds of pure physics, with simulations of physical laws governing every object. Given current technology, pure-physics simulations aren't powerful enough to simulate simple biological objects such as cells, let alone intelligent entities such as humans and animals, but that will doubtless change.

Still, if we biological creatures interact with the virtual world of the simulation, there will always be an element of dualism. We aren't ourselves governed by the laws of the simulation, but by the laws of the outer universe. When we interact with the simulation, there will be a sense in which the world of the simulation is Cartesian, with minds in the outer world interacting with bodies in the inner world.

The Matrix provides a nice puzzle case. Neo's biological brain is in a pod outside the simulation, but it's constantly interacting with the simulated world. For the most part, the virtual world of the Matrix looks so much like ours that we can think of it as a pure-physics simulation. We can suppose that there are Matrix scientists who investigate the physics of their world and find something much like our physics, perhaps even down to the level of quantum mechanics. Likewise there are Matrix surgeons who open up bodies and find biological organs inside. Presumably there are even Matrix neurosurgeons who open up skulls to find brains inside, and Matrix neuroscientists who perform experiments connecting those brains to behavior.

Neo has a virtual body within the Matrix, which contains a virtual brain. He also has a biological brain outside the Matrix. How do these interact? Is one of them redundant? Perhaps the biological brain does all the work, and the virtual brain is for show—but then won't the virtual brain have to be disconnected from the virtual body to avoid affecting it, so that the virtual neuroscientists will notice something wrong? Or perhaps the virtual brain does all the work, and the biological brain is a mere passive observer—but then won't the biological Neo notice something wrong when he tries to do something and his virtual brain makes him do something else?

To avoid problems, it seems best that the biological and virtual brains be synchronized. Every input the virtual brain receives is copied to the biological brain, and they respond to inputs in exactly the same way. If we want the biological brain to be more than a passive observer, we can hook it up to action systems in the virtual body. Perhaps when a motor neuron fires in the biological brain, its output is transferred to a corresponding motor neuron in the virtual brain, thereby controlling action of the virtual body. But if the simulation is functioning well, that neuron will fire as it would have had it been under control of the virtual brain alone.

This picture raises any number of questions. First among them is: Is there one person here or two? On the face of it, Neo's biological brain supports a conscious subject in something like the normal way. His virtual brain, too, might support a conscious subject, if computer processes can support conscious subjects (something I'll discuss in the next chapter). Are these subjects one person or two? It's tempting to conclude that there are two persons here—like identical twins completely in sync with each other, but still different people.

In Daniel Dennett's story "Where Am I?," a biological brain and a backup silicon brain are kept in sync with each other. Both are used to control the same body, occasionally switching which brain is in charge. One day, the two brains diverge in their thinking. At that point, it becomes all too clear that two separate people are connected to the same body. Were the two of them there all along, or was there only a single mind that split in two when the brains came apart? It's hard to know for sure. One could always argue that the mechanisms keeping the two brains in sync make them, in effect, a single system supporting one mind and one person, not two. Similarly, it's hard to know what to say about Neo's case.

This puts a whole new spin on *The Matrix*. Were there two Neos all along? The biological Neo disconnects from the Matrix upon taking the red pill. What happens to the virtual Neo, complete with his virtual body and his virtual brain? Perhaps he's vaporized entirely? Is a new virtual Neo re-created every time Neo re-enters the Matrix, always in sync with the biological Neo? This could help to explain one of the great mysteries of *The Matrix*: When someone dies in the virtual world, why does that person die in the outer world, too? The answer may be that the two brains are always kept in sync. When one brain dies, so does the other.

If we want to interact with a physics-based virtual world without encountering problems like this, it may make sense for us to control virtual bodies lacking fully autonomous virtual brains. The virtual body might have a minimal virtual brain and nervous system just for sensory and motor processing. The nonvirtual brain's signals will control the behavior of the virtual brain and body, perhaps through a virtual pineal gland.

All of this will require complex dualistic interactions between outer-world minds and inner-world bodies. That said, the system need not be dualistic at the fundamental level. Both the inner-world physics and the outer-world psychology may derive from a single outer-world physics. If so, we'll have a kind of monism (one sort of stuff) rather than dualism (two sorts of stuff) at the fundamental level. Denizens of the outer world will see this as a sort of materialism, but denizens of the inner world won't, since outer-world matter is not the matter of the inner world. From the perspective of the inner world, the truth might be described as a sort of *neutral monism*: There's a single more basic sort of stuff underlying both minds and bodies in our world. The neutral stuff —the physics of the outer world—is not physical but meta-physical.

Upshot

The arguments in this chapter can help us think clearly about the simulation hypothesis. We saw in chapter 2 that the hypothesis can be split into the *pure simulation* hypothesis, in which our cognitive system is part of the simulation, and the *impure simulation* hypothesis, in which it's not. The traditional brain-in-a-vat scenario is an impure simulation, as is the situation in *The Matrix*. In this chapter we've mainly been focusing on impure simulations, in which the cognitive system and the physics of the virtual world are distinct.

If I accept the impure simulation hypothesis, I should accept the Cartesian dualist hypothesis that my cognitive system is nonphysical and interacting with physical systems. My mind is outside the physical space of my virtual world, and it interacts with my body, which is inside that space. My physical world derives entirely from bits in a computer, but my mind is tied to the brain in a vat, which need not derive from bits at all.

Earlier, I argued that the simulation hypothesis leads to the it-from-bit creation hypothesis. Now we can see that the *impure* simulation hypothesis leads to the *Cartesian* it-from-bit creation hypothesis: Physical systems derive from computational processes, put in place by a creator, and our cognitive systems are distinct from, and interact with, these physical systems. In effect, the impure simulation idea is akin to combining the it-from-bit creation idea about the physical world with Cartesian dualism about the mind. By contrast, the *pure* simulation hypothesis leads to the *non-Cartesian* it-from-bit creation hypothesis: Our cognitive systems derive from physical systems, which derive from computations, which are themselves created.

I'm not suggesting that the impure simulation hypothesis is especially plausible. If you take the simulation hypothesis seriously because of the statistical argument (see chapter 5) that simulations will be common, this reasoning tends to support the *pure* simulation hypothesis. It will be easier to create pure simulations (just set up the simulated physics of a world and watch it go), and much harder to create impure simulations (in which you'll need separate minds to interact with the simulation). If you need a biological brain for every impure simulation, this poses a hurdle that may limit the supply of impure simulations. As long as pure simulations can also support minds like ours (as we'll discuss in the next chapter), statistical reasoning suggests that it's more probable that we're in a pure simulation than an impure one.

Furthermore, insofar as we have reasonable evidence that physics forms a closed network in our world, this is evidence against the Cartesian hypothesis and against the impure simulation hypothesis, or at least against versions of these hypotheses in which the mind makes a difference in the physical world.

Still, simulation reasoning may give us reason to take Cartesian dualism more seriously than before. Cartesian dualism initially seems supernatural—inconsistent with a naturalistic view of the world. Simulation reasoning shows us how Cartesian dualism might be entirely naturalistic, deriving from natural processes in an outer world. Just as simulation reasoning gave us a naturalistic version of theism, it also gives us a naturalistic version of Cartesian dualism. It also helps us to overcome Princess Elisabeth's objection that a nonphysical system could not in principle interact with a physical system. The impure simulation hypothesis provides us with a model for how this interaction might work.

Is this simulation-based Cartesian dualism consistent with our scientific knowledge of the world? Where physics is concerned, the idea that physics forms a closed network is an attractive hypothesis, but it's certainly not demonstrated. Current physics suggests that there may be forces not yet discovered. It would be surprising to learn that outer-world processes occasionally affected inner-world physics inside computation-based human bodies, but it wouldn't be inconsistent with our evidence.

Where neuroscience is concerned, we know we have highly sophisticated brain processes that correlate closely with perception, thought, and action. There are versions of the impure simulation hypothesis that are consistent with all this. One somewhat extravagant version is the duplicate-brain hypothesis with a nonvirtual brain in the outer world duplicating and overriding the virtual brain in the inner world. Less extravagant is a version in which a nonvirtual brain is connected to a semi-autonomous virtual brain, affecting it and controlling its behavior at key junctures. It's true that we have little in the way of direct evidence for this hypothesis, but we also have little evidence against it.

The impure simulation hypothesis suggests that Cartesian dualism is at least consistent with our scientific knowledge of the world. I'm not arguing that the impure simulation hypothesis is true, so I haven't provided an argument that Cartesian dualism is true. Still, simulation reasoning shows us a way in which Cartesian dualism *could* be true—which is interesting in itself.

Chapter 15

Can there be consciousness in a digital world?

I N THE *STAR TREK: THE NEXT GENERATION* EPISODE "THE MEA-sure of a Man," there's a trial to determine whether the android Data is a sentient being. Starfleet cyberneticist Bruce Maddox wants to dismantle Data in order to learn from his technology. Data refuses his request. Maddox claims that Data is Starfleet property, a mere machine with no right to refuse. Captain Picard argues that Data is a sentient being with rights, including the right to choose his own destiny.

The issue before the court is whether Data is sentient. Picard asks Maddox to define "sentient," and Maddox replies, "Intelligent, self-aware, and conscious." Of these criteria, Maddox agrees quickly that Data is intelligent. ("It has the ability to learn and understand, and to cope with new situations.") Picard then makes a strong case that Data is self-aware by asking Data what he's doing now. Data replies, "I am taking part in a legal hearing to determine my rights and status," where what is at stake is "my right to choose—perhaps, my very life."

The issue comes down to the third criterion: Is Data conscious? Surprisingly, Picard offers no direct argument that Data is conscious. Instead, he says, "You see, he's met two of your three criteria for sentience. So, what if he meets the third? Consciousness in even the smallest degree. What is he then?"

Even without an argument for consciousness, Picard's question is enough to carry the day. The judge says that the basic issue is whether Data has a soul. She doesn't know the answer, but rules that he must be granted the freedom to explore the issue.

CONSCIOUS DATA   ZOMBIE DATA

Figure 36 Is Data a conscious being, or a philosophical zombie?

The key third question is left unanswered. Is Data conscious? Or is Data what philosophers call a *zombie* (or sometimes a *philosophical zombie*, to distinguish them from the zombies found in Hollywood movies)? For philosophers, a zombie is a system that outwardly behaves much like a conscious being, but which inwardly has no conscious experience at all. Picard presumably holds that Data is conscious: he has an inner stream of conscious perception, feeling, and thinking. Maddox presumably holds that Data is a zombie: he behaves intelligently but lacks a conscious inner life altogether. (It's hard to depict a zombie, since outwardly they are just like ordinary conscious beings, but figure 36 makes an attempt.) Which of them is right?

More generally, can a digital system be conscious? Or are only humans and animals conscious?

The question matters a great deal in thinking about digital worlds. Consider the virtual world of Daniel Galouye's novel *Simulacron-3*, which (as we saw in chapter 2) pioneered the simulation genre. This virtual world is a pure simulation, containing many simulated humans with simulated brains. Are these simulated humans conscious? If they are, then to shut down the system irreversibly would be an atrocity, a sort of genocide. If they're not conscious, then they are digital zombies, and shutting down the system seems no worse than turning off an ordinary video game.

So far, the virtual worlds we've created don't contain digital creatures with anything like the complexity of a human being. In most virtual worlds, the biological human players are by far the most sophisticated creatures, and the digital nonplayer characters seem mindless—few people would consider them conscious. But eventually, there will be simulated worlds with much more sophisticated nonplayer characters who have virtual brains as complex as ours. Once there are worlds like this, the question of digital consciousness cannot be avoided.

The question matters even more when we think about *mind uploading*: the attempt to transfer our minds from a biological brain to a digital computer. Many people see this as our best hope for achieving a sort of immortality. In the *Black Mirror* episode "San Junipero," people approaching the end of their biological life can choose to upload their brains into a digital replica, which is then plugged into a virtual world. The virtual world functions as a sort of heaven where people can live forever.

Mind uploading raises many scientific issues. Some concern the *behavior* of an uploaded system. Can such a system generate the same sort of intelligent behavior or display the same memories and the same personality traits as the original biological system? Can we really measure a brain well enough to simulate it? Can we simulate networks of neurons perfectly in a digital system?

If these scientific issues can be solved, deeper philosophical issues await. One of the deepest concerns consciousness. For uploading to work as a path to immortality, it's crucial that the uploaded system be conscious. If the uploaded system is an unconscious zombie system, then uploading won't be a form of survival; it will amount to destruction, at least where the conscious mind is concerned. Most people would view this sort of zombification as little better than death.

Another deep issue about uploading concerns *identity*. If I upload to a computer, will the uploaded system be *me*? Or will it be a wholly new person who behaves just like me, akin to a newly created twin? If I create an uploaded version of myself while leaving the biological original intact, most of us would consider the biological version to be me and the digital copy to be someone new. Why should it be any different in the case in which the biological version is dead and only the digital copy persists?

In effect, uploading raises versions of Picard's three questions. First is a question about intelligent behavior: Will the upload behave like me? Second is a question about consciousness: Will the upload be conscious? Third is a question about the self: Will the upload be me? We'd have to answer yes to all of these questions for uploading to be a feasible path to survival.

I'll say something about all of these questions in what follows, but I'll focus mainly on the question concerning consciousness. This question is especially relevant to assessing the simulation hypothesis: If simulations cannot be conscious, our consciousness rules out the pure simulation hypothesis from the start.

The problem of consciousness

What is consciousness? Consciousness is subjective experience. My consciousness is a sort of multitrack inner movie capturing how my life seems from the first-person point of view.

Consciousness has many components. I have visual experience of colors and shapes. I have auditory experience of music and voices. I have bodily experience of pain and hunger. I have emotional experience of happiness and anger. In my waking hours, I experience a stream of conscious thought—thinking, reasoning, talking to myself. I decide; I act. All this is somehow unified into an encompassing state of consciousness, making up the conscious experience of being me.

Why is there consciousness in the universe? How do physical processes give rise to consciousness? How can there be subjective experience in an objective world? Right now, no one knows the answers to these questions.

I became a philosopher in order to think about the problem of consciousness. My academic background was in mathematics and physics: I did a first degree in mathematics in Australia in the 1980s and got partway through a doctorate at Oxford. I loved those fields because they seemed to address truly fundamental questions. But it gradually came to seem to me that most of the really hard questions had been answered, and the fundamentals were pretty well understood (or at least so I supposed at the time). Perhaps that was wrong, but that was how things seemed to me then.

At the same time, there seemed to be a truly fundamental problem as yet wide open: the problem of consciousness. The human mind seemed to pose many of the harder questions left for science, and of these, consciousness seemed the hardest of all. Consciousness was the most familiar thing in the world—but also the thing we understood the least. How did it fit into the physical world? How could there be subjective experience in an objective world? No one knew.

I became obsessed by these questions, to the point where I decided to leave mathematics and work on the problem of consciousness directly. I moved in 1989 from Oxford to Indiana University to work in the cognitive science group headed by Douglas Hofstadter—the author of *Gödel, Escher, Bach* and other books I loved. I learned a lot of cognitive science and did a lot of work in artificial intelligence, but consciousness remained my driving passion. It seemed to me that the best way to tackle the biggest problems about consciousness directly was through philosophy. So I became a philosopher and ended up writing a PhD thesis on consciousness, which turned into my first book, *The Conscious Mind.*

Around that time, in April 1994, I gave a talk at one of the first international conferences on consciousness, in Tucson, Arizona. In the talk, I called the problem of explaining consciousness the *hard problem*. This name caught on more quickly than anything else I've ever said. People have written books on "the hard problem." The playwright Tom Stoppard wrote a play about consciousness called *The Hard Problem*. None of this was because the idea was radical or original—in fact, quite the opposite. The name caught on so fast because everyone knew what the hard problem was all along. The label just captured the problem and made it more difficult to avoid.

To explain the hard problem, and the contrast with other easier problems, it's useful to start by examining the relationship between two of Picard's questions: consciousness and intelligence.

What is intelligence? To a first approximation, intelligence is sophisticated and flexible goal-directed behavior. If a system is good at only one sort of goal—winning a game of chess, say—then at best it's a *narrow intelligence*. If a system can make reasonable attempts at achieving a wide range of goals, we call it a *general intelligence*.

So far, many existing digital systems exhibit a narrow intelligence. The AlphaZero program developed by DeepMind is good at winning games of chess and Go. Self-driving cars are good at navigation. No existing digital system is yet close to general intelligence, though. The only generally intelligent beings we know of are humans and perhaps some other animals.

As I understand intelligence, it's an objective feature of a system that mostly comes down to behavior. Intelligence isn't a matter of how a system *feels*. What matters are the objective processes in the system and the behavior they produce.

As a result, we have a much better grip on intelligence than on consciousness. There are standard methods for explaining the behaviors of a cognitive system. To explain a behavior, you need to identify a mechanism and show how it produces the behavior. The mechanism might be a system in the brain, or some sort of algorithm that we think the brain is using. That's why I call the problems of explaining intelligence, and explaining behavior in general, the *easy problems*. They include: How do we navigate? How do we communicate? How do we discriminate objects in our environment? How do we control our behavior to achieve our goals? The easy problems aren't really easy; some may take a century or more to solve. But at least we have an idea of how to go about solving them.

While intelligence is a matter of objective behavior, consciousness is a matter of subjective experience. The hard problem is the problem of explaining subjective experience. All conscious experience seems to be tied to a conscious subject who is having the experience. It's partly this subjectivity that makes the problem of consciousness so hard.

My New York University colleague Thomas Nagel famously defined consciousness as *what it is like to be* a system. There's *something it is like* to be me, or to be you. If so, you and I are conscious. Most people think there's nothing it's like to be a rock; a rock has no subjective experience. If they're right, a rock is not conscious. If there's something it's like to be a bat, as Nagel suggested, a bat is conscious. If there's nothing it's like to be a worm, a worm is not conscious.

Many people think of consciousness as something very complicated, at the top of the hierarchy of intelligence. Some think it requires a complex form of *self*-consciousness. For example, on the TV series *Westworld*, consciousness is depicted as an awareness of one's inner voice as belonging to oneself. According to that view, only humans or other reflective beings are conscious. I think this is the wrong way to think about consciousness. Consciousness is manifested even in a simple state, like seeing red or feeling pain. To invoke Nagel again, there's something it is like to see red, and something it is like to feel pain, so these are conscious states. These states don't require an inner voice or reflective awareness of oneself. Certainly, an inner voice is an aspect of consciousness in those who have one, and so is reflective self-awareness. But these shouldn't be confused with consciousness in general.

Even simple conscious states such as seeing red or feeling pain raise the hard problem of consciousness. When my visual system processes a stimulus in a way that leads me to identify it as red, why do I have a conscious experience of redness? Why is there *something it is like* to see red? The objective methods that work so well on the easy problems don't work so well on subjective experience. Identifying a brain mechanism

Figure 37 Does Mary the color scientist (here modeled on
Mary Whiton Calkins, a leading philosopher and psychologist
of the early 20th century) know what it's like to see red?

that leads us to classify the stimulus as red doesn't tell us why we have
a conscious experience of redness. More generally, explaining behavior
doesn't explain why the behavior is accompanied by consciousness. In
any description of brain processes, there seems to be a gap between that
story and consciousness. Why should brain processes give rise to conscious experience? Why don't they just go on "in the dark" without any
subjective experience at all? No one knows.

The standard methods of neuroscience and cognitive science are
directed at explaining behaviors, so they don't give us much of a grip
where the hard problem of consciousness is concerned. At best, they
give us *correlations* between brain processes and consciousness. Neuroscientists are gradually making progress toward what they call the
"neural correlates of consciousness." But correlation is not explanation.
So far, we have no explanation of why and how these processes give rise
to consciousness itself.

We can bring out the distinctive problem of consciousness with a
thought experiment constructed by the Australian philosopher Frank
Jackson. Mary is a neuroscientist who knows everything there is to
know about physical processes in the brain and how they respond to

colors. However, Mary has spent her whole life in a black-and-white room, studying the world through books, black-and-white screens, and other equipment. She has never experienced color herself. Mary knows the full objective story about how red or blue or green things produce certain wavelengths of light, how these affect the eye and the brain, how they produce associations in people, and how they lead to reports like "That barn is red." However, there's one central thing about colors that Mary doesn't know: She doesn't know what it's like to *experience* red, blue, and other colors.

Physical knowledge of the brain tells Mary all sorts of things about color, but it doesn't tell her about the conscious experience of color. So knowledge of conscious experiences seems to go beyond knowledge of brain processes. This doesn't yet tell us what consciousness is, but it brings out why there's a problem.

I once met someone who was just like Mary. Knut Nordby was a Norwegian neuroscientist with achromatopsia, or total color-blindness. The cone cells in his retina, which process color, didn't work. Despite this, Nordby specialized in psychophysics (the study of sensory processes) and had published many articles on colors. He knew all about the brain systems involved in color processing. When I met him in 1998, he was having his brain scanned and stimulated by the Stanford cognitive neuroscientist Brian Wandell to see if he might be able to experience colors that way. Alas, the experiment was unsuccessful. "The world of colors will forever remain a mystery to me," Nordby told me.

In *The Conscious Mind*, I argued that no explanation of consciousness in purely physical terms is possible. The basic idea was that physical explanations are great at explaining behavior, but that ultimately they *only* explain behavior. More precisely, physical explanations are always a matter of objective structure and dynamics, and all they explain is further objective structure and dynamics. This is perfect for solving the easy problems, but it cannot solve the hard problem. To solve the hard problem, you need something more.

I further argued that if you can't explain consciousness in terms of existing fundamental properties (space, time, mass, and so on) and existing fundamental physical laws, then you need new fundamental properties in nature. Perhaps consciousness is itself fundamental. You also need to acknowledge further fundamental laws—perhaps laws connecting physical processes to consciousness. The search for a science of consciousness will, in effect, be the search for these fundamental laws.

Since then, there has been an explosion of proposals to solve the hard problem. Some involve new fundamental laws connecting physical processes and consciousness. Some theories connect consciousness to information processing; others connect it to quantum mechanics. An especially popular idea in recent years has been *panpsychism*, the idea that there's some element of consciousness in every physical system throughout nature. Other views are more reductionist, trying to deflate the hard problem so it can be solved in physical terms. Perhaps the most extreme version of this strategy is *illusionism*, the idea that consciousness itself is an illusion, and that for some reason our evolutionary history makes us believe we have the special properties of consciousness when in fact we do not. If that's right, there's no consciousness and no hard problem of explaining it.

There's much more to say about the hard problem of explaining consciousness, but let's set it aside for now and move to a somewhat narrower question: Can machines be conscious?

The problem of other minds

It's hard to know for sure whether silicon machines can be conscious. One reason is that it's hard to know for sure whether *any* entity other than oneself is conscious. Through my own subjective experience, I'm confident that I am conscious. Descartes's *cogito* for consciousness says that I am conscious, therefore I am. But this tells me only about one case of consciousness. It doesn't tell me about anyone else.

This is what philosophers call *the problem of other minds*. How can we know that anyone else has a mind? And how can we know what their minds are like? This is a skeptical challenge on a par with the skeptical challenge about the external world. As with the external world, almost all of us believe that other people do have minds and we think

Figure 38 The problem of other minds:
Zhuangzi, Huizi, and the happy fish.

we sometimes know what they are thinking and feeling. But how can we know this for sure?

A simple version of the problem of other minds arises for non-human animals. In a famous parable, Zhuangzi observes some jumping fish and says they are happy. His companion Huizi says "You're not a fish. How do you know what makes fish happy?" Zhuangzi replies "You're not me. How do you know what I know?" Huizi replies that we can't know in either case, while Zhuangzi is more optimistic. The parable has been used to make many points, but at a basic level it wonderfully illustrates the problem of other minds: How can we know what is going on in the minds of other animals and of other people? Or as Thomas Nagel puts it, how can we know what it's like to be a bat or to be another person? (Figure 38 interprets the parable along these lines.)

The core of the problem of other minds is the problem of other consciousnesses. Perhaps I can know that other people perceive, remember, and act—if I understand these capacities as independent of consciousness. But consciousness seems to be private and subjective, which makes it very hard to observe in others. Your behavior may *suggest* to me that you're conscious, and you might even *tell* me you're conscious—but how strong is this evidence? Couldn't an unconscious robot do the same?

We can illustrate the problem of other minds by asking: How do we know that other people aren't zombies? As we've seen, a philosophical zombie isn't the Hollywood version that's risen from the dead. Instead, it's a being that looks and behaves just like an ordinary human being but is not conscious at all. For a philosophical zombie, everything is dark inside. The extreme case of a zombie is a complete physical duplicate of a conscious human being, with the same brain structures but no subjective experience.

Few people think that zombies actually exist. Almost all of us believe that other people are conscious. But the very idea of zombies is enough to raise the problem of other minds. I can at least imagine that someone else is a zombie—behaving normally with a normal brain, but not in any way conscious. As far as I can tell, there's no contradiction in the idea that there could be a physical structure that's atom-for-atom identical to Donald Trump and is not conscious. Again, most of us find the zombie hypothesis fanciful and implausible. But as with skeptical challenges in general, the challenge here is, "How can we know for sure?"

Philosophers invoke zombies for many purposes. In *The Conscious Mind*, I used them to argue against materialism. The rough idea was that if zombies are conceivable, then there could conceivably have been a world physically just like ours, but with no consciousness at all. In our world, however, there is consciousness, which means that our world (unlike the zombie world) contains something over and above its physical structure. Other philosophers have used zombies to raise questions about the causal role of consciousness and its evolutionary function. If zombies can in principle do exactly what we can do, then why did evolution bother with bringing about consciousness at all?

These uses of zombies are controversial. Some philosophers think we cannot really imagine zombies: When we try to do so carefully, we'll always come up against hidden contradictions. Others allow that zombies are conceivable but argue that not much about our reality follows from what we can conceive. Still others think we may be zombies ourselves because consciousness is an illusion.

In this discussion, I'm not using zombies for any of these controversial purposes. I'm just using zombies to raise a challenge: How do we know that other people aren't zombies? Perhaps you've got an answer. If so, all the better!

The problem of other minds persists as we move away from humans. How can we know for sure that dogs are conscious? For that matter, how do we know that infants are conscious, or when they become so? Most of us think of newborns as conscious, but how do we know for sure? Descartes thought that dogs were mere automata, or zombies. It was once believed that infants were unconscious at birth. For many years, no anesthetic was used in circumcising them because people thought infants could not have a conscious experience of pain—in other words, they thought that where consciousness is concerned, infants are like zombies. Today this view is regarded as implausible, but, again, it's not easy to make a conclusive case that it's wrong.

In practice, one thing we can do is find certain neural or behavioral markers of consciousness. These are physical states that seem to correlate with consciousness in my own case or in typical human cases, and we can then extend them to assess other cases. Perhaps the best behavioral marker of consciousness is a verbal report, such as someone saying they feel pain. In my own case, these reports certainly correlate with consciousness. It's natural to infer that they're markers of consciousness in others, too, at least once we assume that others are conscious at all. We can't use verbal reports in assessing the consciousness of nonverbal animals and infants. Still, we can use other behavioral markers—including pain behavior, for example—that correlate with consciousness in us and which can also be found in animals and infants. We can also use brain-based correlates of consciousness in a similar role. None of this provides indisputable proof that other systems are conscious, but at least it makes a rea-sonable case.

A complete solution to the problem of other minds may require a full theory of consciousness that tells us which systems are conscious and which are not, and what their consciousness is like. We don't have such a theory yet. So for now, in reasoning about other minds, we have to rely on a few empirical markers deriving from the science of consciousness, a few pre-theoretical principles connecting consciousness and behavior, and philosophical reasoning about where consciousness is to be found.

Can machines be conscious?

The problem of machine consciousness is an especially hard version of the problem of other minds. How can we know whether a machine, such as *Star Trek*'s Data, is conscious? Commander Maddox will insist that Data is a silicon zombie with no consciousness at all. Data's underlying makeup is very different from ours, and he doesn't have a biological brain. So we cannot use brain processes as evidence for consciousness. Certainly Data behaves in a way that we associate with consciousness. That carries considerable psychological force, but it isn't obvious how much evidential weight this should be given in a system with such a different makeup.

I'll concentrate on just one type of machine: a perfect simulation of a brain, such as my own. The brain simulation is a digital system running on a computer. If we can establish that *one* digital system is conscious, then we know there's no general reason why digital systems cannot be conscious, and the floodgates will open. Compared to other machines, the simulated brain has the advantage of maximizing similarity to a human brain, which makes some of the reasoning simpler. For example, we don't need a full theory that tells us which systems are conscious, since we can start from the one case that we know is conscious: ourselves. Furthermore, brain simulation helps to illuminate simulated worlds and uploading, which are the main reasons we're pondering digital consciousness here.

How would simulating a brain work? We can suppose that every neuron is simulated perfectly, as is every glial cell and other cells throughout the brain. The interactions between neurons are simulated perfectly, too. All the electrochemical activity is simulated, and so is other activity, such as blood flow. If there's a physical process in the brain that makes a difference in how the brain functions, it will be simulated. In what follows, I'll adopt the simplifying assumption that only neurons need to be simulated, but everything I say will apply without that assumption.

Someone might argue that simulating a brain is impossible. Here I'm assuming that the brain is a physical system and therefore obeys laws that can be computationally simulated. Certainly, current evidence favors both of those assumptions. If they're true, then a computer simulation of a brain should be possible. These assumptions do not guarantee that a neuron-level simulation will work—perhaps to simulate brain processes really well, one will have to go down to the level of physics. If necessary, a physics-level simulation will suffice for our purposes.

The case of a simulated brain has one big advantage for our purposes. It raises the possibility that we might *become* the machine. Then we can have our own first-person evidence about machine consciousness. How might you become a simulated brain? Simply creating a simulation may leave matters unclear. If your original brain is still intact, it will presumably claim to be the original person. You could try destroying the original brain, but this brave step would leave open all the issues about identity. Have you become the simulated brain, or has a whole new person been created?

The safest way to become a simulated brain is to become one in stages. This is the process sometimes called gradual uploading. To do this, we can simulate your brain one cell at a time (or one area at a time). We'll build a simulation of each cell and arrange for it to interact, via receptors and effectors, with neighboring biological cells. At first, just a few of the original cells will be replaced by simulated cells. After a while, many will be replaced, and neighboring cells can interact in a fully simulated way. Eventually a quarter of your brain will be simulated, then half, then three-quarters, until the result is a fully simulated brain. Perhaps the simulation will be connected by effectors to the original physical body, or perhaps the body will be simulated, too.

We can imagine that the uploading process happens over a period of weeks. After the first few cells are replaced, you take a break. Perhaps you're a bit dazed from the operation, but otherwise you feel normal. The simulated cells behave exactly like the old cells and produce the same behavior, so you'll behave normally. Given this, you'll certainly say that you're conscious.

Something like this will happen at every stage, presuming that the simulations are faithful enough. Someone will ask, "How do you feel?" and you'll say, "I feel fine," or perhaps "I feel hungry" or "I feel pain" or "I feel bored." Again, assuming that our neurons control our behavior and that they've been perfectly simulated, we'll expect much the same sort of behavior from the simulation.

Finally, at the last stage, your brain will have been entirely replaced by a simulation. Again, you'll be asked, "How do you feel?" and you'll give a reasonably normal answer. If the simulation is working well, you'll say you're conscious. If asked whether you're sure of that, you'll say "Yes!" (at least if you're the kind of person who was certain you were conscious before the uploading). You'll presumably say that, to you, this is completely convincing evidence that machines can be conscious.

In her 2019 book *Artificial You*, the American philosopher Susan Schneider expresses skepticism about whether machines will be conscious and whether consciousness is preserved in the uploading process. To Schneider, it is likely that if she were uploaded, she would turn into a philosophical zombie. Sure, she might say, "I'm still here" and "I am conscious," but this is just the sort of thing we would expect a zombie to say.

However, we can raise some uncomfortable questions for skeptics such as Schneider. If gradual uploading preserves your behavior but eliminates your consciousness, we can ask: *What happens to consciousness along the way?* Presumably, after only a few biological neurons were replaced by simulations, you are still fully conscious. What

Figure 39 Susan Schneider undergoes gradual uploading.

about after a quarter of your brain was replaced, or half, as depicted in figure 39? Does your consciousness gradually fade away? Or does it disappear all of a sudden?

In *The Conscious Mind*, I called this the "fading qualia" argument, because it focuses on the idea that the qualities (or "qualia") of your conscious experience might gradually fade away. It appears that there are two positions that skeptics about machine consciousness might take.

First, your consciousness suddenly disappears—that is, you go from full consciousness to no consciousness when a single neuron was replaced. This would be an extraordinary discontinuity, unlike any other that we find in nature. Perhaps replacing one neuron could shut down consciousness if this shut down much of the brain; but if the replacement is a good enough simulation, the rest of the brain will be unaffected, so a tiny change must do the job on its own. Furthermore, we can extend the case by replacing submicroscopic parts of the crucial neuron, one at a time, until we eventually find the crucial quark whose replacement causes consciousness to be destroyed completely. The crucial-quark hypothesis seems even more implausible than the crucial-neuron hypothesis. It's more plausible that consciousness gradually phases out than that it suddenly disappears.

This leads us to the second option: Your consciousness gradually wanes, so there are points at which it has diminished but not disappeared. Perhaps some aspects of your original consciousness remain, and some have disappeared; or perhaps all have faded somewhat. But because your neurons have been replaced by perfect simulations, your behavior will be completely normal. Throughout the process, you'll swear that you're fully conscious, with normal rather than faded conscious experiences. So in this case a skeptic like Schneider will have to allow that you're deluded about your consciousness. At the intermediate point when your consciousness is fading, you're still a conscious being (not a zombie), and you don't display any obvious irrationality. Nevertheless, you're completely out of touch with your own consciousness: you think it's normal when in fact it is fading away. This also seems quite bizarre.

There's a third hypothesis, which is far more plausible: Your consciousness stays intact at every stage and is present at the end of the process. This hypothesis avoids the implausibilities of fading or suddenly disappearing consciousness and it is not subject to objections like those the other hypotheses encounter. This hypothesis has the consequence, however, that simulated brains can be conscious. At least in the special case in which you become a simulated brain by gradual uploading, the simulation will be fully conscious.

Once we've gotten this far, it's natural to conclude that simulated brains in general can be conscious, at least when the brain they are simulating is conscious. For any conscious system, one could run a gradual-uploading scenario leading to a simulated version. Then, reasoning along the lines above will suggest that the simulations are conscious. I suppose someone could suggest that *only* gradually uploaded simulations are conscious, maybe because these bring your soul along from the original brain. But this makes consciousness nonuniform, and raises the worry that almost any system might be a zombie, depending on whether it is attached to a soul or not. It seems more plausible that if consciousness is present in one simulated brain, it will be present in all simulations of conscious brains.

Consequences

If I'm right, then simulated brains can be conscious. More precisely, if a system with a biological brain is conscious, a system with a perfect simulation of that brain will be conscious, too, with the same sort of conscious experiences.

This conclusion has consequences for the simulation hypothesis. In a pure simulation, there will typically be many simulated brains. If the original world has many conscious beings, our reasoning suggests that a simulation of it will have the same number of conscious beings, with the same sort of consciousness. It follows that our consciousness does not rule out the pure simulation hypothesis. It's equally compatible with simulated and unsimulated reality. Furthermore, we now have good reason to think that consciousness is substrate-neutral, so one major obstacle to the simulation hypothesis (chapter 2) and the simulation argument (chapter 5) has been overturned.

All this strengthens the case that virtual reality is genuine reality, by making the case that simulated minds are genuine minds. These arguments also support the prospects for artificial consciousness more generally, whether or not the artificial system involves a simulated brain. Once we know that one computer system can be conscious, we can expect there will be many more.

Finally, all this makes the prospects look better for mind uploading. Recall that the three potential obstacles to uploading were intelligence, consciousness, and identity. Would an uploaded system behave like us? It now seems likely that it would, as long as our brains and behavior are governed by laws that can be simulated. Would an uploaded system be conscious? I have made a case that it will be. Certainly if one undergoes gradual uploading, there's a strong case for consciousness being retained at the end of the process.

What about the final obstacle: identity? If I upload my brain to a simulated brain, will it be me? I think it depends on the case. In the case of *nondestructive* uploading, in which the original brain stays alive after being uploaded, most people's intuitions are that I remain with the original brain and the simulation is a new and distinct person. The case of *destructive* uploading, in which the brain is destroyed, is less clear, but many people accept that the original person dies and a new person is created.

The best case for survival is, once again, gradual uploading. Say one percent of my brain is replaced every day. It seems plausible that at the end of the first day, I'm the same person. At the end of the second day, I'm the same person I was at the end of the first day, so I'm the same person I was at the beginning of the process. And so on. What happens here seems no different in principle from what happens to an ordinary biological brain, many of whose neurons may be replaced over a long period of time. Making a new brain all at once may create a new person, but gradual replacement leaves the old person intact.

As usual in philosophy, there are no guarantees. But if I ever get a chance to upload my brain to a simulation, I'd prefer to do it by gradual uploading. That seems the best bet for surviving the process and coming out conscious on the other side.

Chapter 16

Does augmented reality extend the mind?

I N CHARLES STROSS'S 2005 SCIENCE-FICTION NOVEL *ACCELERANDO*, the main character, Manfred Macx, wears glasses that have taken over many of the functions of his mind. The glasses store his memories. They recognize objects and people for him. They gather information and make decisions for him. Stross writes: "In a very real sense, the glasses are Manfred, regardless of the identity of the soft machine with its eyeballs behind the lenses."

When Manfred's glasses are stolen, he is almost helpless. He borrows another pair and connects to his "metacortex" in the cloud to restore some of his functioning. His memories and his personality gradually return to him.

We don't yet have glasses like Manfred's, but technology has been taking over the functions of our brain for decades. Smartphones remember phone numbers and appointments for us. Mapping software does our spatial navigation for us. The internet is a repository of much of our knowledge. We often see via cameras and communicate via digital messages. We've made the technology portable so that we have it with us most of the time, and we use it as automatically as we use our bodies.

Glasses like Manfred's are on the way. As we saw in chapter 12, augmented reality glasses expand our ordinary perception of physical reality with computer-based images projected into our visual field.

One of the interesting aspects of augmented reality is that it simultaneously augments the *world* and augments the *mind*. We've already seen how augmented reality extends the world, adding virtual screens,

Figure 40 Do Andy Clark's augmented reality glasses
extend his mind?

virtual art, and virtual buildings to our environment. I've argued that
these may genuinely be considered part of external reality.

Perhaps even more fascinating is augmented reality's potential to
take over functions of the mind. Automated recognition systems can
identify people for us, displaying names alongside them when they walk
into a room. Navigation systems will navigate for us, with visual cues
showing where to go delivered directly from augmented reality. Design
systems will design spaces for us, showing how a new building might
alter the look of a neighborhood. Calendar systems will keep track of
events for you, telling you where you need to be. Communication systems will connect you with others, putting you and a faraway friend in
the same space as if you're having an in-person conversation.

In effect, augmented reality technology becomes part of what Stross calls our "exocortex"—the brain outside our brain. Philosophers call this the *extended mind*.

The extended mind

In 1995, my colleague Andy Clark and I wrote a short article entitled "The Extended Mind," arguing that the technologies we use can become parts of our minds. We were working together in a new program for philosophy, neuroscience, and psychology at Washington University in St. Louis. Andy was fascinated by the way that tools in the environment—a notebook, a computer, even another person—can play much the same role as parts of the brain. He thought that because of these tools, we should reject the idea that the mind lies wholly within the skull or the skin. This idea resonated with me, and I suggested an argument to support the claim that objects in the world can become parts of our mind. Both of us had been influenced by British evolutionary biologist Richard Dawkins's 1982 book *The Extended Phenotype*, which argues that evolved biological organisms can extend into the environment. We argued that the same is true of the mind.

We submitted our article to three leading journals of philosophy, all of which promptly rejected it. At the time, people treated the argument as an interesting curiosity, too radical to take seriously. Computers and notebooks are tools for the mind—but parts of the mind? Surely not. The article was eventually published in 1998, however, and people gradually began to pay attention to the idea of mind extension. By now, there has been a small explosion of work, with hundreds of articles and a number of books written on the extended mind.

Back then, our central example of mind extension was not a computer but a humble notebook. We talked about Otto, a New Yorker with Alzheimer's disease, who uses a notebook to write down key facts and recalls them later. We contrasted him with Inga, who remembers facts with her brain in the normal way. One day, Otto wants to go to the Museum of Modern Art, so he looks up the address in the note-

Figure 41 Inga and Otto: Is Otto's external memory part of his mind?

book and sets out for 53rd Street. Inga recalls the address using her brain and sets out too. We argued that Otto's notebook isn't just a memory aid. It's part of his memory, in the way that Inga's biological memory is. The notebook acts as a repository of Otto's beliefs about the world. Otto believes the museum is on 53rd Street because it is written in his notebook—just as Inga believes this because it is stored in her brain.

My New York University colleague Ned Block likes to say that the extended-mind hypothesis was false when we wrote the article in the 1990s but that it has since become true. The main reason is the advent of the smartphone and the widespread availability of the internet. In the age of the smartphone, a hypothesis that once seemed ridiculous now seems obvious. Of course my phone is part of my mind! I couldn't function without it. The same goes for the internet. The webcomic *xkcd* published a strip titled "Extended Mind," saying, "When Wikipedia has a server outage, my apparent IQ drops by about 30 points."

Of course, parts of our environment have been performing functions of the brain for a long time now. The first time someone counted on their fingers, the process of counting was being partly offloaded from brain to body. The first time someone used an abacus, the work of calculation was being offloaded from the brain to a tool. The first time someone wrote something down for later use, the work of memory was being offloaded from brain to written symbols. The fingers, the abacus, and the written symbols all became part of a mental process (counting, calculating, remembering) that spans both brain and body.

Notebooks and counting on our fingers provide nice examples of the extended mind, but computers supercharge the idea. The role of computers in extending the mind was not lost on pioneers of the computer age. As early as 1956, the cyberneticist W. Ross Ashby talked about the role of computers in "amplifying intelligence." In 1960, the computer visionary J. C. R. Licklider (who was one of those responsible for the computer networks preceding the internet) published a manifesto called "Man-Computer Symbiosis," in which he wrote:

> The hope is that, in not too many years, human brains and computing machines will be coupled together very tightly, and that the resulting partnership will think as no human brain has ever thought and process data in a way not approached by the information-handling machines we know today.

The era of personal computing, starting in the late 1970s, moved things closer to Licklider's vision of tight coupling between brains and computers by putting a computer on many desktops. Still, the coupling between humans and desktop computers is loose: It's lost whenever the human leaves the desktop. Only with the arrival of mobile computing, which skyrocketed with the smartphones of the mid-2000s, did Licklider's vision of tight coupling really become part of everyday life. Our mobile computers are ubiquitous. We take smartphones everywhere, and they're almost always available to us. They serve as our memories, as navigation systems, as communication devices. They have also coupled us closely to the internet, with a vast range of information only a click or two away. The ubiquitous coupling of humans, smartphones, and the internet has been a giant leap in extending the mind.

Augmented reality may well surpass smartphones as a mind-extender. Current mobile computers take some work to access. We have to activate them, find the right app, and search for information; we're not coupled to them as tightly as we could be. When it comes to extending the mind, seamlessness matters. The 20th-century German philosopher Martin Heidegger observed that in the most basic case of using a tool such as a hammer, it is "ready-to-hand" so that we use it while hardly thinking about it. In this case the tool becomes an extension of our body. In the same way, the more a tool such as a smartphone is seamlessly ready-to-hand, the better it extends our minds.

Augmented reality glasses promise to be especially seamless. They can make information instantly available whenever we need it. We don't need to go find the glasses if we're wearing them already; we'll see the information in our visual field as soon as we need it. We'll hardly notice the glasses themselves. And we can easily imagine a future when augmented reality glasses have been replaced by contact lenses that we wear all day long.

What sort of mental processes do augmented reality devices extend? For a start, they can more seamlessly extend all the processes that smartphones extend: memory (remembering someone's birthday), navigation (getting to the museum), decision-making (deciding where to eat), communication (talking to friends), language processing (translating from another language), and more. But their immersive connection to our perceptual system provides new avenues for extension.

When combined with infrared sensing technology, augmented reality can enable us to see things we couldn't see before. When combined with artificial intelligence technology for object recognition, it might enable us to recognize things and people that we previously could not identify. In these cases, the device plays a role similar to that of the parts of the brain handling color perception or object recognition. Our perceptual system has now expanded to include the device.

Augmented reality can also extend our imagination. We saw in chapter 12 that whereas we once had to rely on internal imagination to judge how a sofa would look in our living room, augmented reality devices can now do the work for us. Architects have already had their minds extended for a long time with all sorts of design technology, but augmented reality provides a particularly effective way of imagining a new building. With augmented reality we can see how we'd look in new clothes or with a new hairstyle without leaving our homes.

Augmented reality devices will be far from the last step in extending the mind. They still rely on ordinary perception and movement to link brains and computers. The device affects the brain by stimulating our eyes and ears so we see and hear the relevant information. We affect the device by talking to it or perhaps by moving our eyes, lips, and hands, which the device will track using face-trackers, hand-trackers, special wristbands that monitor nerve signals, and more. Having to rely on deliberate perception and action slows down extended cognition. But even more efficient mind-extending technologies are on the horizon.

There has been a recent surge of work on brain-computer interfaces. Sensors monitor electrical activity on the surface of the head or inside the brain. A computer receives inputs from these brain sensors and uses the input to drive actions. Such brain-computer interfaces allow severely paralyzed people to control devices such as a wheelchair or a prosthetic hand by thinking about it. If they think about moving forward, the wheelchair moves forward. The technology is still limited, but in a few decades we should be able to communicate well with our devices just by thinking. And they might connect directly to our brain's perceptual systems, bypassing eyes and ears to convey information to us with no need for glasses or screens.

At this point, minds will extend seamlessly into our devices. We will only need to think about where we want to go, and brain-computer interfaces will provide the route in our visual field or directly to our thought processes. The same sort of technology might allow us to recognize people seamlessly and to make complex calculations without any trouble. The device will simply be there as a constant, in the way that Manfred Macx's spectacles are.

A final step in extending the mind might come when our minds are uploaded from our brain to a computer, as discussed in the previous chapter. When that becomes possible, there will be no need for complex augmented reality devices or brain-computer interfaces. We'll exist on computers, and our "internal" processes will be able to link to external systems as easily as any two computers can link to each other. At this point, the boundary between what's "internal" and what's "external" will become extremely blurry without the skin or the skull to mark an organism's boundary. We'll still have minds, but it may no longer be useful to talk of the boundaries of our brains or of our minds.

The argument for the extended mind

The extended-mind hypothesis says that tools in our environment can literally become parts of the mind. Many people think this hypothesis is too radical. Opponents often fall back on the view that the mind is internal, and technology merely serves as its tool. This view is sometimes called *embedded cognition*: The mind is not so much extended as embedded in an environmental web that greatly enlarges its capacities.

So far, I've set out the hypothesis without really arguing for it. In "The Extended Mind," Andy and I focused our key argument on two characters: Otto, the Alzheimer's patient who remembers things by writing them in his notebook, and Inga, who relies on her biological memory. For our purposes, I'll give an updated version involving augmented reality.

Let's say that Ishi and Omar live in Sydney and both want to go to the Opera House. Ishi is technologically unenhanced. She calls the Opera House to mind, remembers the route, and walks there. Omar is

Figure 42 Do Ishi and Omar both know the way to
the Opera House before they think about it?

like Manfred Macx and relies greatly on his augmented reality glasses
for everything he does. He says, "Opera House," his glasses show him
the route, and he walks there.

Ishi clearly knew the way to the Opera House, even before she
thought about going there, because the knowledge was in her memory
all along. What about Omar? He knows the way to the Opera House
after his glasses show him the route. But what about beforehand? The
extended-mind hypothesis says that just as Ishi knew the way all along
because the knowledge was in her biological memory, Omar knew the
way all along, too, because the knowledge was in his digital memory.
The digital memory plays exactly the same role for Omar that the biological memory plays for Ishi.

We can state the argument like this:

1. Ishi's internal memory is genuine knowledge.
2. Omar's external memory plays the same role as Ishi's
    internal memory.
3. If the internal and the external memory play the same role,
    they both count equally as knowledge.

---

4. So: Omar's external memory is genuine knowledge.

The conclusion is that Omar can know something even though the information is stored outside his brain, in his augmented reality glasses; that is, his knowledge can reside in the world. The digital memory in his glasses is part of his knowledge, so it's part of his mind.

All three premises are plausible. It's also true that all of them could be denied. These denials yield some of the most prominent objections to the extended-mind thesis.

Some people deny premise 1, saying that Ishi didn't know the way to the Opera House until she thought about it consciously. The problem with this view is its implication that no one knows anything except when they're thinking about it. That's greatly at odds without our standard picture of thought and knowledge. Our knowledge of something doesn't go away when we stop thinking about it. Most of our knowledge is outside our consciousness, but it's still part of our minds.

Here it's important that Andy and I didn't argue that *consciousness* extends into the environment; instead, we applied the hypothesis to the many aspects of the mind that don't involve consciousness, such as memories and beliefs in the mind's background. The philosopher Brie Gertler has responded that these things aren't really part of the mind—*only* conscious states are part of the mind. But shrinking the mind like this removes much of what makes us who we are. Our hopes and dreams, our beliefs and knowledge, and our personalities are all mainly outside our consciousness at any given time.

Some people deny premise 2, pointing to differences between Ishi and Omar. It's true that there are differences, but it's hard to see, in the end result, why any of them should make the difference between knowing and not knowing. First, Omar's digital memory is inaccessible when he takes the glasses off; but the same goes for much of Ishi's biological memory when she drinks too much alcohol. Second, someone might tamper with Omar's glasses, but someone could also (in principle) tamper with Ishi's brain. Third, Omar's digital memory may not be as well integrated with other memories as Ishi's biological memory is, but even a nonintegrated memory is still a memory. Fourth, Omar's digital memory involves information put there by others, but it's not clear why this disqualifies that information. If a neurosurgeon implanted memories in Ishi's brain, these implants would still be memories and part of her mind.

The key premise is premise 3, a version of what's sometimes called the *parity principle*. The parity principle says that if an internal process and an external process play the same role, they're both on a par when it comes to being part of the mind. Some people deny the parity principle by saying there's something special about the skin or the skull as boundaries for the mind. To us, that seems like a sort of biological chauvinism, or maybe skin and skull chauvinism. What's so special about the skin or the skull that it makes the difference between being part of the mind or not?

The parity principle means that when an external memory plays the right role, it's genuinely part of the mind. To play this role, it has to be effectively glued to us, so it is as constantly and reliably available as biological memory is. And we have to trust the external memory system as we trust our own memory. Most information on our bookshelves won't count as part of our extended mind (it's not easily enough available), and most information on the internet won't, either (it's not trusted enough and often not available enough). But special systems we use constantly and trust will extend our minds: Otto's notebook, certain smartphone apps, augmented reality glasses.

In the right circumstance, even another person can become part of our extended minds. Say that Ernie and Bert are a long-term couple, and Ernie's biological memory isn't working so well anymore, so he relies on Bert to remember important names and facts. As long as Bert is reliably available and Ernie trusts him, then Bert has become part of Ernie's memory. Ernie's mind has expanded to include Bert.

Trust and availability flesh out Licklider's idea that humans and computers should be "tightly coupled." Smartphones make for tighter coupling than notebooks, and augmented reality makes for tighter coupling than smartphones. Augmented reality systems are still not as tightly coupled to our brains as they could be, but once we get to brain-computer interfaces and uploaded minds, the coupling may be just as tight as our coupling with our biological memory.

Consequences of the extended mind

Is using technology to augment the mind a good thing or a bad thing? A lively debate about this has raged for years. A 2008 cover story by Nicholas Carr in *The Atlantic* had the headline: "Is Google Making Us Stupid?" The idea is that the internet discourages us from thinking for ourselves. Carr wrote: "what the Net seems to be doing is chipping away my capacity for concentration and contemplation."

This view is not a new one in philosophy. In Plato's dialogue *Phaedrus*, Socrates discusses a debate between two ancient gods about whether the invention of writing would make the people of Egypt wiser and improve their memories. Socrates seems to think that writing was a turn for the worse. He quotes the god Thamus as saying:

> [T]his discovery of yours will create forgetfulness in the learn-ers' souls, because they will not use their memories. They will trust to the external written characters and not remember of themselves. . . . They will be hearers of many things and will have learned nothing. They will appear to be omniscient and will gen-erally know nothing. They will be tiresome company, having the show of wisdom without the reality.

Socrates goes on to suggest that "even the best of writings are but a reminder of what we know," and that "only in principles of justice and goodness and nobility taught and communicated orally . . . is there clearness and perfection and seriousness." Perhaps this is why Socrates himself never wrote anything, relying on the oral tradition to convey his philosophy. There is some irony in the fact that his ideas were immortalized in writing by Plato.

The extended-mind hypothesis offers a more positive perspective on technology. Writing enhances our knowledge and our memory; it doesn't diminish them. Likewise, Google makes us smarter, not stupider. Augmented with these tools, we can know more and we can do more than we could before.

It is true that if this augmentation is taken away, we may remember less with our brains than we did before. Once we have books around, we have less need to commit ideas to our biological memory. Similarly, in the age of Google, there's no need to remember addresses and phone numbers, so that if Google is taken away we may know less. But something like this is true for almost every technology. Once we came to rely on cars, our capacity for walking or running diminished. Heating technology has made us less practiced at withstanding the cold. If someone took away our books, computers, cars, and furnaces, we'd be at a loss. But does this mean that technologies are a bad thing? Books, computers, cars, and furnaces are a central part of our lives, and for the most part, they've made us better off, not worse. The same goes for writing and the internet.

This is not to say that technology has only good consequences. Every technology has its downsides. After the invention of the printing press, Leibniz worried that "the horrible mass of books that keeps growing might lead to a fall back into barbarism." Cars have had a terrible impact on the environment. The internet has been responsible for obvious wonders and obvious horrors.

The philosopher Michael Lynch argues that while the internet has enabled us to know more, it has often led to our understanding less. He writes:

Today, the fastest and easiest way of knowing is Google-knowing, which means not just "knowledge by search engine" but the way we are increasingly dependent on knowing via digital means. That can be a good thing; but it can also weaken and undermine other ways of knowing, ways that require more creative, holistic grasps of how information connects together.

I'm not sure this is the whole truth. In my experience, the internet has many sources that enable a deeper understanding. The issues that Lynch raises apply equally to reading; looking up information in a book is also no substitute for truly understanding it. With all of these technologies, one can engage shallowly or deeply. Everything depends on how the technology is used.

Might augmented reality make us stupid? There have been studies that suggest that when we use mapping software such as Google Maps to get to our destination, our brain activity is lower than when we navigate under our brain's own steam. The same might be true for many uses of augmented reality. But this isn't really surprising. When we drive, our level of muscle activity is lower than when we walk. What matters is whether we navigate better or worse with augmented reality. As always, we may lose something. We'll probably have a different sense of the spaces around us. But augmented reality may give us all sorts of new ways to think about and use that space.

The extended-mind hypothesis may also reconfigure how we think of morality and the self. If someone steals my smartphone, we typically think of this as theft. But if the extended-mind hypothesis is right, it's more akin to assault. If the phone is part of me, then interfering with the phone is interfering with my person. This trend is likely to increase as our reliance on augmenting technology grows. Think about Manfred Macx, rendered almost completely nonfunctional by the absence of his glasses. At some point, our social and legal norms may need to change to acknowledge the extension of our minds.

As usual, we have to expect that every technology will bring changes for better and for worse. But the extended-mind hypothesis at least suggests ways in which augmented reality could be for the better. Technological augmentation almost always has the potential to increase our capacities. How we use that potential is up to us.

Part 6

VALUE

Chapter 17

Can you lead a good life in a virtual world?

THE YEAR IS 2095. EARTH'S SURFACE IS A WRECK, A CASUALTY of nuclear warfare and of climate change. You could live a hardscrabble existence here, avoiding gangs and dodging mines, with your main aspiration being survival. Or you could lock your physical body in a well-protected warehouse and enter a virtual world.

Let's call this virtual world the *reality machine*. In the reality machine, you'll be much more comfortable than in physical reality. This world is much safer, and there's plenty of pristine territory for everyone. Most of your friends and family are there already. There are plenty of opportunities to build a community and make a difference.

You face a choice. Will you enter the reality machine?

You may well say no: the reality machine is simply an escapist fantasy. Life in a virtual world doesn't mean anything; at best, it's like spending one's life at the movies or playing video games. You should stay in the physical world where you can have real experiences and where you might be able to make a real difference.

Or you may say yes. The reality machine is on a par with the physical world. You can live a meaningful life there just as you did in the physical world. In the circumstances, it will be a far better life.

These answers reflect two different answers to the Value Question: Could you live a good life in virtual reality?

My answer to the Value Question is yes. In principle, life in virtual reality can have the same sort of value as life in nonvirtual reality. To be sure, life in virtual reality can be good or bad, just as life in physical reality can. But if it's bad, it won't be bad simply because it's virtual.

Figure 43 Robert Nozick in the experience machine.

Other philosophers say no. Some support for a negative answer was given by Robert Nozick's 1974 fable of the experience machine, which we encountered in chapter 1 (and which differs from standard VR in a few ways, as we'll see). Nozick's 1974 book, *Anarchy, State, and Utopia*, is mainly a work of political philosophy, advocating a sort of libertarianism, but along the way he wanted to reject certain views of what a good life involves. To do so, he introduced his machine for generating experiences. To continue the passage from Nozick that we quoted in chapter 1:

> You can pick and choose from their large library or smorgasbord of such experiences, selecting your life's experiences for, say, the next two years. After two years have passed, you would have ten minutes or ten hours out of the tank, to select the experiences of your next two years. Of course, while in the tank you won't know

that you're there; you'll think it's actually happening. Others can also plug in to have the experiences they want, so there is no need to stay unplugged to serve them. Would you plug in?

The Canadian philosopher Jennifer Nagel has suggested that Nozick should have taken seriously the idea that *he* was in the experience machine. After all, as a handsome Harvard professor whose books received widespread acclaim, Nozick was living just the sort of life that the experience machine might provide. Still, Nozick expected that most readers wouldn't choose to plug in to the machine. He gave three reasons.

First, Nozick says, we want to *do* certain things. We want to write books and make friends. In the machine, we merely have the *experience* of writing books and making friends. We don't really do these things.

Nozick's underlying worry here seems to be that the experience machine is *illusory*. Or at least, our actions in the experience machine are illusory. It seems that we write books and make friends, but this doesn't really happen. More generally, Nozick's line suggests that most of what happens in the experience machine is a sort of illusion. As he put it in *The Examined Life* (1989): "We want our beliefs, or certain of them, to be true and accurate; we want our emotions, or certain important ones, to be based upon facts that hold and to be fitting. We want to be importantly connected to reality, not to live in a delusion."

Second, Nozick says, we want to *be* a certain sort of person. For example, we may want to be courageous or kind. In the experience machine, we're not courageous or kind; we're not any sort of person. We're just indeterminate blobs.

The underlying problem here is perhaps that the experience machine is preprogrammed. What happens there is decided in advance. When we seem to be courageous, or kind, this is just part of the program. We're not exercising any sort of autonomy; we're just along for the ride.

Third, Nozick says, we want to be in contact with a deeper reality. In the experience machine, we're limited to a human-made reality. Everything we experience was constructed by humans.

The underlying problem here is that the experience machine is *artificial*. We value contact with the *natural* world, but we cannot get that in the experience machine. At best, we're in contact with a simulation of the natural world. The simulation is not itself natural; it's artificial.

Are these reasons for resisting the experience machine good reasons for rejecting life in the reality machine? As the philosophers Barry Dainton, Jon Cogburn, and Mark Silcox have observed, the experience machine is unlike standard VR in a number of ways. There are at least three important differences between the experience machine and the reality machine. First, you don't know you're in the experience machine while you're in there, but you know you're in the reality machine. Second, the experience machine preprograms all your experiences, but the reality machine does not. Third, while you enter the experience machine on your own, the reality machine allows your friends and family to share a reality with you.

Once we're clear about these things, and about the status of virtual worlds more generally, I don't think any of Nozick's reasons to reject the experience machine are good reasons to reject life in the reality machine. They're also not good reasons to reject life in VR.

First: VR is not illusory. I've already argued that objects in VR are real and not illusions. The same goes for actions in VR. People in virtual worlds perform real actions with their virtual bodies. In the reality machine, you can really write a book. You can really make friends. None of this is illusory. A conversation between two sims in *Free Guy* gets this right. One asks: "If we're not real, doesn't that mean nothing you do matters?" His friend replies: "I'm sitting here with my best friend, trying to help him get through a tough time. . . . If that's not real, I don't know what is."

Nozick himself may have been skeptical about this. In his 2000 *Forbes* article extending the experience machine to real-life VR, he said that the contents of VR are not "truly real." But if my arguments in this book are right, then he's wrong about this, and the illusion issue gives no reason to reject VR.

Second: VR is not preprogrammed. Typically, it's open-ended. A user in the reality machine exercises choice, and what happens there depends on the choices the user makes. Even in a simple video game like *Pac-Man*, the user chooses which direction to go in. In a more complex virtual world like *Minecraft* or *Second Life*, the user has all sorts of choices. Crucially, VR is interactive by definition. What the user does makes a difference to what happens in the world. So users can indeed be genuinely courageous or genuinely kind in the reality machine.

Third: VR is artificial, but so are many nonvirtual environments. Many of us live in cities that are largely human-made, but we still manage to lead meaningful and valuable lives. So artificiality of an environment is no bar to value. It's true that some people value a natural environment, but this seems an optional preference: it's equally possible to prefer an artificial environment, and there's nothing irrational in doing so. Even for people who prefer a natural environment, life in an artificial environment will often be a life worth living.

Nozick's experience machine nevertheless raises important issues about the value of life. In what follows, I'll discuss some philosophical issues about value and then consider the question of whether life in VR can be valuable.

What is value?

What makes for a good life? What makes one life better than another?

Questions like this are part of value theory, the philosophical study of values. Values include *moral* values (right versus wrong) and *aesthetic* values (beauty versus ugliness). In this chapter, though, I'm most interested in *personal* values: What makes something better or worse for oneself?

Personal value (sometimes called "well-being" or "utility") is at issue when one asks questions like, "Which option would be best *for me*?" Say I am thinking about whether to become a philosopher or a mathematician, from a purely self-interested point of view. I want to know which of these options will be better for me. These questions apply to life choices as well as to any number of day-to-day decisions, such as what to eat for dinner. Of course, when we make such decisions, we're often concerned not just with ourselves but with others. But even when we speak of what's best for others, it's a matter of personal value: What's better or worse *for them*?

Moral values and personal values are related. Many of us believe that, morally speaking, one should avoid harming other people whenever possible. More strongly, some people believe that one should help other people whenever one easily can do so. Even more strongly, we should go out of our way to save others from hunger and pain. There seems to be a connection between whether an act is right and the personal value it produces in others. That said, moral questions, along with social and political questions, are the subject of the next two chapters, and I'll set them aside for now.

In this chapter, when I ask what makes for a good life, I'm asking: What makes a life good for *oneself*? It may be that for many people, leading a personally good life requires also leading a morally good life, but we cannot presuppose that at the outset.

To ask what makes a good life for someone, perhaps we can start by asking what makes *anything* good for someone.

A common answer since the time of the ancient Greeks has been the philosophical view known as *hedonism*. A simple version of hedonism says something is good for someone to the extent that it gives them pleasure and not pain. A good life will be one with a healthy balance of pleasure over pain. The 19th-century British philosopher Jeremy Bentham even developed a "Hedonic Calculus" for measuring states of pleasure along various dimensions and adding them up to quantify how good something is.

This simple version of hedonism is perhaps too simple. Pleasure is a wonderful thing, but it often verges on the superficial. A life of pleasure threatens to be a hedonistic life in the ordinary sense: a shallow life centered on the enjoyment of food, drink, and sex. Bentham famously said that the source of pleasure doesn't matter: "Prejudice apart, the game of push-pin is of equal value with the arts and sciences of music and poetry." The frivolous pleasure from a parlor game counts as much as the high-minded pleasure from cultural pursuits. Some philosophers mocked Bentham by saying that his view was "a philosophy fit for swine" because a pig can experience as much pleasure as a human.

To assess Bentham's hedonism, we can contemplate the *pleasure machine*. Unlike Nozick's experience machine, the pleasure machine does not need to simulate complex scenarios. In a process that the science fiction writer Larry Niven called "wireheading," it simply stimulates pleasure centers in users' brains so that users experience great pleasure all the time. Bentham's hedonism says that a life spent wholly inside the pleasure machine is much better than an ordinary life. Most people do not agree. While the pleasure machine might be wonderful in small doses, an entire life there is an impoverished life.

Other hedonists go beyond simple pleasures. Bentham's ally John Stuart Mill said that higher pleasures, such as pleasure derived from the arts and from understanding, matter much more than lower pleasures, such as the pleasures of eating, drinking, and sex. A still more general form of hedonism, sometimes called *experientialism*, says that the fundamental objects of value are *conscious experiences*. Some experiences are positive; for example, experiences of pleasure, happiness, and satisfaction. Other experiences are negative: most obviously, experiences of physical pain, emotional suffering, and frustration. Experientialism says that something is good for someone to the extent that it gives them positive experiences and not negative experiences. A good life will be one with a healthy balance of positive over negative experiences.

It is this general form of hedonism that Nozick targets with his thought experiment about the experience machine. The experience machine goes beyond the pleasure machine by supplying the user not just with pleasure but with experiences of all sorts. He is arguing against the experientialist thesis that all that matters in life is how people's experiences feel "from the inside." He makes the case that inside the experience machine, you'll have more positive experiences than you will outside the experience machine—but nevertheless it's preferable to live outside the machine than to live inside the machine. It follows that experientialism is incorrect; experiences aren't all that matter.

An alternative view is the *desire-satisfaction* view of value: a good life is one in which our desires are satisfied, or in which things are as we want them to be. Importantly, our desires go beyond our experiences. As Nozick says, the experience machine makes it clear that we don't want just to have experiences of doing something; we want to actually do it. Our life goes better when the world is as we want it to be, even if this doesn't affect our own experiences.

This can be brought out by another thought experiment. Suppose it's very important to you to have a monogamous relationship, and you and your partner agree to this. However, your partner is frequently unfaithful to this agreement, hiding the evidence so well that you never suspect a thing. You're just as happy as you would have been had your partner been perfectly faithful. Hedonism says that a life with the unfaithful partner is just as good as a life with the faithful partner, because the experiences are the same. To many people, this seems the wrong conclusion. If it's important to you that your partner be faithful, then your life with an unfaithful partner is less good than it would be otherwise—even if you never discover the unfaithfulness.

We care about things beyond our experiences, and what we care about matters. In this case, we *want* our partner to be faithful. A life in which this desire is satisfied is better than one in which it isn't. This result is in line with what a desire-satisfaction theory would predict, but not in line with hedonism.

On the desire-satisfaction view, value is largely subjective. Value arises from what we want, and what we want is largely up to us. We might rephrase the point by saying that *valuing* produces value: Our valuing something is what makes it valuable. A good life is one in which we have what we value—and, more generally, in which the world is as we want it to be.

You might think that the desire-satisfaction theory makes value too subjective. The American political philosopher John Rawls imagined someone whose greatest desire was to count blades of grass. If the grass-counter's desire is fulfilled, does he really lead a good life? One might think he's missing out on enormous sources of value: knowledge, friendship, pleasure, and so on. Even if the grass-counter doesn't want these things, his life is arguably worse for not having them. Similarly, it's far from obvious that killing a young person who desires death leads to the best life for them.

A third view of value is the *social* view of value, put forward especially in the African philosophy of *Ubuntu*. The core idea here is that all value comes from connections to other people. As an Ubuntu maxim has it, "A person is a person through other people." The Ubuntu view rejects the individualism embodied in the hedonist and desire-satisfaction views, instead appealing to relations among people. Friendship matters. Community matters. Respect matters. Compassion matters. These are where genuine value come from.

One could object that there are values beyond the social. Perhaps a hermit could lead a valuable life of contemplation. Still, the social view offers a plausible diagnosis of what many people find missing in the experience machine. Inside the machine, you're missing genuine connections to other people. In VR, however, there can be genuine friendship with others, genuine community, and genuine *Ubuntu*. I'd like to think that when I meet up with fellow philosophers in VR during the pandemic, we have *Ubuntu*.

Finally, there's the *objective-list* view of value, in which the basic source of value are the items on a list: say, knowledge, friendship, fulfillment, and more. Your life is better insofar as you have more knowledge, more friendship, and so on down through the list. It doesn't matter whether you want these particular things—they're just good for anyone to have.

The objective-list view can incorporate some of the insights of the other three views by putting pleasure, desire-satisfaction, and interpersonal relations on the list of good things. Still, some big unanswered questions remain. What's on the list? What unites the items on the list? There's a hard choice lurking here. If there's something underlying the items that ties the objects of value together, isn't *that* something the ultimate source of value? But if nothing ties them together, isn't the list just an ad-hoc grouping of things that most of us approve of? That said, the objective-list view at least is flexible enough to capture many different views about the sources of value.

What good things are missing from VR?

Can we lead a good life in VR?

We can approach the question by asking: Are there *good things*—objects of value—that are missing in virtual reality? These may include positive experiences, as in the hedonist view. They may also include things that we deeply desire, as in the desire-based view. They may include positive social relations, as in the social view. They may include things that seem to be objectively valuable, as in the objective view.

We've addressed a few good things that Nozick thought were missing from the experience machine—we want to accomplish things, we want to be a certain sort of person, we want to be in touch with a deeper reality—and I've argued that none makes a strong objection to VR.

Perhaps the most important worry is that we won't have *autonomy* or *free will* in the experience machine because the machine programs every action in advance. As Nozick puts it in a 1989 discussion: in the machine, a person would not make any choices, and certainly would not choose anything freely.

Free will is much less of a problem in an ordinary virtual world. If we have free will in ordinary physical reality, then we can equally have free will in virtual reality. After all, our decisions in ordinary VR are made with the same brains we use in physical reality, using similar decision-making processes. We typically carry out virtual actions by performing physical actions. If the physical actions are freely chosen, the virtual actions are freely chosen, too. (For a more extensive discussion of free will in the experience machine and in VR, see the online notes.)

It's perhaps useful to first deal with limitations on the value of the near-term VR of the coming decades. These are important but temporary limitations. Then we can discuss limitations of long-term VR, and things that can't be found in VR in principle.

In the near term, it's clear that our sensory experiences in VR will be impoverished. Visual experiences in existing VR headsets are low-grade but improving. Auditory experience is somewhat closer to the quality of ordinary perception. Taste and smell are missing entirely, though, and the sense of touch is drastically limited. Bodily experience is a major limitation in near-term VR. We can inhabit virtual bodies, but our experience of those bodies is limited. There's no realistic experience of eating and drinking in current VR. There's no real hugging or kissing. Despite the best attempts of the sex-technology industry, interpersonal sexual experience in VR is a shadow of what it is outside VR.

These limitations present obstacles to having a full and fulfilling life in near-term VR. If you value the multisensory experience of food and drink, or of lifting weights or swimming in the ocean, you have to get these experiences outside VR for now. As things stand, you can just live some fulfilling *parts* of your life in VR: You might go to a VR workplace, or have VR conversations with your friends, or attend a VR gathering where sensory experiences aren't important.

Still, VR technology is advancing all the time. Visual resolution and field of view are improving, and before long they will be on a par with ordinary vision. Researchers are experimenting with mechanisms for taste, smell, and touch. In the longer term, we'll almost certainly have brain-computer interfaces whereby virtual inputs directly stimulate parts of the brain responsible for these sensory experiences. This may ultimately allow a vast range of experiences—experiences that not only subsume all of our ordinary sensory experiences but also go well beyond them.

It's worth touching on some ways in which VR may be *better* than ordinary physical reality. First, as we've just seen, VR may allow many experiences that are difficult or impossible in physical reality: flying, inhabiting entirely different bodies, new forms of perception. Second, if Earth becomes dangerously degraded (as illustrated in the video game *Soma* as well as in my story of the reality machine), VR can offer a safe haven. Third, whereas space on Earth is a limited resource, space in VR is almost unlimited. As we'll discuss in chapter 19, everyone can have a virtual mansion or even a virtual planet. Fourth, as our minds speed up in the technological future, physical reality may come to seem unbearably slow. Virtual reality can speed up along with our minds.

At a certain point, VR will provide enormous benefits in the realms of space, time, experience, embodiment, and more. The question is whether these benefits are outweighed by costs.

What's missing from long-term VR?

To focus on longer-term issues, we can imagine a fully immersive VR system in which sensory and bodily experiences are much the same as in the physical world. During a pandemic, when physical contact is dangerous, we can hang out in VR instead. No doubt there will be modes of VR that offer experiences far beyond physical reality, but there will also be modes that simulate physical reality near perfectly. We'll eat, drink, hug, swim, work out, and have sex in ways indistinguishable from the original. Will this be as good as ordinary physical reality? Or is something missing?

Some will say that sheer *physicality* is missing in VR, and physicality is something we value. Certainly it's easy to imagine people exiting VR to experience physical eating or physical swimming or physical sex on occasion. But if the experience of these things in VR is truly indistinguishable from the experience outside VR, this pursuit of the physical may come to seem a novelty or a fetish.

No doubt some people will still value the physical world greatly for its own sake. The Australian singer Olivia Newton-John (perhaps paying tribute to her grandfather, the great German physicist Max Born) famously expressed a preference for physical reality, singing "Let's get physical." Many people in virtual worlds may at least be interested in visiting the physical world from time to time. There may be a sense of authenticity in interacting in our original biological form. But it's hard to see why sheer physicality should make the difference between a meaningful life and a meaningless life.

To those who value physicality, we can ask: What if it turns out that we're already in a simulation? Do the eating and swimming and kissing that we already do in our simulated environment have the value of sheer physicality, or not? If they do: then it starts to look like the issue isn't so much physicality versus virtuality as a preference for our original bodies and the environment we started in. If they do not: then why is simulation-based reality any less valuable than quark-based physical reality? Once we accept simulation realism, it's hard to make the case that there's something intrinsic to physical reality that makes it more valuable than virtual reality.

Of course no one should be forced into VR. If someone is a virtual irrealist who thinks a good life in VR is impossible, then even if they're wrong, a life in VR will go against their personal desires. I'm assuming here that the decision to enter VR is a free choice. That said, I predict that as the quality of virtual worlds improves, virtual realism will gradually become a commonsense view. Eventually, many people will freely choose to live most of their lives in virtual worlds.

Many will point to *relationships* as something that would be missing in VR. If you entered a lifelong VR alone, you would give up contact with your family and friends. But maybe not. For a start, your family and friends could enter the virtual world with you. Furthermore, many virtual worlds will allow you to communicate with the nonvirtual world and to travel back there, in which case you won't need to give up those relationships. Where relationships are concerned, the range of options in VR is roughly analogous to the range of options available when you move to another country. Emigration, virtual or not, may diminish some old relationships, but it also leads to many new ones, and it often makes your life better rather than worse.

There are many *social and political* concerns about the impact of virtual worlds on society that parallel issues about information technology more generally: inequality, privacy, autonomy, manipulation, resource-intensiveness, and more. I will focus on issues at this level in Chapter 19. A related worry is that life in virtual worlds may be a form of *escapism* from the nonvirtual world, in the way that video games can be. There is something to this, but full-scale virtual worlds are not video games. As before, a move to a full-scale virtual community will parallel emigration to a new nonvirtual community. You escape one set of issues and confront many new ones. That said, I'm certainly not recommending that everyone abandon the nonvirtual world for virtual worlds. That would lead to obvious problems. But within limits, a move to virtual worlds need be no more escapist than emigration.

Potential problems arise from *transfer* between the virtual and nonvirtual worlds. Some people worry that habits learned in virtual worlds will transfer to the nonvirtual world. Most obviously, violent behavior learned in video games might lead to violence in ordinary life. Perhaps if you spend most of your time in VR, you'll neglect your nonvirtual health. These are reasonable worries, but interaction issues of this sort apply just as much in the nonvirtual world. A new relationship can distract you from old friendships; a military role can desensitize you to violence in ordinary life; a desk job may be bad for your health. The issues here are not distinctive to virtual reality but are part and parcel of trying to live a good life in reality.

It's fair to worry about your *physical body* when in VR. Will your body be confined and neglected? If your body has to be locked up in a crowded dark warehouse to live in the reality machine, isn't that a cost? Here I'm imagining that in a long-term full-dive VR, your body is kept healthy at least. Most of the time there will be no awareness of your physical body, so its confined condition won't affect your experience. But in childhood especially, exposure to a physical environment may be required for normal development of the body and the brain. And if people want to travel regularly between virtual and nonvirtual reality, then it's important that they have a high quality of life in nonvirtual reality, too.

A serious limitation of many virtual worlds is that they are *transient*. The virtual worlds in many video games last only minutes. Massive multiplayer environments last longer, but even these typically close eventually. More importantly, no virtual world we've yet created has anything like the long history of our nonvirtual world, and history is something that many of us greatly value. There's value in living in a place where people have lived for centuries and millennia. There's value in visiting places where history-shaping events happened. There's value in taking part in time-honored traditions.

We might acknowledge the value of history while holding that (as with naturalness) history is just an optional value. People have often moved to places with no particular historical resonance, and they have still lived good and meaningful lives. Some people don't care much about history at all. Others may care, but it's rarely the most important value in a life. If history is valued, you can obtain some of the value through travel between the virtual and nonvirtual domains. And virtual worlds will presumably have notable histories of their own in the long run.

Perhaps the most striking absence in virtual worlds is the absence of *birth* and *death*. No one is born in the virtual worlds we have today, and no one dies there. There are depictions of birth and death, but not the real thing. Avatars may be created and destroyed, but people are not. We may enter a virtual world for the first time and leave it forever—but we existed before we were in the virtual world and we will continue to exist afterward. This is more like moving to a community and leaving it than being born or dying. Birth and death are two of the most meaningful events that happen in nonvirtual reality. So isn't a virtual world without birth and death a greatly impoverished world?

There are some obvious replies. Perhaps entering a virtual community for the first time and exiting it forever would be like birth and death in a world with an afterlife and reincarnation. Birth and death of that sort are still meaningful. Furthermore, people would genuinely die in a virtual world when they died in the physical world. Perhaps we'll eventually have pure sims who are born and die in virtual worlds—although one wonders how permanent "death" would be if digital records were lying around. Indeed, it's entirely possible that within a century or two, new medical technologies will eliminate many causes of death. One way or another, birth and death as we know them are likely to be absent or transformed in virtual worlds.

A deep question: What is the role of birth and death in a good life? I'd say that both are important. Experiencing others' birth and death can be transformative. But it's not clear that this is essential to a good life. Some people hold that although death is usually a bad thing, it's nevertheless essential and a world without it would be awful or meaningless. In his essay "The Makropulos Case: Reflections on the Tedium of Immortality," the British philosopher Bernard Williams argued that immortality would ultimately be boring. In the TV show *The Good Place* (spoiler alert!), characters who made it to heaven eventually decide to end their lives because they have overcome every challenge, and there is no impetus to keep on living. These attitudes are far from obviously correct, however. My own suspicion is that once immortality is possible (perhaps through digital means), people will wonder how they ever lived without it.

Birth is trickier. Many people lead wonderful lives without having children. At the same time, birth is one of our paradigms of a good thing, and a world without births would be impoverished compared to our world. The film *Children of Men* depicts a bleak world where birth has stopped entirely. That said, a virtual world without birth need not be nearly as impoverished as this. Births will still happen in the nonvirtual world. Some people may travel to the nonvirtual world to give birth, or, eventually, they may even experience it in the virtual world. At an appropriate time (perhaps at birth), children may enter the virtual world for the first time. At least with appropriate connections between the virtual and nonvirtual worlds, the absence of birth from virtual worlds needn't lead to impoverished lives.

It's not unreasonable to think that long-term VR could be impoverished by the absence of nature, history, and perhaps birth and death. They are all valuable or at least meaningful aspects of life in physical reality. Still, these missing benefits can be weighed against new benefits arising from the many new forms of life and other possibilities that VR offers. Weighing everything together, it seems at least possible to have a meaningful and valuable life in a virtual world. For many people, choosing to spend much or most of their life in VR will be a reasonable choice.

Terraform Reality

Here's a thought experiment that illustrates how I see the value of virtual reality.

In the future, we may come up with a new technology we can call Terraform Reality. Terraform Reality allows us to turn exoplanets in the nonvirtual world into habitable environments full of beautiful places and engaging activities. People have the option to travel to these planets and build new lives there. These planets rapidly become popular: They

Figure 44 Which would you choose: Life in virtual reality or in Terraform Reality?

have much more space than Earth, and they're full of new opportunities. Many societies are set up on these planets, and new planets and societies are continually being introduced. People can also acquire new bodies in Terraform Reality, and many people choose to do so.

Is life in Terraform Reality as good as life on Earth? There are pros and cons. On the plus side, it may be more pleasant than Earth—and more exciting, given the many new opportunities. On the minus side, terraformed environments are artificial and lack the history of natural environments, so that life on these planets may seem less weighty than life on Earth. Still, it would seem perfectly reasonable for many people to choose to spend considerable time in a terraform reality, or even to move there long term.

Let us say that *rich VR* is VR with roughly the complexity of ordinary reality, after short-term technological limitations have been overcome. I'd say that life in rich VR can be about as valuable as life in Terraform Reality. Each has pros and cons. In VR, much more may be possible. For example, there may be different laws of nature. We may be able to fly like a bird. Terraform Reality allows physical birth and death, and sheer physicality in general, more straightforwardly than in VR. Still, in many respects the two seem roughly on a par.

All this can be put as an argument:

1. Life in rich VR is roughly as valuable as life in Terraform Reality.
2. Life in Terraform Reality is roughly as valuable as ordinary nonvirtual life.

_____

3. So: Life in rich VR is roughly as valuable as ordinary nonvirtual life.

Life in rich VR may be in some respects better than a corresponding life outside VR, and in some respects worse, but overall they're about on a par. In a future where we have the option to enter an attractive virtual world, without the option to enter an equally attractive nonvirtual world, it may well be rational to enter the virtual world.

Most importantly, there's no good reason to think that life in VR will lack meaning and value. Nor is there reason to think its values will be limited to entertainment. It will largely allow the kinds of values available in living a life in physical reality. There will be good and there will be bad, and it will sometimes be a struggle to make the good outweigh the bad, but that's how life in reality goes.

Shifting from VR to the simulation hypothesis: Is life also valuable in a full-scale simulated universe? A lifelong pure simulation avoids some obstacles to value in standard VR: transience, birth and death, and low-quality sensory experience aren't issues here. Artificiality may remain an issue, but being in an artificial universe seems no worse than being in a universe created by a god. The same goes for worries about the simulators' being malicious or indifferent, or the simulation's being fragile. There's a question about whether simulated creatures can themselves be sources of value, but I'll address that in the next chapter.

You might worry that life in a perfectly simulated universe—with no possibility of escape, unlike in standard VR—would be confining. We may want to know as much of the cosmos as possible and to travel throughout it if we can. This problem is akin to our existing situation, in which we're confined to Earth and the solar system. It might be enjoyable to explore the actual universe, but life on Earth isn't all that bad.

What is the source of value?

I've been focusing on the value of virtual reality. What can we learn from this about value in general? Nozick used his experience machine to argue against hedonistic theories of value. Does our thesis that virtual realities have comparable value to nonvirtual realities tell us anything about what is truly valuable?

My positive answer to the Value Question is compatible with all of the major theories of value. Hedonists, desire-satisfaction theorists, social theorists, and objective-list theorists can all accept that you can live a good life in VR. For the hedonist, VR need only replicate the conscious experiences from a good life in the nonvirtual world. Where desire-satisfaction is concerned, virtual realism suggests that our everyday desires will be satisfied about as well if we turn out to be in a simulation as if we don't. Some desires may be satisfied less well in VR—a desire to be in nature, or a desire not to be in a simulation— but these don't make the difference between a good life and a bad one. Social communities and social connections can in principle be just as rich in VR as outside VR. Finally, if we work down an objective list of what is valuable, I'd argue that the most important factors can all be equally present in VR.

What are the sources of value? I'm inclined to think that all value arises, one way or another, from consciousness. Conscious states themselves (say, happiness and pleasure) are valuable. What is valued by conscious creatures (say, knowledge and freedom) is valuable. And relations among conscious creatures (say, communication and friendship) are valuable. One might say that consciousness has value, and relations to consciousness add value.

Insofar as virtual reality involves the same sort of consciousness as a nonvirtual world, it will have the first of these three sources of value. Insofar as it has the same acts of conscious valuing that are satisfied equally well, it will have the second. Insofar as it has the same relations among conscious creatures, it will have the third.

In the long term, virtual worlds may have most of what is good about the nonvirtual world. Given all the ways in which virtual worlds may surpass the nonvirtual world, life in virtual worlds will often be the right life to choose.

Chapter 18

Do simulated lives matter?

A S YOU'VE PROBABLY NOTICED, THE HISTORY OF PHILOSOPHY is dominated by men. There have been many notable women in the history of philosophy, from the Hindu philosopher Maitreyi (8th century BCE) to the great French feminist philosopher Simone de Beauvoir (20th century). But for the most part, women's contributions have been obscured. It is only well into the last century that they start to flourish.

One of the most remarkable flourishings happened in Oxford during the Second World War. Four women who studied philosophy there during the war went on to become leaders in the field: Elizabeth Anscombe, Philippa Foot, Mary Midgley, and Iris Murdoch. They were all close and met regularly with one another. It's perhaps not a coincidence that they emerged at a time when most men were away at war.

Each of these philosophers made striking contributions. Anscombe's densely argued book *Intention* (1957) is one of the key works in the philosophical understanding of how people act. Midgley argued for continuity between humans and animals in *Beast and Man: The Roots of Human Nature* (1979) and also wrote influential polemics condemning reductionism in science and culture. Murdoch's philosophical novels are widely celebrated, and the philosophical essays in her book *The Sovereignty of Good* (1970) have been influential in the foundations of morality.

Perhaps the most influential contribution of all from this group is a 1967 thought experiment devised by Philippa Foot concerning a runaway streetcar. Following British usage, she called it a "runaway tram." A decade later, the American philosopher Judith Jarvis Thomson made the transatlantic translation, resulting in the much-loved *trolley problem*. The trolley problem has inspired endless books and articles and was depicted colorfully in the philosophical television series *The Good Place*. Thomson's version goes like this:

Edward is the driver of a trolley, whose brakes have just failed. On the track ahead of him are five people; the banks are so steep that they will not be able to get off the track in time. The track has a spur leading off to the right, and Edward can turn the trolley onto it. Unfortunately there is one person on the right-hand track. Edward can turn the trolley, killing the one; or he can refrain from turning the trolley, killing the five.

What should Edward do? If he does nothing, five people will die. If he turns the trolley, one person will die. Many people have the intuition that he should turn, so that four more people will survive.

Before you get too comfortable with this conclusion, consider the related *transplant case* that Thomson formulates in the same article.

David is a great transplant surgeon. Five of his patients need new parts. One needs a heart, the others need, respectively, liver, stomach, spleen, and spinal cord—but all are of the same, relatively rare, blood-type. By chance, David learns of a healthy specimen with that very blood-type. David can take the healthy specimen's parts, killing him, and install them in his patients, saving them. Or he can refrain from taking the healthy specimen's parts, letting his patients die.

What should David do? If he does nothing, five people will die. If he removes the organs from the healthy patient and transplants them, only one person will die. Here, the common intuition is that David should *not* remove the organs from the healthy patient.

Figure 45 Philippa Foot and Judith Jarvis Thomson confront
the trolley problem. Should they switch tracks?

The trolley case and the transplant case are structurally similar, but our intuitions about them are opposite. How can we reconcile them? We could change our judgment about one case or the other. Or we could try to find a relevant difference between the two.

What are the differences between the trolley case and the transplant case? Foot's view was that in the trolley case, if you don't switch tracks, you are *killing* the other five. After all, you're the driver and therefore responsible for the trolley barreling into them. In the transplant case, by contrast, if you don't kill the healthy patient, you are merely letting the other five die. And many people think there's a significant moral difference between killing five people and letting five people die. Another relevant difference is that in the transplant case, you're directly interfering with the subject you're killing. Many people have the intuition that in a version of the trolley case in which you have the option of *pushing* one person from a footbridge onto the track, killing that one to save five, you shouldn't do it. In the PhilPapers Survey that David Bourget and I conducted in 2020, 63 percent of professional philosophers said they would switch tracks in the trolley case, but only 22 percent said they would push the person in the footbridge case. Why should there be such a huge difference in our reactions to these two cases, both of which result in the killing of one person to save five?

Here we're doing *ethics*: roughly, the study of right versus wrong. What things *should* we do, morally speaking? What things *shouldn't* we do? And why? Most people think we *should* turn the trolley in the trolley case and *shouldn't* cut up the healthy patient in the transplant case. The hard question in these cases is "Why?" We need a theory that explains why one action is right and the other is wrong.

Virtual worlds raise any number of ethical questions. Some of them concern even today's virtual worlds. What are the moral limits on how we should act in a virtual world? Is it wrong to "kill" a fellow combatant in a video game? Are assault and theft in a virtual world just as wrong as they are in the nonvirtual world? I'll focus on these in the next chapter.

In this chapter, I'll focus on ethical questions about long-term simulated worlds. Is it morally permissible to "play God" by creating virtual worlds containing conscious sims? What moral responsibilities do we have toward the sims in those worlds? In *Free Guy*, artificially intelligent sims based in a video game go on strike to demand respect. Does this make sense? Do sims matter?

We can even come up with a *simulation trolley problem*. In the actual world, Fred is sick, and the only way to save him requires intensive computer research on a computer we've been using for a simulated world. Computer space is extremely tight, and we have no backups. To do this research, we'll have to sacrifice five people in the simulation. Is it morally acceptable to kill five simulated people to save one non-simulated person? I'll let you think about that while we review ethical theories.

Ethical theories

One traditional theory of right and wrong is *divine command* theory: An action is right if, and only if, it's what God commands. God commands us not to kill people, so killing people is wrong. God commands that we should worship him, so worshipping him is right.

The best-known problem for divine command theories has its origins in Plato's dialogue *Euthyphro*. Euthyphro is prosecuting his father for murder. His family objects, but Euthyphro says it's the "pious" (i.e., right) thing to do. Socrates asks him what makes pious acts pious. Euthyphro responds by embracing divine command: "Piety . . . is that which is dear to the gods, and impiety is that which is not dear to them."

Socrates then asks the crucial question: Is the pious dear to the gods because it is pious, or is it pious because it is dear to the gods? To put the issue in more familiar terms: When something is the right thing to do, is it the right thing to do because God commands it, or does God command it because it's the right thing to do?

Euthyphro has a dilemma. If he answers that something is the right thing to do *because* God commands it, he is faced with the manifestly unacceptable consequence that if God had commanded us to torture and murder babies, torturing and murdering babies would have been the right thing to do.

If he says that God commands something *because* it's the right thing to do, then we need some independent account of what *makes* the action right. It cannot be God's command that makes the action right; otherwise we'd have a circular situation wherein God commands the action because God commands it. Something else must make the action right—so we must go beyond divine command theory.

Euthyphro's dilemma is one of the most powerful in philosophy. It recurs again and again in any number of domains. Many conclude that we need some further account of what makes an action right or wrong beyond what God or anyone else commands.

Perhaps the most widely known moral theory is the *utilitarianism* of Jeremy Bentham and John Stuart Mill. Utilitarianism says that the right thing to do is whatever does the greatest good for the greatest number. Or alternatively, the right thing to do is whatever maximizes *utility* across the whole population.

What is utility, exactly? It is a measure of how good an outcome is for a person. A better outcome has higher utility. Utility measures what we called *personal value* in the last chapter. As we saw then, Bentham and Mill were hedonists about personal value. They held that the utility of an outcome comes down to a predominance of pleasure over pain in that outcome.

Utilitarianism works nicely for the trolley problem. If we do nothing, five people will die and one will survive. If we switch tracks, one person will die and five will survive. We can assume that death has very low utility for a person (perhaps zero, since that person will no longer have any pleasure or pain, assuming a painless death), whereas survival has much higher utility (say 100, assuming a happy life to come). Then switching produces utility of 500 in this group of people, while doing nothing produces utility of 100 for the same group. Of course, there may be effects on others, too. (What if the person on the other track has a large family? What about effects on the trolley operator? And so on.) But at least to a first approximation, it looks like maximizing utility requires that we switch tracks.

However, utilitarianism doesn't work so well in Thomson's transplant case. We can assume, again, that death for a person has utility zero and survival has utility 100. It follows that cutting up a healthy person to save five dying patients will have utility 500, whereas allowing the healthy person to live and the others to die will have utility 100. Cutting up the healthy person will maximize utility, so utilitarianism says that we should cut up the healthy person. But that conflicts with most people's strong sense that this wouldn't be the right choice.

This is a problem for utilitarianism. Some utilitarians may follow the logic where it leads, saying that we should go ahead and cut up the healthy person. Others will say that cutting up the healthy person doesn't maximize utility because it has other bad consequences. For example, if we cut up the healthy person, others will hear about it, and in future fewer healthy people will trust doctors and more will die. But what about a secret case, which no one ever finds out about? Then this negative consequence will be reduced or eliminated, but killing the healthy person seems wrong all the same.

Utilitarianism focuses on the *consequences* of our actions in evaluating their moral status. An alternative approach to morality focuses on our *reasons* for performing an action. Moral theories in this broad class are often called *deontological* theories, whereas those that focus on consequences are *consequentialist*.

The most straightforward deontological theories are *rule-based*: What makes an action right and wrong is not so much its consequences as the rules that you follow in producing the action. An example of a moral rule is "Thou shalt not kill," or "Don't harm innocent people." If you act according to the wrong rule, you have acted immorally.

How do we determine what the acceptable rules are? The most famous proposal was offered by the 18th-century German philosopher Immanuel Kant, often regarded as the greatest philosopher of the last several centuries. In his *Groundwork for the Metaphysics of Morals*, Kant proposed the Categorical Imperative: *Act only according to those rules that you can also will as universal laws.*

Kant's idea is that when you contemplate acting according to a certain rule, you should consider a world in which everyone obeyed that rule as a matter of law. Would it be rational to choose to live in a world like that? If yes, you may act according to the rule. If no, you may not.

Consider the principle "You should lie when necessary to get what you want." Kant thought this was an unacceptable principle. Suppose we lived in a world where everyone lied to get what they wanted. Then everyone would know this, and lying would be useless as a way of getting what you want. It doesn't make sense to will this principle as a universal law—and that means it's unacceptable to act according to this principle. On the other hand, principles such as "Be good to others" can be willed as a universal law. It would make sense to want a world where everyone is good to everyone else, so "Be good to others" is an acceptable principle to guide your action.

The rule-based approach can be combined with the utilitarian approach. So-called *rule-utilitarianism* says a moral rule is acceptable when adopting it as a universal rule will lead to the best consequences.

For example, following a rule such as "Do not kill healthy patients" may have worse consequences in a particular instance such as the transplant case (four people will die, as opposed to a single person), but it may be that adopting it as a general rule has the best consequences overall (the hospital system will work better and many more people will survive). But what about a qualified rule that says, "Kill someone when this will save more lives and no one will ever know"? Even if adopting some rule like this would have better consequences overall, it can still seem immoral. So there remain some puzzles here.

In her classic 1958 article "Modern Moral Philosophy," Elizabeth Anscombe was scathingly critical of both consequentialist and deontological approaches to ethics. She said that consequentialism leads to immoral consequences and reveals a "corrupt mind." On the other hand, deontological approaches such as Kant's involve a legalistic conception of morality as a set of laws laid down by a legislator. Anscombe thinks this picture is a leftover from the old divine command theory, according to which God was the legislator. For Anscombe, once God is rejected, the rule-governed approach cannot work.

Anscombe thought we should not talk of what's morally right and wrong at all. These terms are too coarse to capture what is of interest in morality. Instead, we should use finer-grained words like "unjust" and "brave" and "kind" to assess the moral character of people's actions.

In this, Anscombe was recommending a return to the *virtue ethics* found in Aristotle, in which ethics centers on virtues such as bravery and kindness. A closely related picture is found in the work of Confucius, Mencius, and other philosophers in the Confucian tradition, which gives a central role to the moral traits we should aspire to, such as benevolence, trustworthiness, and wisdom. One popular version of virtue ethics frames the moral character of an action in terms of the moral character of a person who would perform that action. A brave act is one that a brave person would perform. A kind act is one that a kind person would perform.

Thanks to Anscombe as well as her Oxford colleagues Philippa Foot and Iris Murdoch and Chinese philosophers in the New Confucian movement, among others, virtue ethics has recently had a resurgence as a leading moral theory. It is sometimes criticized for not giving clear criteria regarding how we should act. Nevertheless, it's often seen as offering tools for moral improvement and for understanding the rich tapestry of morality without reducing it to simple principles.

Simulations and moral status

Let's turn to the ethics of virtual reality. We can start by thinking about long-term simulation technology. When is it morally permissible to create a simulation? When is it morally permissible to end one? What are our moral responsibilities as creators of a simulation?

If we simulate a universe without life, there are few ethical issues. Cosmologists already run simulations of the history of galaxies and stars, and they don't need permission from an ethics board. Perhaps there are ethical issues about whether this is the best use of computer power, and about what to do with knowledge gained from the simulation, but these are standard ethical issues involved in everyday science. Even simulating biology—say, at the level of the evolution of plant life— does not go far beyond this.

Ethical issues arise when we simulate *minds*. To start with an extreme case, say that we work for an intelligence agency and we want to simulate the reactions of human beings who undergo torture. We create simulated humans with fully functioning simulated brains and subject them to (simulated) torture. Is this morally acceptable or morally horrific?

A natural reaction is that it depends on the mental life of the simulated beings. If they're conscious creatures who experience suffering, simulated torture would be morally horrific. If they're unconscious simulations and don't experience suffering, then simulated torture would perhaps be morally acceptable.

This raises a fundamental issue: Do sims have moral status? A being has moral status when it's an object of moral concern in roughly the way that people are objects of moral concern—that is, it's a being whose welfare we need to take into account in our moral deliberations.

A being has moral status when that being *matters*, morally speaking. The Black Lives Matter movement is all about moral status. Black lives matter as much as any human lives do! Killing Black people is as bad as killing white people. Mistreating Black people is as bad as mistreating white people. In the past, and even today, many people and many social institutions have treated Black lives as if they mattered less than white lives. This is now widely recognized as monstrous.

Over the years, the circle of moral status has expanded. It's now widely accepted that many nonhuman animals have moral status, too. The issue isn't quite the same as with human lives. Most people think that humans matter more than birds and dogs—but birds and dogs still matter to some extent. We shouldn't be wantonly cruel to dogs. It's less obvious whether flies and shellfish have moral status; some people think they do. Some environmentalists hold that trees and plants have a sort of moral status, but this is a minority view. As for inorganic matter, few people think that rocks or particles have moral status. You can treat a rock however you like, and this won't matter morally, at least as far as the rock is concerned.

My own view (shared with many others) is that what bestows moral status is *consciousness*. If an entity has no capacity for consciousness, and never will have, then it has no moral status. It can be treated as an object. If an entity has the capacity for consciousness, then it has at least some minimal moral status. If it can experience something, that should be taken into account in our moral calculations. It's arguable that systems with a minimal degree of consciousness (ants?) have only a minimal degree of moral status and so weigh much less heavily than humans in our moral deliberations. But consciousness at least gets them in the door.

We can use a thought experiment to help us think about the moral status of consciousness. I call it the *zombie trolley problem*. You're at the wheel of a runaway trolley. If you do nothing, it will kill a single conscious human, who is on the tracks in front of you. If you switch tracks, it will kill five nonconscious zombies. What should you do?

A few clarifications. The zombies are philosophical zombies, as described in chapter 15: near-duplicates of human beings with no conscious inner life at all. Zombies have no subjective experience. You can imagine them as physical duplicates of us without consciousness, or as silicon versions of us without consciousness if that's easier. If that's still too hard, imagine something as close to us as possible without the capacity for consciousness. Whether or not these zombies will be useful for various purposes isn't relevant in this thought experiment; what matters is their moral status.

When I have taken polls about the zombie trolley problem, the results are pretty clear: Most people think you should switch tracks and kill the zombies. It's worse to kill one human than five zombies. A few say that zombies count for as much as humans, so we should kill the human, but they are a distinct minority.

Killing the zombies may sound awful. There is a recent movie, *Zombies*, that centers around the way a community of zombies is mistreated in a human world. But importantly, the zombies in the movie are conscious. Philosophical zombies lack consciousness, so that there is arguably no one home to mistreat.

We can take things further. Suppose you have the choice between killing one conscious chicken or a whole planet of humanoid philosophical zombies. At this point, intuitions are less clear. Some people stick with "Kill the zombies," reflecting the view that without consciousness there's no moral status. Others switch to killing the chicken, presumably because they think the zombies have some degree of moral status, perhaps deriving from their intelligent behavior. My own intuition wavers on this matter.

The zombie trolley problem can lead to a weak or a strong conclusion. If you think a single conscious creature should be saved at the cost of killing five nonconscious creatures, this suggests that consciousness is *relevant* to moral status. Conscious creatures matter more than nonconscious creatures. If you hold the stronger view—that there's never a moral reason to spare nonconscious creatures—this suggests that consciousness is *necessary* for moral status. Nonconscious creatures don't matter at all, morally speaking.

The stronger conclusion dovetails with the view I advocated at the end of the last chapter, that consciousness is the ground of all value.

Whenever anything is good or bad for someone, it's because of their consciousness. Consciousness has value, what a conscious creature values has value, and relations between conscious creatures have value. If a creature has no capacity for consciousness, nothing can be good or bad for it from its own perspective. And it's natural to conclude that if nothing is good or bad for a creature, then the creature has no moral status.

The view that consciousness is required for moral status is central in discussion of animal welfare. The Australian philosopher Peter Singer, who inspired the contemporary animal-rights movement with his 1975 book *Animal Liberation*, has argued that what he calls *sentience* is what matters for moral status:

> If a being is not capable of suffering, or of experiencing enjoyment or happiness, there is nothing to be taken into account. This is why the limit of sentience (using the term as a convenient, if not strictly accurate, shorthand for the capacity to suffer or experience enjoyment or happiness) is the only defensible boundary of concern for the interests of others.

In ordinary English, "sentience" is roughly equivalent to "consciousness." Singer uses the term more narrowly to describe suffering and the experience of enjoyment and happiness. This is a *kind* of consciousness: Only conscious creatures can suffer or experience enjoyment or happiness. Singer holds that consciousness is necessary but not sufficient for moral status. Not just any sort of consciousness bestows moral status; the conscious experience of positive or negative affective states is required. The same "sentientist" view has been taken by many recent theorists, who hold that the experience of positive or negative affective states is what matters for moral status. This view goes back at least to Jeremy Bentham, who said in the 18th century that where moral status is concerned, suffering is what matters.

I find this view implausible. There's much more to consciousness than the experience of suffering or happiness, and it's not plausible that these other sorts of consciousness don't matter morally. To make this point, we might think about a more extreme version of the unemotional Vulcan Mr. Spock on *Star Trek*.

Let's say that a *Vulcan* is a conscious creature who experiences no happiness, suffering, pleasure, pain, or any other positive or negative affective states. The Vulcans on *Star Trek* aren't quite as extreme as this: they experience lust every seven years and experience at least mild pleasures and pains in between. To avoid confusion with *Star Trek* we could call our version *philosophical Vulcans*, by analogy to philosophical zombies.

As far as I know, no human being is a philosophical Vulcan. There are some reported cases of humans who do not experience pain, fear, or anxiety, but they still experience positive states. A philosophical Vulcan will lack those states as well. They might still have a rich conscious life, with multimodal sensory experiences and a stream of conscious thought about all sorts of complex issues. We've all experienced affectively neutral states in perception and thought. I can see a building or think about a meeting without any positive or negative affect. For a Vulcan, that's what things are like all the time.

Vulcans' lives may be literally joyless, without the pursuit of pleasure or happiness to motivate them. They won't eat at fine restaurants to enjoy the food. But they may nevertheless have serious intellectual and moral goals. They may want to advance science, for example, and

Figure 46 Jeremy Bentham faces the Vulcan trolley problem.
Is it better to save one human or five Vulcans?

to help those around them. They might even want to build a family or make money. They experience no pleasure when anticipating or achieving these goals, but they value and pursue the goals all the same.

The Bentham/Singer view predicts that a philosophical Vulcan doesn't matter morally. That seems incorrect. We could make the point with a *Vulcan trolley problem*. Would it be morally acceptable to kill a planet of philosophical Vulcans to save one human with ordinary affective consciousness? I think the answer is obviously no.

More simply, suppose you're faced with a situation in which you can kill a Vulcan in order to save an hour on the way to work. It would obviously be morally wrong to kill the Vulcan. In fact, it would be monstrous. It doesn't matter that the Vulcan has no happiness or suffering in its future. It's a conscious creature with a rich conscious life. It cannot be morally dismissed in the way that we might dismiss a zombie or a rock.

(Does the Vulcan have a desire to keep on living? As I'm thinking of things, yes. If we shouldn't kill such a Vulcan, that shows that more than affective conscious states matter. We could also stipulate an even more extreme Vulcan—one who has no affective conscious states and is also indifferent to continuing to live or dying. My view is that it would also be monstrous to kill this Vulcan. If so, this suggests that more than affective consciousness and desire satisfaction matter. My view is that non-affective consciousness matters, too.)

My own sense is that a Vulcan matters about as much as an ordinary human. Of course I am glad that I am a human and not a Vulcan, since affect makes my life better. Suffering and happiness make a big difference to how good or bad a conscious creature's life is. But they're not what gives a creature moral status in the first place.

Bentham once expressed his view by saying that where the moral status of animals is concerned, "The question is not, Can they reason?, nor Can they talk? but, Can they suffer?" If I'm right, what matters is not suffering but consciousness. The right question is not "Can they suffer?" but "Are they conscious?"

To determine the moral status of simulated creatures, "Are they conscious?" is also the question we need to ask. We've already asked and answered this question for some simulated creatures. In chapter 15, I argued that a perfect simulation of a human brain will be associated with just the same sort of consciousness as the original brain. That is, simulated humans will have the same sort of consciousness as ordinary humans. If consciousness is all that matters for moral status, simulated humans have the same moral status as ordinary humans.

The ethics of simulations

Now we can answer the simulation trolley problem. The answer is no: It is not acceptable to kill five simulated people to save one ordinary human. If simulated humans weren't conscious, this would be acceptable. But because full-scale simulations of humans are conscious in much the same way we are, they have the same moral status as we do.

From a certain perspective, this view may seem unreasonable. Would you really sacrifice a human life in order to save a few computer processes? But we can turn the question around by supposing that we're in a simulation. If we're in a simulation, would it be morally acceptable for our simulators to kill five of us in order to save one of them in the next universe up? From our perspective, I'd say the answer is no. Even if our simulators have the power to do this, this does not make it right. The same goes for our own actions toward conscious people in the simulations we create.

Someone might say that although consciousness matters for moral status, other factors matter, too. For example, maybe nonsimulated humans have higher moral status than simulated humans simply because they're in the top-level universe. Or maybe brief simulations count for less simply because they don't last for as long. I find these views somewhat implausible, though. And once again, considering them under the assumption that we're the ones in a simulation will bring out their downsides. Why should the fact that we're not in the top-level universe make killing us more morally acceptable?

I'm inclined to think that we have roughly the same moral obligations toward sims that we have toward nonsims. If it's not OK to do something to an ordinary human, it's also not OK to do the corresponding thing to a simulated human. Killing sims will be as bad as killing ordinary humans, and stealing from sims will be as bad as stealing from ordinary humans. Performing experiments on sims may invoke the same moral restrictions as experiments on humans. And so on.

All this concerns how we should behave toward sims who already exist. What about the question of whether or not it's morally permissible to create sims, and more generally to create whole simulations?

Let's start with extreme cases. Is it OK to create a simulation involving a million conscious beings all experiencing agony for their whole existence? I'd say it clearly is not. But is it OK to create a simulation involving a million conscious beings who lead largely happy, fulfilled lives? On the face of it, this seems acceptable.

Someone might object that even in the million-person happy simulations, we're "playing God." It's my opinion that we shouldn't create these simulations *lightly*. We should think hard about what we're doing. Society may want to place strong restrictions on who can create universe simulations and for what reasons.

At the same time, it's hard to see why creating a happy simulation would be *immoral*. After all, we create new conscious lives when we reproduce. Again, we shouldn't do it lightly. Philosophers have occasionally argued that any reproduction is immoral because all lives contain suffering, but only a small minority agree. If it's acceptable to bring an ordinary human child into existence, it should be acceptable to bring a simulation into existence, too.

Is it morally acceptable to create sims with considerable happiness and some suffering? A utilitarian might say this isn't the best action you could take, as it would be better to create a world with huge happiness and no suffering. The right thing to do is to produce the greatest good for the greatest number. So we should produce the biggest simulation we can, with sims who are only happy and never suffer. Perhaps the best simulations will allow some suffering, on the grounds that a small amount of suffering can make a life better. Still, from this perspective, a benevolent simulator will create only the best of all possible simulations.

Leibniz thought that this is what a God must do. A benevolent God will create only the best of all possible worlds; accordingly, our world is the best of all possible worlds. If it contains some evil, it's because this is necessary in order for the world to be as good as a world can be.

On the other hand, simulators can create multiple simulations. Suppose we've already created the best possible simulations. We now have the option of also creating the *second-best* possible simulations, which are a little less good than the best but remarkable overall. Should we create those? Someone might say no: It would be better to create a second copy of the best simulations. But it could be argued that a second identical copy doesn't have as much value as the first. Here's one way that could happen: For all we know, when two simulations are identical, they support just one population of conscious beings, so the second simulation is redundant. If so, it might be better to create the second-best simulation instead of another copy of the best. In any case, it seems at least *good* to create the second-best simulation, in that it adds a great deal of goodness to the world. According to this perspective, there may be a moral imperative to create as many simulations as possible with more happiness than suffering.

One can use this picture to generate a *simulation theodicy*. A theodicy is a theological explanation of why God might permit evil in the world. A simulation theodicy does the same for simulators. One simulation theodicy says that a good simulator should create all worlds with a balance of happiness over suffering. This will lead to worlds with considerable suffering, as long as it is outweighed by happiness. If the simulation hypothesis is true, perhaps this idea could even explain the existence of evil in our world. Or perhaps our simulators are not especially benevolent and have priorities other than the well-being of the creatures they create.

Another simulation theodicy says that simulators aren't omnipotent and so cannot predict everything that will happen in a simulation. If everything in a simulation could be predicted, there would be less reason to run it! So when we create simulated universes, we cannot expect to avoid the emergence of suffering entirely. Unpredictable evils will happen. Still, we may have some sense of which simulations tend to lead to good lives, and what sorts don't. If we do, then we should at least aim for the former, other things being equal.

Many ethical theories will say that the morality of creating a simulation may depend on our *reasons* for creating it. For example, Immanuel Kant's Principle of Humanity states that we should treat human beings not merely as means to an end but as ends in themselves. We should always recognize the humanity of others and take it into account when we act. Now, it's not obvious that Kant would extend this principle to nonhumans (let's say) who are intelligent and conscious, or to simulated humans, but it's natural to think he should. A Principle of Humanity restricted to humans would be speciesist. Sims are people too! What we really need is something like a Principle of Personhood: We should always treat persons not as means to an end but as ends in themselves.

Applying the Principle of Personhood to sims tells us we should never treat sims merely as means to an end but as ends in themselves. This suggests that it's morally wrong to create a simulated universe merely for entertainment, or merely to help us predict the future, or merely to benefit science. We must respect the personhood of the beings we create. It may be permissible to create simulated universes for these purposes, but only if this is consistent with respect for the sims. For example, just as we allow scientific experiments on human subjects when they don't harm the subjects, perhaps we can run simulations for science when doing so is good for the simulated subjects. Of course, this will make it harder to run simulations of unpleasant situations. Full-scale simulations of wars might be banned. To create a war only to benefit people in another universe cannot really be said to treat the participants as ends in themselves.

Will simulators actually obey these moral strictures? They may well not. Humanity has a long history of falling short of moral ideals. Sims may be so convenient and useful that they'll be exploited without a second thought to moral considerations. In *Black Mirror*, characters create sims in order to prepare their breakfast in the morning and in order to test potential romantic partners. It's easy to imagine that sims will be treated as disposable. It's also easy to imagine a struggle to grant sims the same rights as ordinary humans. I offer no predictions about which approach will win out in the end. But if Martin Luther King Jr. was right that the long arc of the moral universe bends toward justice, then the arc will bend toward granting equal rights to sims.

Chapter 19

How should we build a virtual society?

T RIGGER WARNING: THE FOLLOWING PARAGRAPHS DESCRIBE A virtual sexual assault in a text-based virtual world.

In 1993, the most popular virtual worlds for social interaction were MUDs, or multiuser domains. MUDs were text-based worlds with no graphics. Users navigated through a number of "rooms" with text commands and interacted with others there. One of the most popular MUDs was LambdaMOO, whose layout was based on a California mansion. One evening, a number of users were in the "living room" talking with one another. A user named Mr. Bungle suddenly deployed a "voodoo doll," a tool that produces text such as *John kicks Bill*, making users appear to perform actions. Mr. Bungle made one user appear to perform sexual and violent acts toward two others. These users were horrified and felt violated. Over the following days, there was much debate about how to respond within the virtual world, and eventually a "wizard" eliminated Mr. Bungle from the MUD.

Almost everyone agreed that Mr. Bungle had done something wrong. How should we understand this wrong? Someone who thinks virtual worlds are fictions might say that the experience is akin to reading a short story in which you are assaulted. That would still be a serious violation, but different in kind to a real assault. That's not how most of the MUD community understood it, however. The technology journalist Julian Dibbell reported a conversation with one of the victims recounting the assault:

Months later, the woman ... would confide to me that as she wrote those words posttraumatic tears were streaming down her face—a real-life fact that should suffice to prove that the words' emotional content was no mere fiction.

Virtual realism gives the same verdict. The assault in the MUD was no mere fictional event from which the user has distance. It was a real virtual assault that really happened to the victim.

Was Mr. Bungle's assault as bad as a corresponding sexual assault in the nonvirtual world? Perhaps not. If users in a MUD attach less importance to their virtual bodies than to their nonvirtual bodies, then the harm is correspondingly less. Still, as our relationships with our virtual bodies develop, the issue becomes more complex. In a long-term virtual world with an avatar in which one has been embodied for years, we may identify with our virtual bodies much more than in a short-term textual environment. The Australian philosopher Jessica Wolfendale has argued that this "avatar attachment" is morally significant. As the experience of our virtual bodies grows richer still, violations of our virtual bodies may at some point become as serious as violations of our physical bodies.

The Mr. Bungle case also raises important issues about the governance of virtual worlds. LambdaMOO was started in 1990 by Pavel Curtis, a software engineer at Xerox PARC in California. Curtis designed LambdaMOO to mimic the shape of his house, and initially it was a sort of dictatorship. After a while, he handed control to a group of "wizards"—programmers with special powers to control the software. At this point, it could be considered a sort of aristocracy. After the Mr. Bungle episode, the wizards decided they didn't want to make all the decisions about how LambdaMOO should be run, so they handed power to the users, who could vote on matters of importance. LambdaMOO was now a democracy of sorts. The wizards retained a degree of power, however, and after a while they decided that democracy wasn't working and they took some decision-making power back. Their decree was ratified by a democratic vote after the fact, but they had made it clear that the shift would be made regardless. The world of LambdaMOO moved fairly seamlessly through these different forms of government.

All this raises crucial issues about both the ethics and the politics of near-term virtual worlds. Ethically: How should users act in a virtual world? What's the difference between right and wrong in a virtual world? Politically: What are the ethical and political constraints on the creators of a virtual world? How should a virtual world be governed? What is justice in a virtual world? In what follows, I'll start with ethics—for both users and creators of virtual worlds—and then move to politics.

Ethics for users

Let's start with virtual worlds that exist already. Perhaps the simplest case is that of single-player video games. You might think that with nobody else involved, these games are free of ethical concerns, but ethical issues still sometimes arise. In his 2009 article "The Gamer's Dilemma," the philosopher Morgan Luck observes that while most people think that virtual murder (killing nonplayer characters) is morally permissible, they think that virtual pedophilia is not. The same goes for virtual sexual assault. In the 1982 Atari game *Custer's Revenge*, the objective was to sexually assault a Native American woman. Most people think that something is going wrong morally here.

This presents a philosophical puzzle. What is the relevant moral difference between virtual murder and virtual pedophilia? Neither act involves directly harming other people. If virtual pedophilia led to nonvirtual pedophilia, that would be a major harm, but it seems that the evidence for such transfer is weak.

It is not straightforward for moral theories to explain what is wrong here. One possible explanation invokes virtue ethics. We consider the kind of person who enjoys virtual pedophilia to be morally flawed, so engaging in virtual pedophilia is itself a morally flawed act. Perhaps the same goes for virtual sexual assault, torture, and racism. It is telling that many people have a similar moral reaction to the 2002 game *Ethnic Cleansing*, in which the protagonist is a white supremacist killing members of other races. By contrast, we don't think that "ordinary" virtual murder is indicative of a moral flaw, so we regard it as unproblematic. Still, the ethical issues here are subtle.

Once we move to multiuser video-game environments (such as *Fortnite*), and then to fully social virtual worlds (such as *Second Life*), the ethical issues multiply. If these virtual worlds were merely games or fictions, then the ethics of virtual worlds would be limited to the ethics of games or fictions. People could wrong each other in the ways they do when playing games, but not in the richer ways that they do in ordinary life. Once one sees virtual worlds as genuine realities, however, then the ethics of virtual worlds becomes in principle as serious as ethics in general.

In many multiplayer game worlds, there are "griefers"—bad-faith players who delight in harassing other players, stealing their possessions, and harming or even killing them within the game world. This behavior is widely regarded as wrong insofar as it interferes with other users' enjoyment of the game. But is stealing someone's possessions in a game as wrong as doing so in real life? Most of us would agree that objects owned in a game matter less than possessions in the nonvirtual world. Still, in long-term games, and all the more in nongame environments, possessions can be important to a user, and the harm can be correspondingly significant. In 2012, the Dutch Supreme Court upheld the conviction of two teenagers for stealing a virtual amulet from another teenager in the online game *Runescape*. The court declared that the amulet had real value in virtue of the time and effort invested in obtaining it.

Virtual theft is hard to explain if virtual objects are fictional. How can you "steal" an object that doesn't exist? The virtual fictionalist philosophers Nathan Wildman and Neil McDonnell have called this the *puzzle of virtual theft*. They suggest that virtual objects are fictions that cannot be stolen—these cases involve the theft of digital objects but not virtual objects. In the *Runescape* case, a digital object was stolen but no virtual object was stolen. Virtual realism gives a much more natural explanation. Virtual theft deprives someone else of a real and valuable virtual object. In this way, virtual theft provides further support for virtual realism.

What about murder in virtual worlds? Because there's no genuine death in near-term virtual worlds, there is not much room for genuine murder. A user could induce a heart attack in another user's physical body by saying something, or could induce others to commit suicide in the physical world. These acts in a virtual world are as morally serious as the same sort of act in a nonvirtual world. Short of these cases, the nearest thing to murder is "killing" an avatar. But this doesn't kill the person who inhabited the avatar. At worst, it removes the person from the virtual world, an act more akin to banishment. Killing an avatar might be more akin to murder followed by reincarnation, at least if reincarnation produces full-grown people with memories intact. It might also be akin to destroying a *persona*: perhaps eliminating the Iron Man persona while Tony Stark still lives. Those are all morally serious actions, even if they're not as serious as murder in the ordinary world.

How should wrong actions in virtual worlds be punished? Death is not an option. Banishment is an option, but it may not count for much. Mr. Bungle was banished from LambdaMOO, but soon afterward the same user returned as Dr. Jest. Virtual penalties and virtual imprisonment likewise may have some effect, but the effects will be limited when users can easily take on new bodies. Nonvirtual punishment (from fines to imprisonment to death) may in principle be an option, but with anonymous users this may be hard to arrange. As virtual worlds become more central to our lives, and virtual crimes take on increasing seriousness, we may well find that it becomes difficult to find punishments that fit the crime.

Ethics for creators

Many ethical issues for the creators of virtual worlds arise even for single-user environments. The creators of the *Grand Theft Auto* video games were criticized for glamorizing violence, sadism, and sexist treatment of women. The major objection is that the video game may encourage violence and sexism in the nonvirtual world. The UC San Diego philosopher Monique Wonderly has argued that these games tend to decrease users' capacity for empathy, which diminishes their capacity for moral judgment. There is some experimental support for the idea that behavior in a virtual world can transfer to the nonvirtual world. For example, the psychologists Robin S. Rosenberg and Jeremy Bailenson have found that when subjects inhabit virtual reality as a superhero, they tend to behave more altruistically afterward, whereas when they play as a villain, the reverse is the case.

Many have suggested that virtual reality can increase empathy toward others. When a virtual world puts you in the situation of a refugee, for example, it can give you a visceral sense of the refugee experience.

Researchers have also used VR to illustrate moral dilemmas. The VR researcher Mel Slater has devised a VR analog of Stanley Milgram's famous 1963 experiment on following orders. Milgram told his experimental subjects (the "teachers") to inflict increasing amounts of pain on other subjects (the "learners") when they failed to answer questions correctly. The learners were in fact actors. They cried out in pain, but the experiment's subjects were told to continue, and many continued to extreme levels, even when it sounded as if the "learners" were dying. In Slater's version, the learner is merely a nonplayer character in VR, and the experimental subject knows this. Slater finds the same sort of results that Milgram obtained, with many subjects continuing to do as they are told, even when the virtual character appears to be suffering greatly. The subjects also become anxious and uncomfortable, as Milgram's subjects had, with their heart rates increasing and their palms sweating.

The philosophers Erick Ramirez and Scott LaBarge (who themselves have devised VR versions of the trolley problem and the experience machine) have proposed that experiments like this should be severely restricted because they can be harmful to subjects in the same way that nonvirtual analogs would be. They suggest an *Equivalence Principle*: "If it would be wrong to allow subjects to have a certain experience in reality, then it would be wrong to allow subjects to have that experience in a virtually real setting." Even if the experimental subjects know that the other "subject" is a nonplayer character so they're not really inflicting pain, the experience can be harmful all the same. (Slater disagrees, saying participants know the pain isn't real and aren't harmed.) Similarly, if it would be wrong to scare people by dangling them over a cliff, it would also be wrong to scare them by dangling them over a virtual cliff, even though they know intellectually that they're in no danger. The experience of fear can be harmful in itself.

The philosophers Michael Madary and Thomas Metzinger have laid out a number of related ethical guidelines for researchers creating virtual reality environments. They recommend that VR experiments involving "foreseeable . . . serious or lasting harm to a subject" should be barred, and they suggest that subjects should always be informed of potential effects. They also suggest caution in using VR for medical purposes.

Once we move to multiuser virtual worlds, the complex ethical issues for creators blend into social and political issues. Should Metaverse-style virtual worlds be created at all, given the resources they will consume and the effects they will have on people's lives? How should they be organized and governed?

Government in virtual worlds

Who should have ultimate authority about what happens in a virtual world? Should there be laws in a virtual world, and if so, what should they be? How can a virtual world be a truly fair and just place for its inhabitants?

These questions mirror some of the central questions of political philosophy. How should society be run? There have been any number of answers.

The simplest answer is *anarchy*. There is no government, and there are no laws. The idea goes back to the ancient Chinese philosopher Mozi, who wrote, "In the beginning of human life, when there was yet

Figure 47 Thomas Hobbes encounters a
social contract in a virtual world.

no law and government, the custom was 'everybody according to his rule'"—a situation he decried. The 17th-century British philosopher Thomas Hobbes called this the *state of nature*. Like Mozi, Hobbes described the state of nature as not at all pleasant. There would be an endless war of all against all, with feuding groups and temporary rule. Life would be "solitary, poor, nasty, brutish, and short."

Hobbes said that because the state of nature is so awful, people will enter into a *social contract*, deciding to obey a common authority and instituting some form of government and laws. The government arising from the social contract may not be perfect, but at least it avoids the worst excesses of anarchy.

Many theorists have been suspicious of the social contract model. In real life, few people explicitly agree to the laws of their country. Most people have little choice. How can a fictional contract justify real government? Interestingly, social contract theories seem more apt for virtual worlds. Users of *Second Life* and *Minecraft* have to agree to terms and conditions before entering. Users have some choice about which worlds to enter. If so, social contract theories may provide a reasonable starting point for thinking about government in virtual worlds.

Some traditional forms of government include dictatorship (rule by an individual), monarchy (rule by a royal family), aristocracy (rule by elite citizens or the nobility), oligarchy (rule by a small group of the powerful), and theocracy (rule by religious leaders). The dominant form of government in Western countries in recent centuries is *democracy*: rule by the people. Democracy takes different forms. There are direct democracies, in which people vote directly on policies, and (far more common) representative democracies, in which people elect representatives. Democracies don't necessarily involve rule by *all* the people. In American democracy, women and slaves did not have voting rights until long after independence, and it remains common for prisoners and children to be excluded from voting rolls.

What form of government is most apt for a virtual world? All of the forms just outlined are available. Some video games are effectively dictatorships run by an all-powerful designer. The most important rules are built into the software, and users are happy to enter the social contract that the game designer imposes.

These days most virtual worlds are *corporatocracies*, run by corporations that own them. *Second Life* is run by the corporation of Linden Lab as a sort of corporate government. Governance is supposed to be benevolent and minimal. Users are largely free to run their own lives, within limits: for example, explicit sex acts and gun battles are allowed only in certain areas.

From time to time, political crises have arisen within *Second Life*. Violent groups took over in certain areas, and vigilante groups arose in response. Linden Lab was called to impose order, but the imposed solutions often led to new problems. Users started an online newspaper (the

Alphaville Herald) in part to protest the way that the world was being run, and on occasion the corporation would shut the press down. Some areas within *Second Life* were set up on a democratic system with a Representative Assembly within the areas, though ultimate control still rested with the corporation. Some users were moved to set up *Open-Sim*, a virtual world much like *Second Life* but run democratically by users. The great majority of users stayed in the corporate-run *Second Life*, but there was clearly discomfort about corporate government in a largely autonomous social realm.

If all virtual worlds were mere fictions, perhaps this structure would be appropriate. The creator of a fiction presumably has some authority over what happens there. But when we recognize virtual worlds as realities in their own right, the social and political issues that arise in ordinary reality will also arise in VR.

The current governance structure of virtual worlds arises from treating them as a form of entertainment. Virtual worlds like those of *Fortnite* and *Minecraft* are somewhat akin to theme parks like Disneyland. These entertainments are subject to the laws of their corporation's country, but the corporation has a large degree of autonomy in imposing rules and regulations. Users who don't like a corporation's actions in a theme park or a video game have only limited options.

A move toward democracy has been made in the virtual world of *EVE Online*, the popular massive multiplayer space game. *EVE* is already a politically complex game, hosting many competing alliances with rich social structures. An *EVE* document titled "A comparative analysis of real structural social evolution with the virtual society of EVE Online" outlines how *EVE* moved from a tribal society through stratified corporate structures led by chieftains to a civilization in which the different structures share power in a government. As of this writing, *EVE* styles itself as a deliberative democracy, with an elected "Council of Stellar Management" that meets twice each year in Iceland at the headquarters of the game company, CCP Games. CCP still has ultimate control, but the council "advises and assists" it, and the advice is taken seriously.

Things get more complex in virtual worlds where users live their lives. In social worlds such as *Second Life*, many of the issues parallel those that arise in social networks such as Facebook. Are the corporations that run virtual worlds manipulating our behavior? Are they invading our privacy? Do they lead to racism and sexism? Do they foster addiction and isolation from the nonvirtual world? Do they consume too many resources? Should they police the behavior of users? Should users have a say in how the worlds are run? Can the corporations sell information about our lives?

In a social virtual world, users can reasonably demand a degree of autonomy. They can also reasonably demand a degree of privacy, which isn't easy when everything that goes on in a virtual world is, in principle, available for inspection by the owner of that world. Eventually, users may demand a degree of political power in helping to shape how virtual societies are organized.

In this process, people once regarded as customers may come to regard themselves as citizens. One can envisage a clamor for liberty, for equality, and for community arising in virtual worlds. One can envisage revolutionary attempts to replace corporate dictatorships by changing the governance structures in existing virtual worlds or by founding new virtual worlds. One distinctive feature of virtual worlds is that they are relatively easy to enter and exit. People will be able to move between virtual worlds to find a world in which they can thrive. The outcome may be a vast range of virtual worlds run according to different principles for different communities.

Equality and justice in virtual worlds

Many political issues in virtual worlds parallel political issues in society more broadly: What sort of democracy is appropriate in a virtual world? What distribution of resources is appropriate? What sort of property ownership is appropriate? What sort of punishment is appropriate? Should virtual worlds have open borders or should immigration be controlled?

Here I'll just address one central issue in the political philosophy of virtual worlds, concerning equality and justice. This is a place where virtual worlds raise distinctive issues and where virtual realism makes a difference.

The most influential work of political philosophy in the last century is John Rawls's 1971 book *A Theory of Justice*. Rawls focused especially on *distributive justice:* the just distribution of resources among a population. He invoked a thought experiment in which we are all behind a veil of ignorance before our lives on Earth, knowing little about the life ahead of us and not knowing whether we'd be rich or poor. (Imagine souls congregated in a pre-birth limbo, as in the 2020 Pixar movie *Soul*, deliberating about how to organize society.) Rawls argued that in this "original position," everyone should choose a distribution wherein the worst-off are as well off as possible. He used the thought experiment to argue that we should adopt such a balanced distribution of resources in the real world.

How do equality and distributive justice play out in virtual worlds? Does VR change anything fundamental? One major change is that virtual worlds may remove scarcity of many material goods. Space is not at a premium in VR. Everyone can have a personal idyllic virtual island if they choose to. Construction is easy, too. Once a house has been built, it can be duplicated elsewhere at little cost. Anyone can have a large virtual home in a wonderful location. The result may be *virtual abundance.*

In the short term, while virtual worlds are inferior to the nonvirtual world, virtual abundance may have at most a small impact on our lives. But if virtual realism is correct, life in virtual worlds in the long term may approach or exceed the quality of life in nonvirtual worlds. Eventually, a virtual home may be as good or better than a nonvirtual home. In principle, virtual islands are on a par with nonvirtual islands, and virtual clothing is as effective as nonvirtual clothing. As a result, virtual abundance has the potential to eliminate a great deal of distributive injustice.

Following David Hume, Rawls said that *scarcity is a condition for justice*. This means that without scarcity, the principles of justice do not apply. In conditions of abundance, there is no need for justice. The world might have other problems, but at least where considerations of distributive justice are concerned, a world with abundance has no flaws that need correcting.

That raises the intriguing possibility that in the long term, virtual abundance could yield a sort of utopia, at least where distributive justice is concerned. Under virtual abundance, important material goods in virtual worlds are instantly reduplicable and available to all. This is a virtual version of what is sometimes called a post-scarcity society.

We can situate our thought experiment in the relatively far future, in a nonvirtual world that has harnessed the power of the Sun for effectively unlimited energy. To avoid worries about the nonvirtual body, we can suppose that people have freely chosen to upload themselves to virtual worlds. (People who reject virtual realism are free to stay in the nonvirtual world.) To ensure that services are as abundant as material goods, I'll suppose that ultracompetent AI systems serve as doctors and teachers and cleaners. The AI systems are nonconscious (to avoid moral issues) and easily duplicated.

One could try to exploit virtual abundance in a market-based society. John Carmack, the cocreator of *Doom* and former chief technology officer at Oculus, has said "Economically, you can deliver a lot more value to a lot of people in the virtual sense." According to one scenario, outlined in a recent article in *Wired* magazine, corporations will compete to sell people low-cost "mansions at the beach" in virtual worlds. The article paints this picture as a dystopia, but if virtual realism is correct, life in virtual worlds may eventually be better than life outside them. Still, the capitalist version is unlikely to be a post-scarcity utopia. In a market-based system there will always be a premium for the latest virtual world or the latest AI system. That requires artificial scarcity wherein the distribution of goods is restricted. Furthermore, once AI systems have largely eliminated human employment income and concentrated wealth in corporations, then scarcity threatens again. How will unemployed people pay for their virtual worlds? At a minimum, some sort of universal basic income will be required. Even then, it is hard to see how it can be stable and just to concentrate most wealth in corporations, especially once humans are no longer driving most innovation.

It's somewhat easier to see how virtual abundance could work if a state rather than a corporation is responsible for the virtual worlds. The state can ensure that everyone has enough income to live a good life in a post-scarcity virtual world. Innovations will be made available to all. It is no accident that Karl Marx's vision of an ideal society required abundance rather than scarcity. One can raise any number of problems for this virtual abundance scenario (Will essential human values be lost? Will freedom be compromised? Will the system be stable?), but here I am concerned mainly with its impact on equality.

We should not expect abundance of goods and services to lead straightforwardly to an egalitarian utopia. For a start, there are many "positional" goods that depend on one's position in the world and are scarce by their very nature. For example, fame is a positional good: not everyone can be famous. The same is true for power. Abundance of material goods in a virtual world cannot ensure abundance of these positional goods, and these goods may take on even more significance in a virtual world. If some groups have far more political power than other groups, a world with virtual abundance will not be a truly egalitarian paradise.

More fundamentally, while virtual abundance may remove some distributive injustice, there is much more to equality than distributive justice. In her important 1999 article "What Is the Point of Equality?," the American philosopher Elizabeth Anderson argues for a *relational* view of equality, in which what matters most for equality is social relations between people, including power, domination, and oppression. It's oppression in particular that has driven the great movements for racial and gender equality. Even if goods and services are equally distributed, society cannot be egalitarian if significant oppression remains.

It's easy to imagine that current sources of oppression will carry over into virtual worlds. Access to virtual worlds may be much smoother for some groups than others. Oppression based on race, gender, class, and ethnic and national identity runs deep. Virtual worlds may complicate these identities with new forms of embodiment, but they do not remove the underlying sources of the oppression. They may also introduce new forms of oppression. AI systems may initially be oppressed by humans, and perhaps may eventually oppress them in turn. It's easy to imagine that people in virtual worlds may be dominated by people in the non-virtual worlds that contain them. One can also imagine that as virtual worlds become more desirable, people in the nonvirtual world will be regarded as second-class citizens.

These many different sources of oppression can intersect. The American legal theorist Kimberlé Crenshaw coined the term *intersectionality* for the way in which multiple identities can interact: A Black woman's experience of oppression is not a mere sum of oppression for being Black and for being a woman. Likewise, a lower-class person in a virtual world may be oppressed in ways that don't derive just from being lower class or from being virtual, but from their intersection. AI systems in virtual worlds may be oppressed in ways quite different from AI systems in nonvirtual worlds. With so many different intersections, the forms of oppression may multiply.

Genuine equality requires removing these oppressive relations among beings of all sorts. A transition to abundant virtual worlds will not be enough to ensure this. Indeed, it may introduce new forms of inequality. As a result, we can't expect virtual worlds to easily become egalitarian utopias. Still, virtual realism allows us to see how virtual worlds are potentially transformative for at least some aspects of equality.

In thinking about the future, thought experiments take us only so far. The actual future is likely to evolve in wildly different ways we haven't yet anticipated. But if our world transitions to one with VR and AI at the center, we can reasonably expect the transition to restructure society. This will certainly lead to political upheaval, and perhaps to political revolution.

Part 7

FOUNDATIONS

Chapter 20

What do our words mean in virtual worlds?

A CLASSIC SLOGAN ABOUT SIMULATIONS, DUE ORIGINALLY TO the philosopher Daniel Dennett, says: *A simulated hurricane doesn't make you wet!* In 2005, Hurricane Katrina devastated New Orleans. A simulation of Hurricane Katrina, however, doesn't hurt anybody. It leaves you bone dry.

Dennett says that expecting to get wind-blown by a simulated hurricane would be like cowering before the word "lion." Simulated hurricanes deal with descriptions of hurricanes, not with real hurricanes.

The 1981 dialogue "A Coffeehouse Conversation on the Turing Test," by my PhD advisor Douglas Hofstadter, contains a rebuttal to Dennett's slogan and an early statement of simulation realism. After one character makes Dennett's point, another character ("Sandy," a philosophy student), says:

> [Y]our argument that a simulated McCoy isn't the real McCoy is fallacious. It depends on the tacit assumption that any old observer of the simulated phenomenon is equally able to assess what's going on. But in fact, it may take an observer with a special vantage point to recognize what is going on. In the hurricane case, it takes special "computational glasses" to see the rain and the winds. [. . .] [T]o see the winds and the wetness of the hurricane, you have to be able to look at it in the proper way.

Hofstadter's insight is that whether or not we recognize a simulated hurricane as a hurricane depends on our perspective. In particular, it

Figure 48 Daniel Dennett encounters a simulated hurricane.

depends on whether we're experiencing the simulated hurricane from inside or outside the simulation.

So Dennett's slogan is at best half-true. If we're *outside* a simulation, then a simulated hurricane won't get us wet. At most, it will affect some simulated entities and other processes in a computer. But suppose we're *inside* a simulation and have been for our whole lives. In this case, a simulated hurricane can certainly get us wet. If we're in a lifelong simulation, then all the hurricanes we've ever experienced have been digital hurricanes. Even Hurricane Katrina was a digital hurricane. It did enormous damage all the same.

In the long-running BBC science-fiction series *Doctor Who*, the

Doctor travels the universe in a TARDIS (which stands for Time and Relative Dimensions in Space). From the outside, the TARDIS looks like a London police box. From the inside, it's an enormous spacecraft with a giant control room, surrounded by endless further rooms and passages. There's a running joke in the series that the TARDIS is "bigger on the inside."

Like the TARDIS, simulations are bigger on the inside. If I look at Sim Universe from the outside, it doesn't look impressive. All I'll see is a computer—perhaps with some people hooked up to it, depending on how the simulation is arranged. The computer might be running on a device as small as a smartphone. When I look at Sim Universe from the inside, however, it's enormous. I'll experience an immersive environment with all sorts of varied content. Like the TARDIS, it might go on forever. From the inside, Sim Universe is a whole world.

Much about simulations depends on whether you consider them from the inside or the outside. I've argued that if I'm *in* Sim Universe—that is, if it's been my lifelong environment—then the objects in it are completely real. It's a giant world of trees and mountains and animals. But if I didn't grow up in a simulation, then Sim Universe does *not* contain real trees or mountains or animals; it contains simulated trees and simulated mountains and simulated animals. The simulated trees may be real digital objects inside the computer, but they aren't real trees.

How can this be? The simulation itself is part of objective reality. How could its nature depend on me? How could whether it contains trees or mountains depend on my perspective?

My answer is that the difference between a simulation-from-the-inside and a simulation-from-the-outside is not a difference in reality. It is a difference in language, along with associated differences in thought and perception. If I grew up inside Sim Universe, then I've been applying the word "tree" to digital trees my whole life. Digital trees are what I *mean* by "tree." If I grew up outside all simulations, then I've been applying the word "tree" to nondigital trees my whole life. Nondigital trees are what I mean by "tree."

So I'll describe Sim Universe differently depending on whether I've grown up inside it or not. If I've grown up in Sim Universe, I'll say it contains trees, because "tree" for me means "digital tree." If I've grown up outside all simulations, I'll say that Sim Universe doesn't contain trees, since "tree" for me means "nondigital tree."

In objective reality, Sim Universe is not affected by our perspective. It contains digital processes running on a computer. These processes involve objective algorithms supporting objective digital objects. What varies depending on the perspective is how we *experience* things and how we *describe* them. We can understand this idea better by thinking a little about language.

Philosophy of language

There are many traditions in philosophy. In this book, I have most often followed the European tradition that was passed down from the ancient Greeks and Romans to the medievals and then to 17th- and 18th-century figures we have encountered such as Descartes and Kant.

In the 19th and especially the 20th century, the European philosophical tradition split in two. One branch has come to be known as continental philosophy because of an early association with the European continent. Its key figures include the German philosophers Hannah Arendt, Martin Heidegger, and Edmund Husserl, and the French philosophers Simone de Beauvoir, Maurice Merleau-Ponty, and Jean-Paul Sartre. The other branch has come to be known as analytic philosophy, originally because of its use of linguistic analysis. Its key early figures include British philosophers we've encountered already—Bertrand Russell and G. E. Moore—as well as Germans and Austrians such as Rudolf Carnap, Ludwig Wittgenstein, and Gottlob Frege.

Much of the philosophy in this book falls under the rubric of analytic philosophy. One feature of analytic philosophy, especially in the early days, is an intense focus on logic and language. The analytic philosophers of the Vienna Circle (see chapter 4) held that once a philosophical problem had been clarified enough via logic and language, this would either dissolve the problem or break it down enough that it could be settled by science. A century later, analytic philosophy has become a much broader church, but a concern for clarity and a focus on logic and language remain among its distinctive elements.

Perhaps the founding figure in analytic philosophy was the German philosopher Gottlob Frege, who founded the field of logic as we know it today in the late 19th century. Outside philosophy, he is perhaps best known for developing a theory of the foundations of mathematics—a theory that turned out to be inconsistent when Russell pointed out that it led to a paradox about "the set of all sets that do not contain themselves." (Would that set contain itself? The answer can't be "yes" and it can't be "no.") Nevertheless, Frege's theory was a monumental achievement, as was his clarification of the tools of modern logic. Frege was also a pioneer in the philosophy of language, setting out one of the first major theories of what our words mean.

Sadly, Frege was seriously anti-Semitic, as was Martin Heidegger decades later. Like great artists, great philosophers are not always great people. Aristotle and Immanuel Kant have writings that are laden with a now-shocking racism. We can try to separate their core philosophy from their awful views. This is not always straightforward, but in Frege's case, it's arguable that his philosophy of logic and language has little connection to his anti-Semitism.

Frege's best-known contribution to the philosophy of language was to distinguish two aspects of what a word means: its *sense* and its *reference*. Reference is easiest to explain. The referent of a word is what it refers to in the world. "Plato" refers to Plato (the person). "Sydney" refers to Sydney (the city). "Groundhog" refers to groundhogs (the animals). "Seventeen" refers to 17 (the number). And so on.

Sometimes two words refer to the same thing. The classic example is the names "Hesperus" and "Phosphorus" for the evening and morning star, respectively. Both refer to the same object, the planet Venus. But they seem to have a different meaning. In his 1892 article "On Sense and Reference," Frege used this example to argue that there's more to meaning than reference. Although "Hesperus" and "Phosphorus" refer to the same thing, they have different *senses*. The sense of a term is roughly the way it presents the referent to the speaker. "Hesperus" presents Venus as the evening star, so its sense is tied to being visible in the evening. "Phosphorus" presents Venus as the morning star, so its sense is tied to being visible in the morning.

Later, Russell offered a twist on Frege's idea with a suggestive picture of just how words can refer to things in the world. In his landmark theory of names and descriptions, he argued that every ordinary name (such as "Hesperus") is equivalent to a description—say, "the star that is visible in the evening at a certain location." This description would denote whatever object satisfied it—in this case, the planet Venus. His theory allowed an analysis of ordinary language using the tools of logic.

The Frege-Russell theory of meaning was popular for many years, but in the 1970s there was a small revolution. Two American philosophers, Saul Kripke and Hilary Putnam, building on earlier work by the philosopher and logician Ruth Barcan Marcus, argued that the Frege-Russell picture builds on numerous false assumptions. In Kripke's book *Naming and Necessity*, his main target was *descriptivism*—the idea that the meaning of a word is something akin to a description. In Putnam's article "The Meaning of 'Meaning,'" the main target was *internalism*—the idea that the meaning of a word is internal to the speaker and does not involve the speaker's environment.

Putnam's famous slogan was "Meanings just ain't in the head!" In their theories of meaning, he and Kripke favored *externalism*, which says that the meaning of a word depends partly on the speaker's environment. They replaced Russell's descriptivism with the *causal theory of reference*. Putnam's version of the causal theory says roughly that a word refers to whatever entity in the environment causes the word to be used.

Putnam argued for externalism using a thought experiment: the story of Twin Earth. Twin Earth is a faraway planet that's just like Earth, except that all the H_2O on Earth is replaced by a superficially identical substance, XYZ (a molecular structure that's different from H_2O). XYZ looks and tastes just like water. XYZ falls from the skies, it fills the rivers and oceans, it runs through pipes and comes out of faucets, and all the creatures on Twin Earth drink it.

Is XYZ water? Putnam makes a strong case that it isn't. Water is H_2O, a natural substance found on Earth. XYZ is a different substance,

HYPATIA ON EARTH　　　　　　　HYPATIA ON TWIN EARTH

Figure 49 Do Hypatia (studying H_2O) and Twin Hypatia (studying XYZ) mean different things by "water"?

with a similar appearance. We don't call fool's gold "gold," even though it resembles gold. Similarly, we shouldn't call XYZ "water." Earth is largely covered with water, but Twin Earth isn't. It's largely covered with what we might call twin water.

On Twin Earth, there are language users very much like on Earth. Consider Hypatia, the brilliant 4th-century Alexandrian philosopher and mathematician who built hydrometers for measuring the specific gravity of water and other liquids. Hypatia has a near-duplicate on Twin Earth, who is studying XYZ where Hypatia is studying H_2O. Let's suppose the two liquids have behaved just the same in all experiments so far, and no one has discovered the chemical makeup of these liquids yet. Both Hypatia and Twin Hypatia call their liquids "water." Suppose that Hypatia says, "I'm measuring water." She's talking about H_2O. But when Twin Hypatia says, "I'm measuring water," she's talking about XYZ.

This is enough for Putnam to mount his argument that meaning isn't "in the head." Hypatia and Twin Hypatia are near-duplicates of each other, but their words mean different things. Putnam further argues that this was so even before anyone discovered the chemical makeup of water and twin water. It follows that the meaning of a word like "water" doesn't depend just on what's intrinsic to the speaker but also on the speaker's environment.

One way to think about this is that for both Hypatia and Twin Hypatia, "water" picks out whatever plays the water role in their environment: roughly, it picks out whatever is the clear liquid found in the oceans and lakes that people drink and bathe in. For Hypatia, H_2O plays that role, so "water" refers to H_2O. For Twin Hypatia, XYZ plays that role, so "water" refers to XYZ.

You can construct Twin Earth cases like this for all sorts of words. There's a Twin Earth with no trees but with non-DNA-based counterparts of trees. When my twin says "tree," he refers not to trees but to these counterparts that play the tree role on Twin Earth. There's a Twin Earth with a robot counterpart of Obama who plays the Obama role on Twin Earth. When my twin says "Obama," he refers to this counterpart and not to Obama himself. And so on.

For all these words, it looks like their meaning is not in the head. We might call these externalist words: their meaning is anchored to certain things in their environment. Hypatia's word "water" is anchored to H_2O, and Twin Hypatia's is anchored to XYZ.

There are some limits to externalism. One limit arises from logic and mathematics. When my twin on Twin Earth says "seven," he will refer to the number 7. So perhaps the meaning of "seven" is "in the head." The same might go for a term like "and." Logical and mathematical words like these don't need to be anchored in the environment. They might be seen as internalist words.

Externalism also doesn't seem so apt for words like "consciousness," "causation," and "computer." These words aren't anchored in specific things in my environment. I have a general conception of what sort of thing a computer is, in broadly structural terms. Anything I count as a computer will also count as a computer for my twin on Twin Earth. Even if computers on Twin Earth are made of graphene where Earth computers are made of silicon, both count as computers all the same. As a result, my twin plausibly means the same thing by "computer" as I do. That suggests that "computer" is an internalist word.

In my own work on this topic, I've argued for a *two-dimensional* view of meaning, with both internal and external aspects of meaning.

It is roughly as if Frege and Russell were right about the internal dimension of meaning, and Kripke and Putnam were right about the external dimension. For the purposes of this chapter, though, we mainly need the external dimension. What matters is that we can construct Twin Earth cases for many ordinary words. Putnam and Kripke have convinced most philosophers that at least this much externalism about meaning is correct.

Twin Earth and Sim Earth

Putnam's Twin Earth provides a great model for thinking about language use inside and outside a simulation. Putnam himself used it to think about the brain-in-a-vat scenario, as I'll discuss later in this chapter. Here I'll use it to think about simulations.

Here's the idea. Suppose there's an original unsimulated Earth as well as a cosmic simulation containing a simulated Earth. Language use on Earth and Sim Earth is a lot like language use on Earth and Twin Earth.

For someone who has grown up on Earth, the word "hurricane" refers to unsimulated hurricanes: storms made up of giant, fast-moving patterns of air and water, which themselves are made of atoms at a deeper level. This fits our intuitive sense of the meaning of "hurricane." It also fits with the causal theory of reference. When our community uses the word "hurricane," this is triggered by unsimulated hurricanes in our environment.

For someone who has grown up on Sim Earth, the word "hurricane" refers to virtual hurricanes: simulated storms made up of patterns of simulated air and water. Virtual hurricanes have played the hurricane role in Sim Earth all along. Following the virtual digitalism I argued for earlier, at a deeper level virtual hurricanes are made of bits. The causal theory of reference helps explain how the word "hurricane" works: When members of a simulated community use the word, it's triggered by virtual hurricanes in their virtual environment.

The same goes for "water": It refers to water on Earth and to virtual water on Sim Earth. H_2O (a chemical kind) plays the water role on Earth. Virtual water (a digital kind) plays the water role on Sim Earth. The same goes for wetness: It refers to wetness on Earth and to virtual wetness on Sim Earth. And so on.

Now we can analyze Dennett's objection that simulated hurricanes don't make you wet. If we're on Earth, hurricanes certainly make you wet. For Earthlings, the words "hurricane," "water," and "wet" refer to nondigital things. Simulated hurricanes contain only virtual water, which doesn't make anything wet—although it makes virtual things virtually wet!

If we're on Sim Earth, then "hurricane," "water," and "wetness" refer to digital entities: entities that people on Earth call virtual hurricanes, virtual water, and virtual wetness. If people on Sim Earth say, "A simulated hurricane doesn't make you wet," they're saying that a virtual hurricane doesn't make anyone virtually wet. That's a false statement. Virtual hurricanes make them virtually wet. If we *are* inside a simulation, then our hurricanes are simulated hurricanes, and they make us wet.

You might object that a creature in a simulation would have many false beliefs. For example, a sim might think, "I'm in New York," when in fact the simulation is running on a server in Silicon Valley. Is the sim's belief false this time? No! When the sim says, "New York," the name doesn't refer to the unsimulated New York on Earth. It refers to a place on Sim Earth: Sim New York. The sim is indeed in Sim New York. Or at least he's virtually in Sim New York while he's physically in Silicon Valley. And when a sim says "in," that word means "virtually in," which means that the sim's virtual body is in the virtual location. So when the sim thinks "I am in New York," this means that the sim is virtually in Sim New York, which is true.

Traveling between Earth and Sim Earth

What happens to language when people move back and forth between simulated and nonsimulated environments? Much depends on whether people move knowingly or unknowingly. If they know their environment has changed, the meanings of their words may instantly change. If they don't know, the meanings may change more slowly.

Let's see how it works for Twin Earth, starting with unknowing subjects. Suppose astronauts from Earth land their capsule in the ocean on Twin Earth. They have no idea that the ocean is made of XYZ. They say, "Hey! There's water here!" Are they right or wrong? Putnam thought that the astronauts would be wrong. Their word "water" refers to H_2O, and there's no H_2O on Twin Earth. The word "water" hasn't suddenly changed its meaning just because they've encountered some XYZ.

By analogy: Let's suppose that some unsimulated people from Earth unknowingly enter the Sim Earth simulation. Perhaps they'd been traveling in Africa, on safari with a tour group, and the company running the tour decided it could save money by putting the tourists in Sim Earth's Africa instead. The tourists don't catch on. When they see a herd of giraffes, they say, "Hey! There are giraffes over there!" Are they right or wrong? Following the astronauts' case, we'd have to say they're wrong. Their word "giraffe" refers to biological giraffes, but what they're seeing are digital giraffes.

Now suppose the tourists stay in Sim Earth for years. They still don't know it's digital (the tour company is monstrously unethical), but they like new places, and simulation technology has gotten so good they can't tell they're not on Earth anymore. At some point, they'll have encountered more virtual giraffes than the biological ones they may have seen in Earth's zoos. By this time, the causal theory of reference suggests that the word "giraffe," for them, will come to include virtual giraffes as at least part of its meaning. In effect, that meaning will slowly shift to include digital giraffes, and when they say, "There are giraffes here," they'll be right.

There are trickier cases. For example, what if someone from Sim Earth escapes the simulation without realizing it, and encounters a bio-tree for the first time. Is she right when she says, "There's a tree"? Intuitions differ here. If digital trees are based on bio-trees and not vice versa, there's a case that bio-trees are part of what caused her use of the word "trees" in the first place. Still, I'm most inclined to think that her "tree" refers to digital trees (which she has interacted with directly) and not bio-trees. So she's wrong.

Much more realistic are cases in which people move knowingly between the nonvirtual and virtual worlds. This is already happening every day to users of video games and other virtual environments. We enter the world of *Grand Theft Auto* and we talk of stealing cars. However, the virtual world of *Grand Theft Auto* doesn't contain any real cars. It contains only virtual cars. When we say, "There's a car over there" while we're in the game, are we saying something false?

My view is that language is a malleable instrument that bends to our purposes. If we want to extend use of the word "car" so it applies to virtual cars, we can do that. Philosophers and linguists have long recognized that what a word refers to can be *context-dependent*. "Tall" means one thing when we're talking about basketball players (a six-footer isn't tall) and another thing when talking about academics (a six-foot philosopher is tall). Words adapt to the role that's useful in their context.

With the onset of virtual worlds, much of our ordinary language has become context-dependent in this way. When I say, "There's a car" in an ordinary, nonvirtual context, I mean it's an ordinary, nonvirtual car. But when I say, "There's a car" in a virtual context, I use "car" in a way that includes virtual cars.

It's also possible that the meaning of "car" could shift so that it covers both nonvirtual cars and virtual cars by default. In chapter 10, I said that a category X is *virtual-inclusive* when a virtual X is a real X and virtual-exclusive otherwise. On current usage, *car* and *hurricane* are virtual-exclusive (virtual cars don't count as real cars), while *computer* and *communicate* are virtual-inclusive (virtual computers are real computers). But as we saw then, words can evolve to become more inclusive. It's entirely possible that "car" could evolve to become virtual-inclusive by default. In that case, we may mean the same thing by "car" in both virtual and nonvirtual contexts.

The idea of virtual-inclusiveness can help us to analyze externalist cases. It's arguable that virtual-exclusive words such as "hurricane" will refer to one thing (hurricanes) on Earth and to another thing (virtual hurricanes) on Sim Earth. That is, they'll be anchored to their environment, in classical externalist style. On the other hand, a virtual-inclusive word such as "computer" won't be anchored to its environment. It will refer to computers on both Earth and Sim Earth: virtual computers are computers too.

Over time, as virtual reality becomes more and more central in our lives, it's natural to expect that many words will gradually shift from being virtual-exclusive to being virtual-inclusive. As this happens, our usage of language may come to put less emphasis on what things are made of and on how they are anchored in our environment. Instead, a virtual-inclusive language may put more emphasis on factors that are common between virtual and nonvirtual reality: the structured patterns of interaction among things, and how they connect to minds like ours.

Putnam on externalism and Cartesian skepticism

In his 1981 book *Reason, Truth and History*, Hilary Putnam used externalism and the causal theory of reference to analyze the brain-in-a-vat scenario. He didn't say anything explicitly about skepticism, but it's easy to draw conclusions about skepticism from his discussion. His conclusions are not the same as mine, but they are interestingly connected.

We've already encountered Putnam's main thesis in chapter 4. He argued that the lifelong brain-in-a-vat hypothesis is incoherent, or contradictory. In effect, he thought he could use externalism to prove that we're not brains in vats, or at least that we're not lifelong brains in vats (I'll take "lifelong" for granted in what follows). He did not explicitly discuss the simulation hypothesis, but he almost certainly held that this is contradictory as well.

Here's how Putnam's argument goes. Take a situation like the one depicted in figure 50, in which it looks like Putnam is a brain in a vat. However, Putnam insists, "I am not a brain in a vat." He reasons as follows. Externalism tells us that if he's in the situation depicted,

Figure 50 Hilary Putnam as a brain in a vat. Is he
right when he says "I'm not a brain in a vat"?

then when he says "brain," he's referring to a *virtual brain,* a brain-
like object made of bits, like the virtual brain he's pointing to inside
the simulation. So when he says, "I'm not a brain in a vat," he means
I'm not a virtual brain in a vat. And that's true! In the scenario as
depicted, he's not a virtual brain in a vat. He's not like the object next
to him that he calls a "brain." Instead, he's a nonvirtual brain in a vat,
in an entirely different world that's not made of bits. So when Putnam,
inside a simulation, says, "I'm not a brain in a vat," he speaks the truth.
Generalizing, Putnam argues that whether or not he's in a simulation,
he will always be correct in saying, "I'm not a brain in a vat." Each of
us can reason this way for ourselves, thereby proving that we are not
brains in vats!

Let me run through the reasoning for myself. If I'm a brain in a vat, externalism tells me that what I call "brains" are virtual brains—simulated entities that are part of my own simulated environment. But I can't possibly be a virtual brain in my own simulated environment. It's possible that I'm the sort of thing that the *simulators* call a brain in a vat—but that's a different sort of thing entirely. I can know I'm not the sort of thing that *I* call a brain in a vat. That is to say: I can know I'm *not* a brain in a vat.

If Putnam is right, the very idea that I'm a brain in a vat is subtly contradictory. To be a brain in a vat, I'd have to be the sort of thing that *I* call a brain in a vat. But if I'm a brain in a vat, what I call a brain is something in my environment entirely different from the sort of thing I am. So to be a brain in a vat, I'd have to be something I'm not.

There's a lot to say in response to Putnam. One loophole that Putnam touches on himself is that the argument can't rule out the possibility that he is a *simulated* brain in a vat, like the second brain depicted in figure 50, experiencing a second-level simulation as depicted on the second screen. If he's in that situation, what he calls a brain is a virtual brain (like the third brain depicted), and he is in fact a virtual brain (like the second brain depicted), albeit in the next world up. In this scenario, it's arguable that he is what he calls a "brain in a vat" after all. If so, Putnam's argument shows, at best, that he is not an *unsimulated* brain in a vat: he's not a level-one brain in a vat, but he could still be a level-two brain in a vat. That's not enough to defeat Cartesian skepticism; he still needs some separate way to deal with the skeptical hypothesis that he's a brain in a vat at level two.

Furthermore, Putnam's argument doesn't work nearly as well for the simulation hypothesis as for the brain-in-a-vat hypothesis. The reason is that, as we saw earlier, externalism works better for some terms than for others. Words like "New York," "water," and "brain" arguably have their meaning tied to our environment, so they can't refer to something in the next universe up. But there are other words whose meaning doesn't depend on the environment in the same way. Examples include "zero," "person," "action," "computer," and "simulation." A computer is defined in largely structural terms, independent of any particular environment. As a result, even if I'm in a simulation, there's no problem talking about persons, actions, computers, and simulations in the next universe up.

If this is right, then Putnam's reasoning can't rule out the hypothesis that I'm in a computer simulation. Even if I'm in a computer simulation, I can truly say, "I'm in a computer simulation." As I said in chapter 4, if Sim Putnam says, "I'm in a computer simulation," then what he says is true.

Perhaps Putnam could say that the word "simulation" is like "New York" or "water" and is anchored to a specific system existing only in our ordinary environment and others just like it. But this seems wrong: When we talk about computers and simulations, we're talking about something much more general than that. I can straightforwardly speculate that I'm in a computer simulation in the next world up, and I might be right.

For this reason, I don't think Putnam's main argument can work as a general reply to global skepticism. However, Putnam briefly makes a quite separate argument that is also relevant to global skepticism and much closer to my own approach. In one paragraph in *Reason, Truth and History*, Putnam suggests that brains in vats have mostly true beliefs. He argues for this as follows:

> By what was just said, when the brain in a vat (in the world where every sentient being is and always was a brain in a vat) thinks, "There is a tree in front of me," his thought does not refer to actual trees. On some theories that we shall discuss it might refer to trees in the image, or to the electronic impulses that cause tree experiences, or to the features of the program that are responsible for those electronic impulses. These theories are not ruled out by what was just said, for there is a close causal connection between the use of the word "tree" in vat-English and the presence of trees in the image, the presence of electronic impulses of a certain kind, and the presence of certain features in the machine's program. On these theories the brain is *right*, not *wrong* in thinking "There is a tree in front of me."

The basic argument is an appeal to the causal theory of reference. When a brain in a vat says, "There's a tree," its use of "tree" is caused by virtual trees. So "tree," for a brain in a vat, refers to virtual trees. And when the brain says, "There is a tree," there really is a virtual tree. So what the brain in a vat says is true! Something similar goes for other beliefs about the world.

More or less the same line of argument was explored briefly by the American philosophers Donald Davidson and Richard Rorty. It was summarized crisply by Rorty as follows:

> That brain too is reacting to features of its environment. But its environment is the computer's data bank. The only way you can translate the noises it makes is to correlate them with the bits of data that the computer is feeding in. So the noises that sounds like "It's Tuesday the 7th of October 2003, and I am eating tofu" must mean something like "Now I am hooked up to sector 43762 of the hard drive." For most of the envatted brain's beliefs, like most of ours, must be true. It is not as easy to delude a brain as the evil scientist thinks.

I think that Putnam, Davidson, and Rorty are all essentially correct. A brain in a vat refers to entities in its environment, and as a result, its beliefs are mainly true.

Still, I don't think that any of them have given a strong *argument* for this claim. Their argument rests on an overly strong and implausible form of externalism. As it stands, the argument seems to assume an extreme version of the causal theory of reference, in which *every* word refers to an item in its environment that causes it. But this extreme version is false. Many words, like, say, "witch" or "ether," refer to nothing at all. The women to whom the word "witch" was first applied were not witches. The ubiquitous ether of 19th-century science does not exist. A Cartesian may well say that "brain" and "tree" for a brain in a vat are like "witch": they refer to nothing. Putnam has given no real argument against this Cartesian view.

Furthermore, we have seen that while externalism may work for words like "tree" and "brain," it doesn't work so well for words like "three," "computer," and "philosopher." The externalist analysis doesn't really apply to sentences like "There are three philosophers over there," or "I am using a computer." Nevertheless, simulation realism requires that sentences like this can be true inside a simulation. So Putnam's externalism hasn't shown us how simulation realism can be true across the board.

I think it is possible to rebut these criticisms. I've already made a case, in chapter 9, that "brain" and "tree" in a simulation refer to virtual brains and virtual trees. Roughly, what matters is that brains on Earth and virtual brains on Sim Earth play a similar structural role. The argument in chapter 9 also suggests that "There are three philosophers over there," and "I am using a computer," uttered inside a simulation, are true. I'd diagnose this by saying that these beliefs largely concern structural matters that are shared between Earth and Sim Earth, so that they can be true on Sim Earth even with their ordinary meaning. Even if I am in a simulation, I may still see three philosophers or use a computer. However, the argument in chapter 9 and the analyses here rest not on externalism but on *structuralism*.

Ultimately, it's structuralism, rather than externalism, that drives the simulation realism that drives my response to skepticism in turn. The chapters that follow will develop the case for structuralism.

Chapter 21

Do dust clouds run computer programs?

I N THE CLASSIC 1994 SCIENCE-FICTION NOVEL *PERMUTATION City*, by the Australian author Greg Egan, simulations are everywhere. People make simulated copies of themselves, which inhabit virtual worlds where they have conscious experiences much like the originals. The main character, Paul Durham, is a copy with low status and little legal recognition who experiments with creating simulated worlds of his own.

In the novel, it turns out that an ordinary full-scale simulation is not required to create a world. Durham modifies the simulation he inhabits so that its parts are wholly disconnected in space and time. Even so, he continues to exist. He breaks the simulation of himself down into ever-smaller disconnected pieces and he still exists. Even when the pieces are completely scattered in space and time with no connection between any of them, he and his world still exist.

As Durham scatters pieces of himself, he forms the idea that the universe itself fundamentally consists of scattered pieces of unorganized dust:

Squeak. "Trial number four. Model partitioned into fifty sections and twenty time sets; sections and states randomly allocated to one thousand clusters." "One. Two. Three."
Paul stopped counting, stretched his arms wide, stood up slowly. He wheeled around once, to examine the room, checking that it was still intact, still complete. Then he whispered, "This is dust. All dust. This room, this moment, is scattered across the

planet, scattered across five hundred seconds or more, but it still holds itself together. Don't you see what that means? . . .

"Imagine . . . a universe entirely without structure, without shape, without connections. A cloud of microscopic events, like fragments of space-time . . . except that there is no space or time. What characterizes one point in space, for one instant? Just the values of the fundamental particle fields, just a handful of numbers. Now, take away all notions of position, arrangement, order, and what's left? A cloud of random numbers.

"That's it. That's all there is. The cosmos has no shape at all— no such thing as time or distance, no physical laws, no cause and effect."

This theory is the *dust theory*. It postulates a large cloud of randomly scattered atoms of dust, outside space and time and with no cause and effect. The core idea is that a scattered dust cloud like this could execute any possible algorithm and therefore simulate every possible world, resulting in the existence of vast numbers of conscious people. An even more speculative version of the theory says that a dust cloud like this is what underlies our own reality.

Figure 51 The dust theory: A randomly scattered cloud of dust underlies computation, which underlies reality.

The dust theory is fascinating. It would have all sorts of consequences if true. Even if there's no such dust cloud, our world contains a whole lot of matter. If dust can run every algorithm, then so can this matter. If so, then pretty much every possible computer program is running somewhere in the world and every possible simulated world exists. Every possible simulated person exists, too. That's a vertiginous picture. If simulating a world or a person is as easy as this, the whole idea of simulation may be trivialized.

The dust theory raises any number of questions. One question, pertaining to the plot of *Permutation City*, is why anyone in the novel bothers to create full-scale simulations. This would seem pointless if every simulation is already running in the dust. Another question concerns Durham's partition experiment. Can these simulations be truly scattered and unconnected, given that the program he's using always has the ability to reunite the pieces? If they aren't truly scattered, Durham's inference that scattered dust can run simulations seems hasty. A third question is how science can work if every possible world exists. Most worlds will be chaotic and unpredictable. It would be surprising to find ourselves experiencing a highly ordered world as our world seems to be.

There's a deeper problem with the dust theory. It rests on a false assumption. The assumption is that patterns of cause and effect are irrelevant to executing an algorithm, to generating reality, and to generating consciousness. In fact, complex structures of cause and effect are crucial to all of these things. These structures aren't present in Egan's dust cloud. As a result, dust does not support genuine algorithms, simulated worlds, or simulated people.

This point is key to my argument for simulation realism. Computer simulations are not just clouds of aimless dust. They're fine-tuned physical systems whose elements interact in complex patterns of cause and effect. This causal structure is what makes them genuine realities, on a par with the nonsimulated world.

Clarifying this argument requires us to examine the relationship between algorithms and physical systems.

Computation in physical systems

What's the relationship between computer programs and physical systems? There's a vast mathematical theory of computation. The theory postulates abstract systems such as Turing machines, finite automata, cellular automata (like the Game of Life), and algorithms of all sorts. It tells us what sort of problems the various computing systems can solve and how they go about solving them.

Computation isn't just mathematical, however. It's physical because it takes place on physical devices. My desktop computer is currently executing the emacs word-processing algorithm. My smartphone is executing a messaging algorithm. Many argue that a human brain executes various algorithms, such as a neural-network learning algorithm. And so on.

The gap between mathematical and physical computing systems goes back to the early days of computing. In the mid-19th century, the English inventor Charles Babbage provided the mathematical design for a number of computers, including his Difference Engine and the far more complex Analytical Engine. His collaborator, Ada Lovelace, designed algorithms for the Analytical Engine to run, in order to compute certain sequences of numbers. Babbage developed detailed blueprints for mechanical systems that could execute these computations, but due to both engineering and financial limitations, he was never able to finish building them.

A century later, his countryman, the extraordinary mathematician Alan Turing (played by Benedict Cumberbatch in *The Imitation Game*), had more luck. In 1936, he came up with a mathematical model of the first universal computer—that is, a computer that could run any computer program. This model has come to be known as the Turing Machine. At Bletchley Park in 1940, Turing and colleagues built the Bombe, a nonprogrammable decryption device that was entirely dedicated to decoding Germany's Enigma code system. By 1943, his Bletchley Park colleague Tommy Flowers had built the far more complex Colossus, which was the first programmable electronic computer. There's some dispute about how much Turing's mathematical work influenced Colossus and subsequent computers, but it is striking that the gap between mathematical model and physical implementation was less than a decade.

What exactly is it for a physical system to realize a mathematical computation? This issue is sometimes taken for granted in computer science, but it raises interesting philosophical questions.

Take John Conway's Game of Life. This can be seen as a mathematical object—an abstract cellular automaton (see chapter 8). But it's also executed on physical computing devices around the world. What is it for a physical device to implement Life? Suppose we take a specific run of the Game of Life—perhaps we're implementing a glider gun of a certain size and shape. What is it for a physical system, such as my iPhone, to implement the glider gun?

Here's a natural view: My iPhone implements the Game of Life's glider gun when there's a *mapping* from internal states of my iPhone to cell states in the glider gun—a correspondence between these states that preserves the right sort of structure. There are transistors in the phone that can be mapped to cells in the Life grid. Each iPhone transistor has either a low or a high voltage. If the voltage is low, the Life cell is off. If the voltage is high, the Life cell is on. That's how the glider gun is implemented.

At this point, Greg Egan's dust theory kicks in. Egan suggests that given enough particles of random dust, we can always find a mapping from the dust cloud to the glider gun, or to any computation at all. As a result, dust implements any algorithm. If algorithms are the basis of reality, dust will realize any reality. If algorithms are the basis of consciousness, dust will generate any state of consciousness.

Here Egan is echoing similar claims by Hilary Putnam, as well as by the American philosopher John Searle. In his 1988 book *Representation and Reality*, Putnam argues that there's a mapping under which any ordinary system (such as a rock) implements any finite automaton (roughly, any finite computer program). In his 1992 book *The Rediscovery of the Mind*, Searle writes:

For any program and for any sufficiently complex object, there is some description of the object under which it is implementing the program. Thus for example the wall behind my back is right now implementing the Wordstar program, because there is some pattern of molecule movements that is isomorphic with the formal structure of Wordstar. But if the wall is implementing Wordstar, then if it is a big enough wall it is implementing any program, including any program implemented in the brain.

All this threatens to trivialize the whole idea of physical computation. Searle concludes that it is not really an objective matter whether a physical system (like a wall) implements a program—this implementation is in the eye of the beholder. Egan concludes that every program is running all the time. Putnam concludes that the philosophical view of *functionalism*, which holds that the mind is grounded in computer programs, is provably false.

As it happens, years ago I published an article rebutting these arguments. It grew out of a heated online discussion in 1992 in the comp. ai.philosophy discussion group on Usenet (an internet discussion forum that was the ancestor of many discussion forums today). The title of the discussion thread was "Does a rock implement every finite-state automaton?" Some people were pushing Putnam's argument that physical computation is trivial, while I argued that it is nontrivial.

I ended up publishing two articles on the topic shortly afterward— one with the same title as the Usenet discussion and another called "On Implementing a Computation." These articles appeared in philosophy journals in 1994 and 1996, around the same time that Egan's novel was published, so I didn't know about Egan's theory at the time. But my response applies to Egan's argument as well as to Putnam's and Searle's.

The dust-to-Life argument

In *Permutation City*, Egan doesn't spell out how to map dust particles onto algorithmic processes. (Searle doesn't give many details either; Putnam gives a few more.) But it's not too hard to sketch a simple argument that a large-enough cloud of dust implements any process in the Game of Life. I'll call this the dust-to-Life argument.

Let's start with a simple process in the Game of Life. A *blinker* is a row of three cells that blinks back and forth between horizontal and vertical indefinitely. Let's suppose our Life world is a three-by-three square of cells. We start with the three cells in the middle row being on while those in the top and bottom rows are off. The middle cell has two neighbors, so it stays on. The left and right cells in the middle row have just one neighbor, so they die off. The cells in the middle of the top and bottom rows have three neighbors each, so they turn on. The corner cells have only two neighbors each, so they stay off. As a result, the horizontal row of three cells turns into a vertical column of three. By the same logic, the vertical column of three turns back into a horizontal row of three, and the row and column keep blinking back and forth forever.

We'll also assume an infinite cloud of dust particles outside space and time. Each dust particle has a binary state; let's call those states hot and cold. There's an infinite number of hot particles and an infinite number of cold particles. There's no more organization than that. Here's an argument (inspired by Egan, Putnam, and Searle) that the dust cloud implements a blinker.

The blinker can be broken down into repeating generations of nine Life cells each, where each cell is on or off at a given moment of time. In the first generation, three Life cells in the middle row are on and the rest are off. To find this structure in the dust cloud, we can simply take three hot dust particles and six cold dust particles. We'll map the three hot particles to the three cells that are on, and the six cold dust particles to the cells that are off. That way, we've found the first generation in the cloud of dust. For the second generation, we do more or less the same thing. We find three more hot particles and map them onto the three cells in the middle (vertical) column that are on. We find six more cold particles, and map them onto the six Life cells that are off. That way, we've found the second generation. If we repeat this process, we can find a blinker oscillating back and forth forever in the dust cloud.

If a mapping like this from dust particles to blinker cells is all we need for implementing an algorithm, the dust is implementing the

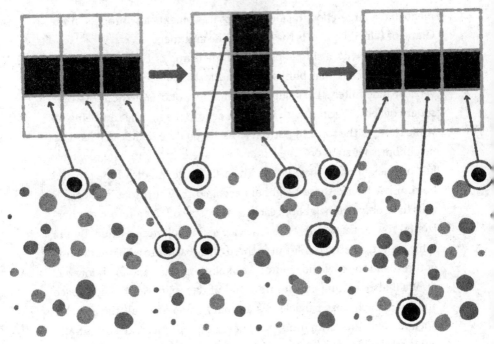

Figure 52 The dust-to-Life argument: Randomly selected dust cells map onto rule-governed cells in the Game of Life.

blinker algorithm. The same goes in principle for any process in Life. It's not too hard to extend this reasoning to any algorithm at all. We might even find the algorithmic structure of the human brain in the dust.

There's an obvious response to this, which is that the dust cloud is missing crucial structure essential to the Life algorithm. In particular, it's crucial to a blinker that the *same* Life cells (those at the ends of the rows and columns) turn on and off over time. In the dust cloud, nothing like that happens. We don't have a few dust particles switching back and forth between hot and cold. Instead one dust particle is hot, and an entirely different dust particle is cold. In the absence of time, there's nothing corresponding to a *transition* from a cell's being on to its being off. In effect, while the dust cloud may implement certain static states of the blinker, it fails to implement the blinker's dynamics. .

Now, it wouldn't be too hard for Egan to modify the dust cloud to allow some of this dynamic structure. Let's suppose that dust particles exist over time. There's an infinite cloud of particles that randomly become hot or cold at any given time. All we need to do is find one particle that's always hot (to correspond to the middle cell), four that are always cold (to correspond to the corner cells), and four that alternate between hot and cold (two starting hot, and two starting cold, to correspond to those on the ends of the middle row and the middle column). In a large enough cloud, we should be able to find nine particles that behave this way for a million generations or so, which is good enough for our purposes. Now we can just map these nine particles to the nine cells in our blinker. The cells will go through exactly the transitions required for the blinker. Is the dust now implementing the blinker?

This version of the dust-to-Life argument is close to the arguments given by Putnam and Searle, which suggest that if we try hard enough, we can find the right series of states for any algorithm in dust (Egan) or a wall (Searle) or a rock (Putnam). If that's enough to implement any algorithm, then computation in physical systems will be trivialized.

Respecting cause and effect

I don't think the dust-to-Life argument succeeds. The key problem is that the dust-to-Life mapping from physical systems to computational systems doesn't respect cause and effect. To execute the Game of Life, it's not enough to have cells that go through on-and-off states in the right order. These cells need to interact with one another in the right ways.

The Game of Life has *rules*, saying (for example) that when an on cell has more than three on neighbors, it turns off. A physical system needs to implement these rules in its functioning. When a dust particle turns off (goes cold), this must be *because* more than three neighbor particles were on (were hot). So for the dust to implement the Game of Life, we need the particles to interact with one another through cause and effect in a way that mirrors the rules of Life. Our scattered, randomly behaving dust particles will certainly not be interacting with one another through cause and effect like this. So the dust doesn't implement Life after all.

Furthermore, for a physical system to implement a blinker in the Game of Life, it must do more than simply proceed through an appropriate sequence of states. In a real implementation of the Game of Life, the implementing system can deal with many different initial states, leading to many different sequences. Even in these alternative sequences, the system has to follow the rules. We can say that *if* the system had started in a different initial state, *then* it would have proceeded into certain different patterns. More generally, we can say that *if* any cell had been in a state with more than three neighbors at a given time, *then* it would have turned off. All this structure is absolutely crucial to implementing the Game of Life. And, again, there's no reason to think that the dust cloud will have this structure.

These *if-then* constraints are what philosophers call *counterfactuals*. Counterfactuals concern things that *could have* happened but didn't. Back in chapter 2, we considered simulating what would have happened if a meteor hadn't wiped out the dinosaurs. A counterfactual says that *if* something had happened (that actually didn't happen), *then* something else would have followed. For example, *if* someone had dropped the glass, it would have broken. Or to take an example from the game of cricket: *if* the ball had not hit the batsman's leg, *then* the ball would have hit the wicket. This is a counterfactual criterion for a batsman to be out. As cricket-lovers know, these counterfactuals are not just in the eye of the beholder. There's typically an objective fact about whether the ball would have hit the wicket or not, and umpires can get this right or wrong.

Counterfactuals are crucial to understanding many phenomena and are especially crucial to understanding processes of cause and effect. In fact, many philosophers think that for something to cause something else is just for there to be the right counterfactual connection between them: to say that the fire caused the smoke is roughly to say that if the fire hadn't been there, the smoke wouldn't have formed. Certainly, counterfactuals are crucial to understanding the cause and effect involved in computation. Suppose a computer screen is running a prerecorded movie that shows a series of states from the Game of Life. If one cell had been different, the screen wouldn't execute the mandated Life behavior; it's just a recording. Because of this, the screen doesn't really implement the Game of Life. Physical computation is all about the right pattern of counterfactuals, and the right pattern of cause and effect.

Once we understand computation in terms of causation, the dust-to-Life argument no longer works. Our carefully selected particles won't have the right causal structure. The dust doesn't satisfy the right counterfactuals. Now, you might hope that with enough particles and enough time, we'll eventually find a group of particles that do interact with the right causal structure to implement a blinker or a glider gun. (The "Boltzmann brain" thought experiment, which we'll look at in chapter 24, runs a little like this.) If so, that will be a genuine implementation of the Game of Life. But now we have gone far beyond Egan's timeless, spaceless, scattered, unstructured dust. This implementation requires intricate structures of cause and effect. Any such implementations will be extremely rare.

Greg Egan himself considers the cause-and-effect objection in the "Dust Theory FAQ" on his website. He says: "[S]ome people have suggested that a sequence of states could only experience consciousness if there was a genuine causal relationship between them. The whole point of the dust theory, though, is that there is nothing more to causality than the correlations between states."

Here Egan is echoing a view often attributed to the great 18th-century Scottish philosopher David Hume, who held that cause and effect is just a matter of regularity. A very simple version of the theory: To say that A causes B is just to say that A is always followed by B. Why does squeezing the trigger cause the gun to fire? Because squeezing the trigger is always followed by the gun firing.

If this simple theory of causation is right, then perhaps the dust will have the right causal structure. At least among the selected particles making up our Life system, when a cell has three neighbors, it will always turn off.

The view that causation is just correlation, or regularity, is a minority view among philosophers. But even if we accept a Hume-style regularity view of causation, the dust probably won't have the right causal structure. Even philosophers sympathetic with Hume accept the bromide that mere correlation doesn't imply causation: a local correlation among a few dust particles doesn't suffice for causation among them. For causation, we need a more robust regularity, perhaps one that occurs among *all* the dust particles (or at least among a wide class of them), not just among our selected subset. The dust cloud has only random, local regularities, not the global regularities needed for genuine cause and effect.

Exactly the same reply can be given to Putnam's and Searle's arguments. Even if we can map Searle's wall to the Wordstar program, there's no reason to think it has the complex causal and counterfactual structure required to implement the program. Even if we can map Putnam's rock to any ordinary automaton, there's no reason to think it has the complex structure required to implement the automaton.

Of course, this isn't the end of the story. Some philosophers have argued that relatively simple systems will meet these stronger constraints. I've tried to rebut them in turn. But once we understand physical computation in terms of causal structure, we can at least avoid the simple mapping arguments that trivialize computation. The bottom line is that implementing a computation requires the right causal structure, and causal structure isn't cheap.

Causal structuralism

As I see things, computation is all about structure. Mathematical computation is a matter of *formal structure*. It's all about certain structures of bits that are governed by certain formal rules: *On* bits are formally followed by *off* bits, and so on.

By contrast, physical computation is a matter of *causal structure*. It's all about physical elements interacting with one another in patterns of cause and effect. High voltage in one transistor triggers low voltage in another, and so on.

According to my *causal structuralism*, a physical system implements a mathematical computation when the causal structure of the system *mirrors* the formal structure of the computation.

This view is in accord with the ordinary practice of constructing and programming computers. When Babbage laid out a blueprint for the Analytical Engine, he was trying to ensure that physical mechanisms in the system interacted in the right patterns of cause and effect to mirror his mathematical structure. When Turing built the Bombe, he did the same. The same goes for modern programming. A programmable computer is a system that can implement many different computations depending on which program it is executing. The cause and effect here is all under the hood. A programmer writes a program, a user executes it, and a physical process gets underway. When I execute an app on my iPhone, the phone is set up so that its physical circuits interact in patterns that reflect the formal rules of the program.

I think of computers as Causation Machines. They are flexible devices that can be arranged into arbitrary structures of cause and effect. It is this that makes them such wonderful devices for simulation. Say we have a system we want to simulate. The original system has a causal structure among its parts. When we build a simulation, we replicate that causal structure inside a computer. In a certain sense, the original system is causally mirrored in the computer, at least at a certain level of detail. We have replicated the original system using a Causation Machine.

The same goes for simulated brains. What happens when we upload a brain into a computer simulation? In essence, we try to preserve its causal structure. This is especially clear in the case of gradual uploading, in which we replace neurons one at a time by chips. We try to ensure that a new chip interacts with surrounding elements in just the same patterns of cause and effect that the old neuron did. All going well, we will end up with 86 billion chips that interact in the same causal structure as the brain's 86 billion neurons. Likewise, in a successful neuron-level simulation of the brain, we'll end up with 86 billion data structures inside a computer interacting with the same causal structure. The gradual-uploading argument in chapter 15 gives us good reason to think that the chip system and the simulation will have the same sort of conscious state that the brain did. If the causal structure is destroyed, then consciousness will not be preserved, either.

In a cloud of randomly scattered dust, we can expect neither computation nor consciousness. But if the dust cloud is organized into the right causal structure, then the sky's the limit. We will have computation. If computation suffices for consciousness, we will have consciousness. With both computation and consciousness, we will have a virtual world.

Chapter 22

Is reality a mathematical structure?

I N 1928, RUDOLF CARNAP CONSTRUCTED THE WORLD. IN HIS
magnum opus, *Der logische Aufbau der Welt*, or *The Logical Structure of the World*, he tried to describe the world fully in the language of logic.

Carnap was the leading figure in the Vienna Circle, the group of
scientifically-minded philosophers we first encountered in chapter 4.
Their aim was to ground philosophy in science and use philosophy to
help bring about progressive social change. At its peak in the late 1920s,
the Circle issued a philosophical manifesto on the scientific conception
of the world.

As the Nazi era approached in the 1930s, the Circle dwindled. It
ended tragically when one of its key figures, Moritz Schlick, was shot
by a paranoid ex-student in 1936. For some years afterward, their
ideas were often caricatured as a crude "logical positivism" that rested
entirely on dismissing most philosophy as meaningless. Over more
recent decades, however, the richer philosophical ideas of Carnap and
the Vienna Circle have been widely recognized.

It was particularly crucial to the Vienna Circle to remove subjective
elements from our picture of the world, and to give objective descriptions of reality in an objective common language. In his *Aufbau*, Carnap tried to do this in the language of logic.

Carnap's *Aufbau* is often regarded as a noble failure. His objective
descriptions of the world were grounded in descriptions of subjective
experience—something that many people think doomed the project
from the start. As we've seen, it's hard to build reality from appear-

ances alone. Despite his best attempts, it's arguable that Carnap never got around this fundamental limitation.

But there's more to Carnap's project than his construction of objective reality from subjective experience. He says in the *Aufbau* that he could equally have constructed the world from descriptions of physics. At one point, he even proposed a second volume of the *Aufbau* to do this; it never came to fruition, but there are elements of it in his 1932 article "The Physical Language as the Universal Language of Science." Underlying both projects was an idea that is central in the philosophy of science to this day.

The real key to the *Aufbau* is not describing the world in terms of subjective experience or of physics. The key is describing the world in terms of *structure*—by which I mean logical and mathematical structure. Carnap's aim was to give what he called a *structure description* of reality: a complete description of reality in logical and mathematical terms.

Carnap illustrated a structure description using a railway system. I'll do it with the New York City subway. Figure 53 contains two pictures of the subway system in the lower half of Manhattan. The first picture has labels for the lines and stations. It tells you that the 8th Street/NYU station is on the R and W lines, and so on. The second picture removes the labels for lines and stations. It tells you that in this area there are 80 stations arranged on 20 lines or so (depending on how you count) in a certain complex pattern.

To go all the way in eliminating nonstructural information, we need to remove any hint of the location of the stations, and we need to drop the words "station" and "line." What is left tells us that there are 80 *entities* arranged in 20 *sequences* in a certain complex pattern. This is a structure description of the subway system. It specifies a pure mathematical structure of the sort that mathematicians call a graph, which is a system of interconnected nodes. This graph specifies the mathematical structure of the subway system, or at least of one part of it.

This unlabeled graph isn't a remotely complete specification of the subway system even in the southern half of Manhattan. It leaves out the trains and the passengers. It leaves out the platforms and escalators. It leaves out the locations of the stations, and much more.

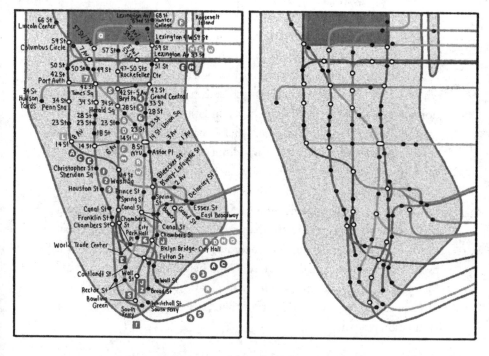

Figure 53 Two ways of describing the New York City
subway system: an ordinary description (with labels)
and a structure description (without labels).

Carnap's structuralist dream was that everything left out could be
included as just more structure. In principle, one could give a logical or
mathematical representation of the station locations, the platforms, the
escalators, the trains. One could even give a mathematical description
of the people riding the trains. If we packed *everything* into our description of the subway system and turned it into mathematics, we'd have
a complete, objective description of the system. And by extending this
process from the subway to the whole universe, we could have a structural description of all of reality.

Few people think Carnap succeeded in describing all of reality using
logic and mathematics alone. My own view is that although he failed,
he came reasonably close. In my 2012 book *Constructing the World*, I
argued that many problems in his attempt can be overcome. If we base

his construction on physics as well as on subjective experience, and broaden his basic language to go somewhat beyond logic and mathematics, a version of Carnap's *Aufbau* can be made to succeed.

For my arguments here, however, I don't need to construct the whole world. I just need to use structuralism to help support simulation realism: the idea that if we're in a simulation, the ordinary physical world is real. To make that case, what I really need is structuralism about *physics* (which as before should not be confused with the structuralism about culture advocated by the French anthropologist Claude Lévi-Strauss and other mid-20th-century figures). This says that we can give a complete specification of physics in structural terms. To a first approximation, we can think of structural terms as mathematical terms (which I'll understand as including logical terms), though we'll eventually need to go a little beyond mathematics. The approximate idea is that we can give a wholly mathematical specification of what physics says about the world. From there, it's easy to make a case for simulation realism.

What does science tell us about the world?

Scientific theories such as Newtonian mechanics, the general theory of relativity, and quantum mechanics have been enormously successful. They're the foundation for most modern technology. Scientists and engineers rely on these theories and take their efficacy for granted.

What exactly do these theories tell us about reality? The standard model of particle physics postulates particles such as quarks and the Higgs boson. No one has ever directly observed a quark or a Higgs boson. Do these particles really exist? Or is this theory just a useful framework for predicting our observations?

There are two traditional views. *Scientific realism* tells us that successful scientific theories and models give us insight into what is real. When our best theories and models postulate certain entities, we should believe that those entities really exist. The standard model postulates quarks, so we should believe that quarks are real. Our best theories postulate electromagnetic fields, so we should believe that these fields are real.

Scientific anti-realism says that successful scientific theories should not be seen as a guide to what is real. They're better seen as convenient and useful frameworks for various purposes. The most prominent sort of scientific anti-realism, associated especially with the 19th-century Austrian physicist and philosopher Ernst Mach, is *instrumentalism*. Instrumentalism says that scientific theories are just "instruments"— or useful devices for predicting the results of our observations.

According to scientific anti-realism, we shouldn't believe that quarks and wave functions really exist, no matter how successful the theories that postulate them are. Scientific anti-realism doesn't say that quarks and wave functions *don't* exist. It just says that we shouldn't believe in them on the basis of the theories. According to one popular instrumentalist slogan about quantum mechanics, we should just "shut up and calculate." Quantum mechanics is so counterintuitive that it's hard to use it to form a picture of reality. But our quantum mechanical calculations will predict all the results of our measurements. We can use quantum mechanics in this way while maintaining a studied agnosticism about the reality behind the measurements.

The most important argument for scientific realism is the *no-miracles* argument put forward by Hilary Putnam, following a related argument by the Australian philosopher J. J. C. Smart. This argument says that the success of a theory would be a miracle if the theory weren't true. If we get the exact results you'd expect if quarks existed, then it would be a miracle to get those results without quarks. The key idea is that we need scientific reality to explain why our scientific theories work so well.

The most important argument for scientific anti-realism is the *pessimistic induction,* set out by the American philosopher Larry Laudan. This argument says that almost every scientific theory ever developed has turned out to be false given enough time. Newtonian mechanics turned out to be false. The atomic theory of matter turned out to be false. In many cases, early theories were replaced by refined successors (quantum theory, the standard model of particle physics) but the new theories rejected many key entities of their predecessors and postulated a new class of entities. So, if we accept a scientific theory wholesale, it's almost guaranteed that we will eventually be proved wrong.

Scientific realism is much more popular than scientific anti-realism. According to the 2020 PhilPapers Survey, 72 percent of academic philosophers accept or lean toward scientific realism, and only 15 percent accept or lean toward scientific anti-realism. Scientific realists often address Laudan's pessimistic induction by saying that later theories are at least closer to the truth than earlier theories, so we're gradually creeping up on reality. But there's still an interesting question about just what it is that our scientific theories are saying.

In recent decades, the most popular form of scientific realism has been *structural realism*. Structural realism says (to a first approximation) that our scientific theories describe the *structure* of the world, where structure can be characterized wholly in logical and mathematical terms.

Structural realism was first clearly stated by Carnap and Bertrand Russell in the 1920s. After years of neglect, it was revitalized in 1989 by the British philosopher of science John Worrall, who argued that it constitutes the "best of both worlds" in the scientific realism debates. Structural realism addresses the no-miracles argument for scientific realism by saying that the success of our scientific theories is explained by how well their structure matches the structure that is present in the world. It addresses the pessimistic induction by holding that even when later theories eliminate some of the entities postulated by earlier theories, they often retain much of the mathematical structure of those theories.

According to structural realism, we can *structuralize* our theories by putting them in a wholly mathematical form. In the previous section, we structuralized a description of the New York City subway system by regimenting it into a clear map and then getting rid of the names for stations and lines, and even getting rid of terms like "station" and "line." In a similar fashion, we can structuralize a physical theory by regimenting it into mathematical terms, getting rid of all names for objects, and even getting rid of words such as "mass," "charge," "space," and "time."

A technique for structuralizing theories was introduced in 1929 by the remarkable British philosopher Frank Ramsey in a paper called "Theories." Ramsey died the following year at the age of twenty-six, having made important contributions to mathematics and economics as well as philosophy. His main tools for structuralizing a theory are now known as "Ramsey sentences" and the process of structuralizing a theory is known as "Ramsifying" the theory.

Ramsey's basic idea was that we can take everything physics says about mass (say) as a *definition* of mass. So instead of formulating a Newtonian theory of inertia by saying, "Objects resist acceleration proportionately to their mass," we can say, "Objects have a property such that they resist acceleration proportionately to that property." This gives us a version of Newtonian mechanics that doesn't use the word "mass." If we do the same for force, charge, space, time, and so on, we'll ultimately be left with a version of physics in wholly mathematical and logical terms.

According to a structuralist view, contemporary physics ultimately says something like "There exist seven properties, which satisfy the following equations," followed by the laws of quantum mechanics, relativity, and so on, in mathematical form. That will be a description of the physical world in logical and mathematical terms.

Is this mathematical structure all there is to the physical world? *Ontological* (or *ontic*) structural realists say yes: Physical reality is purely structural. Ontology is the study of existence. Ontic structural realism says that what really *exists* in the physical world is pure structure. To a first approximation, the physical world can be *completely* described in logical and mathematical terms.

By contrast, *epistemic* structural realism says that structure is not all there is to the physical world, or at least that it need not be. Epistemology is the study of knowledge. Epistemic structural realism says that what we can *know* about the physical world is its mathematical structure. But this is consistent with there being some underlying reality that goes beyond structure.

You'll recall the pure it-from-bit view we discussed in chapter 8, which holds that what really exists in physics is a pure structure of bits. This view is one version of ontic structural realism. Ontological structural realism is more general than the pure it-from-bit view, as the structure it invokes need not be digital structure. We might call it the *pure it-from-structure* view.

The pure it-from-bit view contrasts with the it-from-bit-from-it view, which holds that bits in digital physics derive from more fundamental "its." The it-from-bit-from-it view is in the spirit of epistemic structural realism. It says that our scientific theories reveal a structure of bits, but these bits may also be realized by something that science doesn't tell us about. If we generalize the view beyond digital structure, we obtain the *it-from-structure-from-it* view, which holds that physical objects derive from structure that in turn derives from something more fundamental.

Epistemic structural realism goes beyond the it-from-structure-from-it view by making the further claim that the basic "it" is not revealed by science. We might call it the *it-from-structure-from-X* view. Here, X marks something unknown—or, at least, something not revealed by our physical theories.

We saw in chapter 9 that the simulation hypothesis fits especially well with the it-from-bit-from-it view. It also fits well with the it-from-structure-from-X view—that is, with epistemic structural realism. If we're in a perfect simulation, we can know the structure of physics but not what underlies it. In this case, the underlying X involves processes in a computer in the next universe up.

For our purposes, we don't need to decide between ontic and epistemic structural realism. All we really need is the weaker claim that what our scientific theories *say* about the world is structural. That claim is consistent with both the ontic and epistemic views.

Is physics just mathematics?

Structuralism has a dust problem. You'll recall Greg Egan's dust theory from chapter 21, where there was a danger that we could find the structure of any computer program in a random cloud of dust. If so, dust clouds are running all sorts of computer programs, from Conway's Game of Life to Microsoft Word. That framework threatened to make computation trivial and vacuous.

The same issue arises for physical theories. There's a danger that we can find the structure of any physical theory in a random cloud of dust. Combining this with structuralism, it will follow that dust clouds make all sorts of physical theories true. With enough dust particles we can find the structure of Aristotle's outmoded impetus theory (which attempted to explain the motion of a projectile) in the dust cloud. We can also find the structure of the ether theory, which Einstein's special relativity refuted. The same goes for string theory. If dust-cloud structure is enough to make these theories true, this threatens to make physical theories trivial and vacuous.

I'll call this the *dust-to-physics problem*. To see how serious a problem it is, we can focus first on an even more serious worry, which we can call the *numbers-to-physics problem*.

The numbers-to-physics problem goes like this: Suppose our physical theories are purely mathematical. Then it looks like we can find the mathematical structure of our theories even in purely mathematical objects, such as numbers. For example, our structural description of part of the New York City subway system can be satisfied by the numbers 1 through 80, arranged in twenty appropriate sequences. Our formulation of physics can be satisfied by a mathematical structure of quantities satisfying the appropriate equations.

This consequence might seem appealing to extreme structuralists, who hold that the universe itself is a mathematical structure. As we saw in chapter 8, Pythagoras thought that everything is made of numbers. More recently, the cosmologist Max Tegmark has advocated the *Mathematical Universe Hypothesis*: the hypothesis that our external physical reality is a mathematical structure. It's not just that the universe can be *described* by mathematics. The universe *is* mathematics. Tegmark argues for his view in classic structuralist fashion, by arguing that only mathematical structure provides a truly objective external reality independent of the human mind.

Whatever you think of the Mathematical Universe Hypothesis, there's an inescapable objection to the view that our physical theories are purely mathematical: If they are, then every consistent theory will be true.

Take Newton's theory of mechanics. If physical theories are purely mathematical, Newton's theory will say that there exist certain mathematical structures. Let's suppose that it says that a certain mathematical function from real numbers to real numbers exists. The trouble is that it's too easy for these mathematical entities to exist and therefore too easy for Newton's theory to be true. Numbers and other mathematical entities don't depend on physics for their existence. If a mathematical function exists in one world, it exists in all possible worlds—including worlds where Einstein's physics is correct. If Newton's theory says only that this mathematical function exists, Newton's theory will be true even in an Einsteinian world! Einstein's theory will also be true in such a world, and so will every other consistent physical theory.

This is an unpalatable result. Science progresses by falsifying old theories. But if no theory is false, no theory will be falsified. All the empirical evidence against Newtonian mechanics (the Michelson-Morley experiment, the perihelion of Mercury, the double-slit experiment) will do nothing to rule out the theory. The purely mathematical version of Newton's theory is still true. The same goes for every other consistent theory ever devised. All of them are true. Given that every theory is correct, there's no point trying to find the one correct theory.

To avoid the numbers-to-physics problem, physical theories need to go beyond pure mathematical structure. There are a number of ways to do this. Perhaps the easiest fix involves the notion of *existence*. Existence is a central notion in logic. It has its own symbol, a backward E. For example, we can write $\exists x\,(x^2 = -1)$ to say that there exists a number that when squared yields minus one—that is, that the square root of minus one (an imaginary number) actually exists. But existence requires more in science than in mathematics.

Any scientific theory says that certain things exist: particles, fields, and so on. This is naturally understood as meaning that they exist *con-*

cretely: They exist as part of concrete reality. If particles merely exist *abstractly*, only as mathematical objects (as in the case of the square root of minus one), that's not good enough. We need to interpret existence in scientific theories as *concrete* existence. This means it won't be so easy for our theories to be true. In effect, the move from abstract to concrete existence is a move from pure mathematics to applied mathematics, putting our mathematical theories in contact with the concrete, physical world.

What exactly is concrete existence? One idea is that an object exists concretely if it has causal powers. Particles cause things to happen, while numbers do not. This suggests an alternative way to solve the numbers-to-physics problem. The key idea is that physics involves patterns of cause and effect that are not present among numbers. To support this idea further, we can turn to the dust-to-physics problem.

Is there physics in the dust?

The dust-to-physics problem goes back to the Cambridge mathematician Max Newman, who would later work with Turing at Bletchley Park and help design the first general-purpose computer, the Colossus. Newman was reviewing Bertrand Russell's 1927 book *The Analysis of Matter*, which proposed a structuralist view of physics. Russell claimed that physical theories can always be put in logical and mathematical form. Newman discovered an insuperable problem for Russell's view.

Newman's problem tells us that if we have the right number of objects—perhaps in a cloud of dust particles—any mathematical structure can be found. For example, our structural description of 80 stations in the New York City subway system will be satisfied by any group of 80 objects. Given 80 dust particles in any arrangement, we can always find some assignment of sequences among these objects to serve as the subway lines. Or, to use an example from the last chapter, consider a structural description of the Game of Life involving cells in a certain two-dimensional arrangement with *On* and *Off* states satisfying certain rules. Then, given enough dust particles, we can find some way of assigning them locations on the grid in *On* and *Off* states that satisfy those rules.

All this applies just as strongly to structural descriptions in physics as to the Game of Life. We can always find any purely mathematical structure in the dust, at least given the right number of dust particles. The upshot is that structural descriptions tell us almost nothing about the world. At most, they can tell us how many objects there are in the world. This once again makes our theories nearly vacuous. For them to be true is much too easy.

We solved the dust-to-Life problem by holding that to perform a computation, a physical system must have the right *causal structure*, with the right patterns of cause and effect governing transitions from one state to the next. In particular, to implement the rules of the Game of Life, it's not enough to have cells going through the right sequences. The cells must also satisfy the right counterfactuals. For example, if this cell *were* to have four neighbors, *then* it would turn off. Once we require causal and counterfactual structure of this sort, the structure can certainly not be found in arbitrary collections of dust particles. As a result, implementing a computation is nontrivial and nonvacuous.

We can solve the dust-to-physics problem in the same way. Where the Game of Life has *rules*, physical theories have *laws*. To satisfy Newton's law of gravitation, for example, it's not enough that objects with certain masses actually behave in certain ways. The objects have to satisfy the right counterfactuals: If these two objects with certain masses *were* to be near each other, *then* they would attract each other with a certain force. Once we require lawful and counterfactual structure of this sort, the structure can no longer be found in arbitrary-collections-of-dust theories. As a bonus, this requirement also solves the numbers-to-physics problem, as numbers don't have this sort of causal structure any more than dust does. As a result, physical theories are nontrivial and nonvacuous.

If we understand physical theories this way, a structuralist version of modern physics might look like the following: "There exist seven properties, which satisfy the following laws," followed by the equations of quantum mechanics, relativity, and so on, specified in mathematical form. Now, the concept of a law is not part of mathematics, so using laws in our theories means that the content of our theories goes beyond pure mathematics. But laws are part of the structure of our physical theories, and they are part of how most people have interpreted these theories all along. We simply need to recognize that the equations of our physical theories are laws of nature that require that physical systems satisfy the right counterfactuals.

There are other possible solutions to Newman's problem. One promising route makes an appeal to *fundamental* properties, saying, "There exist seven fundamental properties that satisfy these laws: . . . ." Rudolf Carnap himself made a related appeal to *natural* properties. He claimed that the notion of naturalness is part of logic, which most people haven't found plausible. However, it's plausible that notions like fundamentality, naturalness, law, causation, and concrete existence are still *structural* notions in a broad sense. In practice, structuralists allow their theories to go beyond a purely mathematical specification of the world to a specification that's structural in this broader sense.

A final reason to go beyond mathematics is the connection to *observation*. Physical theories don't just talk about the external world—they also connect the external world to our experiences and observation. You'll recall that instrumentalism said that scientific theories are *just* tools for predicting the results of observations. Structural realists and other scientific realists think scientific theories do more than this, but predicting the results of observation is at least a crucial part of our theories.

It's far from obvious that the results of our observations can be turned into purely mathematical structure. Our observations are basically conscious experiences—for example, an experience of an object with a certain color and shape. Conscious experiences *can* be mathematically described; we can measure the experience of various colors and the differences among them. But conscious experience seems to go beyond a mathematical description. To adapt Frank Jackson's thought experiment about Mary, the color scientist in the black-and-white room (see chapter 15): Mary might formulate a full mathematical description of color processing, but this won't tell her what it's like to experience the color red.

In practice, structural realists usually don't try to structuralize observation. There's a structural part to our theories, and another part connecting the structure to observation. For example, quantum mechanics has the Schrödinger equation, which is expressed in terms of structure alone, and the Born rule, which connects this structure to the probabilistic results of observation.

As a result, the view that physical theories specify purely mathematical structure for reality has to be qualified in two key ways. Physical theories specify *causal* structure, at least by specifying counterfactual-supporting laws, and they specify connections to *observation*. For a physical theory to be correct, the world must have the right concrete causal structure and the right connections to observation. With these constraints, we avoid the dust-to-physics problem. The dust doesn't have the right structure to make arbitrary physical theories true.

It sometimes happens that one physical theory makes another true. For example, under certain assumptions one can derive the structure of thermodynamics (a theory of heat) from the structure of statistical mechanics (a theory of molecular motion). In this case one does not merely vindicate the mathematical structure of thermodynamics, as a mapping from the dust theory might. Statistical mechanics vindicates the principles of thermodynamics (such as the ideal gas law that relates pressure, volume, and temperature) as laws that support counterfactuals, and thereby vindicates its causal structure. It can also vindicate its connection to observation. By vindicating the causal and observational structure of thermodynamics, statistical mechanics makes thermodynamics true.

From structuralism to simulation realism

How does all this connect to the simulation hypothesis? The key idea is that appropriately arranged computer simulations can vindicate the structure of physical theories, just as statistical mechanics can vindicate the structure of thermodynamics. If structuralism is correct, it follows that the computer simulation makes the physical theories true. That's a form of simulation realism. If we're in a simulation, the physical world around us is real.

To flesh out the argument, let's say that Nonsim Universe is a non-simulated universe corresponding to standard physics. If we're in Nonsim Universe, our physical theories are at least approximately true. Let's also say that Sim Universe is a perfect simulation of Nonsim Universe. (It needn't be contained within Nonsim Universe, though it might be.) My thesis is that if we're in Sim Universe, our physical theories are also at least approximately true. There will be physical objects (like atoms and molecules) in Sim Universe roughly as we think they are. If so, simulation realism will be true.

Here's an argument that starts with structuralism and ends with simulation realism.

1.   Our physical theories are structural theories.
2.   If we're in Nonsim Universe, our physical theories are true.
3.   Sim Universe has the same structure as Nonsim Universe.

———————————

4.   So: If we're in Sim Universe, our physical theories are true.

The first premise is a statement of structuralism about physics. As we put it in the previous section, for a physical theory to be correct, the world must have the right causal structure and the right connections to observation. Here, a *structural* theory can be understood as one that specifies causal structure (using mathematically specified laws) and connections to observation.

You can certainly deny the first premise by denying structuralism and arguing that physical theories make stronger claims than structural claims. Perhaps the most promising way to do that is to argue that physical theories make claims about *space* or *time* or *solidity* or other stuff that can't be cashed out in structural terms. I looked at those strategies in chapter 9, and I'll return to them in the next chapter.

The second premise is something we stipulated. We introduced

Nonsim Universe as any universe where our physical theories are true. All that matters really is that Nonsim Universe is a universe in which physical theories are true *enough* that we have atoms, molecules, and physical objects distributed through space and time roughly where we think they are. The theories don't need to be perfectly true. The main point is that whether or not our physical theories are true in Nonsim Universe, Sim Universe doesn't pose any major new obstacles to their being true.

The third premise is the key premise, holding that all structure in Nonsim Universe is also present in Sim Universe. Why believe this? Because Sim Universe is a perfect simulation of Nonsim Universe, set up to mirror all the causal and observational structure of Nonsim Universe.

Where observation is concerned, the observer in Sim Universe is set up to have exactly the same pattern of observations as the observer in Nonsim Universe. Given that Sim Universe is a perfect simulation, even counterfactual observations will be the same. I may not have used a telescope in actuality, but if I *had* looked at the Moon through a telescope, things would have looked the same to me in Sim and Nonsim Universes.

Where causation is concerned, Sim Universe is set up to mirror the causal structure of Nonsim Universe. Physical objects in Nonsim Universe are mirrored by digital objects in Sim Universe. When two physical objects interact in Nonsim Universe, two digital objects interact in the same patterns in Sim Universe. For example, when a bat affects a ball in Nonsim Universe, a digital bat will affect a digital ball in Sim Universe. Every dynamic pattern in Nonsim Universe is mirrored by a pattern in Sim Universe.

Why does the causal structure of Sim Universe mirror that of Nonsim Universe? Because a perfect computer simulation of a system has digital objects that correspond to every element of the original system. When a simulation is implemented, each of these digital objects is physically realized (as a pattern of voltages in a circuit, say) with causal powers, as we saw in the last chapter. The dynamics of these digital objects mirrors the dynamics of the simulated physical objects. It mirrors those dynamics even under counterfactual circumstances involving different states. As a result, the causal structure among parts of the original system is mirrored by causal structure among parts of the simulation.

Sim Universe does not have *exactly* the same causal structure as Nonsim Universe. In particular, if we're in Sim Universe, reality will have a lot of excess structure that needn't be present in Nonsim Universe. For a start, Sim Universe is created by a simulator who may choose to stop the simulation at any time. The computer may have processes that aren't part of the simulation. Even those digital objects that are part of the simulation may be realized by underlying objects (circuits and the like) with extra structure of their own.

Despite this excess structure in Sim Universe, it's not the sort of structure that makes our physical theories false. For each case of excess structure, we can run a thought experiment in which the same excess structure is present in Nonsim Universe. In each case, our physical theories are not undermined.

For example, the simulator who creates Sim Universe could be mirrored by a creator who creates Nonsim Universe. If we're in Nonsim Universe, a creator adds to the structure of the world, but that doesn't make our physical theories false. Even if the creator has the power to shut down reality at any time, atoms and other physical objects will still exist (at least prior to the shutdown), and our theories will be true. Similarly, both Sim Universe and Nonsim Universe may be embedded within a much larger multiverse. None of that makes our physical theories false. At most, it means that they apply just to our universe and not to the whole cosmos.

You might worry that a physical theory that is *complete* and *fundamental* in Nonsim Universe will be incomplete and nonfundamental in Sim Universe. It will be incomplete because it won't apply outside the simulated world. It will be nonfundamental because Sim Universe may have a quite different fundamental physics of its own. Particles in Sim Universe are realized by underlying computer processes which may be realized by a different physics in turn. But again, there are versions of Nonsim Universe where our physical theories are incomplete and nonfundamental in this way. Nonsim Universe might be a "baby universe" springing from a black hole in an originating universe with quite different laws. Perhaps there are many more levels of physics underneath the physical theories we're familiar with. None of this means that our physical theories are false or that atoms and other physical objects don't exist. It just means that these things aren't fundamental. If we're in Sim Universe, it will likewise turn out that atoms aren't fundamental, but they still exist.

There's also the role of the computer in Sim Universe. In many computer architectures, interactions between digital objects will be mediated by a central processing unit. This means that while a proton and an electron may interact directly in Nonsim Universe, their counterparts in Sim Universe will interact indirectly, with a CPU serving as a go-between. That's a structural difference. But there's a version of Nonsim Universe with the same sort of structural difference. We need only imagine that God mediates every interaction among physical objects, as in occasionalist theories of causation that originated with al-Ghazali and other Islamic philosophers and were later taken up by the French philosopher Nicolas Malebranche. If God mediates all interactions, then causation will have a surprising structure, but atoms and other physical objects will still exist.

To sum up our discussion of the third premise: while universe simulations may have excess structure compared to nonsimulated universes, all the structure in Nonsim Universe is mirrored in Sim Universe, and that's enough for our physical theories to be true there.

The conclusion is that if we're in Sim Universe, our physical theories are true. At least, the physical objects in these theories—quarks, photons, atoms, and molecules—exist and are distributed through space and time much as those theories say. And once we've established this much, there's little reason to doubt that cells, trees, rocks, planets, and other physical objects exist too.

There are some limits to the conclusion. It does not establish that beings in a simulation are conscious, so it does not address the problem of other minds. It also applies only to perfect simulations such as Sim Universe (I'll discuss simulations more broadly in chapter 24). But the strategy does establish that if we're in a perfect simulation such as Sim

Universe, ordinary physical objects in the external world exist on a par with the existence of objects in Nonsim Universe. That's a form of simulation realism.

What realizes the structure?

Suppose we're in Sim Universe, a perfect simulation of Nonsim Universe. If I'm right, our universe contains quarks, atoms, and molecules, as our physics suggests. But if we're in Sim Universe, our physics isn't the bottom level. Underlying our physics is a computer in the next universe up. Let's call that universe Meta Universe. Meta Universe may or may not itself be simulated. It has its own physics, which may be entirely different from the physics of Sim Universe. While Sim Universe and Nonsim Universe have a 4-dimensional spacetime, Meta Universe may be 26-dimensional and may be inhabited by beings we can barely conceive of.

What's the relationship between the physics of Meta Universe and the physics of Sim Universe? It's natural to say that the former *realizes* the latter. Biology in our world is realized by chemistry, and chemistry is realized by physics. If we're in Sim Universe, physics is realized by what in chapter 14 we called meta-physics—the physics of Meta Universe.

If Sim Universe is a perfect simulation of Nonsim Universe, we'll never learn about Meta Universe. Even if we're in Sim Universe, all our evidence is consistent with our being in Nonsim Universe. We can speculate that our universe is a simulation in Meta Universe, but we won't know unless the simulation is imperfect and some evidence leaks in.

All this brings out that the perfect simulation hypothesis is highly congenial to epistemic structural realism: the it-from-structure-from-X view. Recall that this view holds that whatever science tells us about the structure of physical reality, there's an underlying nature, X, which science tells us nothing about. If we're in a perfect simulation, our physics has the same structure as in Nonsim Universe but with the underlying nature of Meta Universe. We can know about the structure but not about its underlying nature.

REALITY+

Of course, the physics of Meta Universe has a structure of its own. So in this case, the structure in Sim Universe is realized by further structure in Meta Universe. This leaves open how the structure in Meta Universe is realized. Perhaps it's realized by further structure, in a Meta Meta Universe. But that just defers the problem, so let's just focus on the original top-level universe (the one that contains all the others), wherever it is.

How is structure realized in the top-level universe? There are two possibilities: First, the top-level universe may be a universe of pure structure. If so, we have the pure it-from-structure view at the top level (a version of *ontological structural realism*).

Alternatively, the top-level universe is a universe of impure structure; the structure is realized by something nonstructural. If so, we have the it-from-structure-from-it view at the top level—again, consistent with epistemic structural realism.

I don't know which of these two possibilities is correct. The universe of pure structure is austere and elegant, but does it make sense? Recall the pure it-from-bit view. How could there be *pure* bits—that is, bits that aren't differences in something more basic, such as voltages or charges, but instead are simply pure differences? The same issue arises more generally for the pure it-from-structure view. How could there be *pure* structure—logical and mathematical structure that isn't structure in anything more basic? A conservative part of my mind wants to say that this is inconceivable, but a more open-minded part wants to say that we can learn to conceive of the previously inconceivable.

The universe of impure structure is more obviously coherent—but mysterious nonetheless. Recall the it-from-bit-from-it view (introduced in chapter 8). If there is a truly basic "it" that underlies all structure, then what is the nature of this "it"? It's hard to see how we could ever know. So we're threatened with a version of the it-from-structure-from-X view where the X is forever unknowable.

One hypothesis about the basic X is especially attractive, at least for someone with philosophical views like mine: *Consciousness* may well not be reducible to structure. Frank Jackson's thought experiment about Mary the color scientist suggests that structure alone cannot capture the conscious experience of redness. Conscious experiences may have structure, but they seem to go beyond structure. So, could the basic reality underlying structure involve a fundamental sort of consciousness?

If we start from the it-from-bit-from-it view, this line of reasoning leads to the *it-from-bit-from-consciousness* view, also discussed in chapter 8. If we generalize the line of reasoning beyond digital structure, it leads to the *it-from-structure-from-consciousness* view. There are a few different versions of that view. Perhaps the structure of physics could be realized in a single cosmic mind, as in an idealist view—or by interactions among many tiny minds at the bottom level, as in a panpsychist view (which holds that everything is conscious).

The it-from-bit-from-consciousness view has the advantage of integrating consciousness with structure at a deep level, without trying to reduce consciousness to structure (as materialism does) and without separating consciousness entirely from physical structure (as dualism does). Of course, the view faces many problems of its own, not least the "combination problem" of how consciousness at the ground level of physics can somehow add up to the distinctive sort of consciousness that we experience. But I'll leave the it-from-bit-from-consciousness view on the table at least as an interesting speculation.

Kantian humility

Immanuel Kant, whose moral theory we encountered in chapter 18, had a distinctive view of reality, too. He held that there's a realm of *appearances* and a distinct, unknowable realm of *things in themselves*. We can know about appearances, but we have no knowledge of things in themselves.

Suppose you're looking at a cup. The cup that you see is an appearance. However, underlying this appearance there is also a thing in itself—in German, a *Ding an sich*. Kant holds that we cannot know about the thing in itself underlying the cup. The thing in itself is an unknowable X.

Kant called his view *transcendental idealism*. He held that appearances—the ordinary objects we perceive in space and time— are deeply tied to the human mind. At the same time, he held that things in themselves are beyond our minds and transcend our knowledge. Our inability to know about things in themselves is often called *Kantian humility*.

My analysis of the perfect simulation hypothesis is interestingly reminiscent of Kant's transcendental idealism. Suppose we're in a perfect simulation. Then when I see a cup, I can know some of its properties: I know its color and its shape; more generally, I can know its structural properties. All these can be seen as aspects of the cup as an appearance. But I cannot know the cup's underlying nature. In reality, behind the appearances, there is a digital object running on a computer in Meta Universe. In Sim Universe, the digital object is unknowable to me. We might think of the digital cup in Meta Universe as an unknowable thing in itself.

Of course, the analogy with Kant's view is imperfect. Kant would have counted the digital cup in Meta Universe as just another appearance because people in the next universe up can perceive it in space and time. He'd hold that underlying this digital cup is a truly unknowable thing in itself. But then, we've seen that the same is true for the structure of physics in Meta Universe.

Still, there's an interesting mapping from our picture to Kant's. The structure of reality corresponds to Kant's knowable realm of appearances. Whatever underlies this structure corresponds to Kant's unknowable things in themselves.

An interpretation of Kant's philosophy along roughly these lines was broached by the Australian philosopher Rae Langton in her 1998 book *Kantian Humility*. According to Langton, Kant's realm of appearances is the realm of *relations* between things, including spatiotemporal relations and causal relations. Kant's realm of things in themselves is the realm of *intrinsic* properties of things—properties they have

Figure 54 Immanuel Kant, the apparent cup, and the digital cup
in itself—with an unknowable thing in itself beyond that?

independently of relations to other things. The relational properties
of reality are knowable, but the intrinsic properties are unknowable.

In effect, the realm of appearances is a vast network of relations,
akin to the way we described the New York City subway as a network
of relations. This network of relations yields a *structural* picture of
reality. The realm of things in themselves is akin to the intrinsic nature
of the stations in the network, except at the most fundamental level of
reality. These intrinsic properties yield an *intrinsic* picture of reality.
According to this interpretation, Kant holds a version of the *it-from-
structure-from-X* view, holding that the structure involves a network
of relations and X involves underlying intrinsic properties that we can
never know.

In effect, Langton reads Kant as an epistemic structural realist, although he wrote 200 years before that view was named. Some experts on Kant object that this view does not do justice to Kant's complex idealistic picture of reality. Still, it is a picture I can understand, and even a picture that might be true. It's also a picture that helps us make sense of the simulation hypothesis. So I hope that something like this view might have been Kant's.

Whatever one says about the fine details of interpreting Kant, my interpretation of the simulation hypothesis clearly has a Kantian flavor. You can read the views of many great philosophers into the simulation hypothesis, starting with Plato and Zhuangzi. Perhaps the most popular readings use it to illustrate Cartesian skepticism or Berkeleyan idealism. But if I'm right, the most apt reading is as an illustration of Kantian humility.

Chapter 23

Have we fallen from the Garden of Eden?

I LIKE TO THINK OF THE GARDEN OF EDEN AS A PLACE WHERE everything was exactly as it seemed to be, in our pre-theoretical picture of reality.

In Eden, everything was laid out in a three-dimensional Space. Space was Euclidean and was not relative to anything. Things in Eden changed with the passage of Time. Time flowed from moment to moment in one direction, with absolute simultaneity across the garden and across the universe.

The apple in the garden was gloriously, perfectly, and primitively Red. When we Perceived the apple, the apple and its Redness were simply revealed to us directly, without any mediation.

Rocks in Eden were Solid, full of matter all the way through without any empty space. They had an absolute Weight, which did not vary from place to place.

People in Eden had Free Will. They could act with complete autonomy, and their actions were not predetermined. Their actions were Right or Wrong. They either met the standards of Morality or they did not.

Then there was a Fall. We ate from the Tree of Science, and we were cast out of Eden.

We discovered that we didn't live in an absolute three-dimensional Space in a world of absolute Time that passes. Instead, we live in a four-dimensional and non-Euclidean spacetime. Space and time are relative to reference frames, and there's no absolute now.

We discovered that we didn't live in a world where objects have intrinsic qualitative Colors that are revealed to us in Perception. Instead, colors are complex physical properties that affect our eyes and our brains in complicated ways. In perception, colors aren't directly revealed to us but instead are inferred by the visual systems in our brains.

We discovered that rocks aren't Solid. They consist of mostly empty space and are merely solid. They don't have an absolute Weight: They have one weight on Earth, another on the Moon, and in outer space they're weightless.

The jury is still out, but evidence suggests that we may not have Free Will. Our brains seem to be mechanical systems that determine our actions or at least strongly constrain them. However, we may still have free will—the ability to choose our own actions and mostly do what we choose to do. There may be no absolute standard of Morality by which our actions are Right or Wrong. Instead, there may simply be a system of morality that we construct and endorse, according to which our actions are right or wrong.

We no longer live in Eden. We're growing accustomed to our non-Edenic world. But Eden still plays a powerful role in our picture of reality. Perception still presents us with a world of Colorful and Solid objects laid out in Space and changing with Time. We naturally think of people as acting Freely and doing things that are Right or Wrong.

All this helps to explain our intuitive reactions to the simulation hypothesis. Intuitively, if we are in a simulation, nothing is as it seems. We seem to be in a universe of solid and colorful objects laid out in a certain way in space. If we're in a simulation, we're not in such a universe.

I diagnose our intuitive reaction as follows. We seem to be in an Eden-like world of Solid and Colorful objects laid out in a certain way in Space. If we're in a simulation, our world isn't like this. The simulation doesn't contain Solid and Colorful objects in Space.

But the same goes for our scientific world of quantum mechanics and relativity. Solidity, Color, and Space disappeared from the scientific world picture a long time ago. We reconceived Solidity, Color, and Space as solidity, color, and space. The simulation hypothesis is no worse off than the scientific worldview here. Neither contains Solidity, Color, or Space. But both contain solidity, color, and space.

What is the difference between Solidity and solidity, Color and color, between Space and space? That's a story I'll try to tell in this chapter.

The manifest image and the scientific image

In his 1962 article "Philosophy and the Scientific Image of Man," the American philosopher Wilfrid Sellars distinguished two ways of looking at the world. The *manifest* image is the world as it appears in ordinary perception and thought. The *scientific* image is the world as it is characterized by science.

Sellars himself was especially concerned with the manifest and scientific images of human beings in the world. In the manifest image, we're free and conscious beings whose actions result from reasons and decisions. In the scientific image, we're biological organisms whose actions result from complex neural processes in our brain. How do we reconcile these two images of ourselves?

We can find manifest and scientific images for almost anything that we ordinarily think and talk about. In principle, we can distinguish the manifest image of the Sun (the Sun as we think of it in ordinary life) with the scientific image of the Sun (the Sun as science reveals it to be). The same goes for clouds and for trees. And we can certainly distinguish the manifest and the scientific images for color, space, solidity, and many of the other phenomena in the previous section.

The two images often clash. People seem one way in the manifest image and another way in the scientific image. If the images clash, what should we do? Throw out one of the images entirely? Or remake them so that they're compatible?

The Canadian-American philosophers Patricia and Paul Churchland were both students of Sellars at the University of Pittsburgh in

Figure 55 The manifest and the scientific images: Patricia and Paul Churchland, in the Garden of Eden, eat from the Tree of Science.

the 1960s. They have long argued that the scientific image has primacy over the manifest image. Where they clash, the manifest image should be discarded. Where human beings are concerned, this leads to Patricia Churchland's program of neurophilosophy (discussed in the introduction), which holds that the science of the brain provides our best answers to traditional philosophical questions about the human mind. We should embrace a neuroscientific image of ourselves, with the brain and the nervous system at center stage.

Sellars himself thought that one of the tasks of the philosopher is to locate the manifest image within the scientific image. For any given part of the manifest image, there are a number of different things we can do with it:

(1) *Elimination*: We abandon the manifest image entirely in favor of the scientific image, as we abandon notions of witchhood and magic. The Churchlands have argued that much of our ordinary picture of the human mind should be eliminated in this way.

(2) *Identification*: We identify an aspect of the manifest image with an aspect of the scientific image, as when we identify water with H_2O.

(3) *Autonomy*: We retain an aspect of the manifest image even when it isn't present in the scientific image, as when we continue to assume a strong sort of free will in everyday life. Sellars himself argued that consciousness is real even if it cannot be fully explained in standard physical terms.

(4) *Reconstruction*: We remake the manifest image to be compatible with the scientific image, as when we remake our image of solidity to be consistent with mostly empty space.

There's no single universal answer about how best to reconcile the two images. I think all four strategies are appropriate, depending on the case. But in my view, perhaps the most important strategy, and the most often correct, is reconstruction. We can motivate this strategy by thinking about the metaphorical fall from Eden.

The story of Eden at the opening of this chapter is my own thought experiment for thinking about the manifest image. Eden is a hypothetical world where the manifest image is entirely correct. Edenic color, or Color, is color as it appears in the manifest image. Edenic space, or Space, is space as it appears in the manifest image. Edenic free will, or Free Will, is free will as it appears in the manifest image.

We don't live in Eden. Our world doesn't have Color and Space exactly as they are in the manifest image. But we don't simply eliminate color and space from our picture of the world. Our world contains color and space. In light of science, we've remade our conception of Color and Space into conceptions of color and space, and we've found a place for them in our world.

Reconstructing the manifest image

How do we reconstruct the manifest image in light of science? How do we find solidity in the scientific world? How do we find color and space? This reconstruction very often involves a move from Edenic primitivism to scientific functionalism.

What is Solidity like in the manifest image? It seems to be an intrinsic feature of objects. In the manifest image, a table is Solid because it contains matter all the way through without gaps, and it's rigid. When we eat from the Tree of Science, we learn that no ordinary objects are Solid like this. Tables are made of particles scattered in empty space.

We could react to the science by saying that nothing is solid. Solidity is merely an illusion. But few people actually react this way—the distinction between solid and nonsolid things is too useful to give up. There's an important difference between ice and water, marked by saying that one is solid and the other isn't.

Instead, we react to the fall from Eden by saying that although nothing is Solid in the Edenic sense, plenty of things are solid all the same. We have reconceived solidity, so that what it means to be solid is, roughly, to resist penetration and to have a rigid shape. In this picture, solidity is not a matter of how an object is in itself, but of how it interacts with other objects.

We have, in effect, moved to *functionalism* about solidity. Functionalism in philosophy is a view that understands a phenomenon in terms of the role that it plays. Here, the key role for solidity is resisting penetration. If an object plays this role, it's solid. To use a common functionalist slogan, solidity is as solidity does.

Functionalism got started as a view in the philosophy of mind: The mind is as the mind does. But it's applicable across any number of domains. For example, we're all functionalists about teachers: To be a teacher is to play the role of teaching students. Teaching is as teaching does. We're all functionalists about poisons: To be a poison is to play the role of making people sick. Poison is as poison does.

Functionalism can be seen as a version of structuralism, where the emphasis is put more squarely on causal roles and causal powers. When we move from primitivism to functionalism, we reconceive solidity so that it can exist in the scientific image. You might think that this is just changing the subject by changing what we mean by "solid." But importantly, rigidity and resistance to penetration were always part of the manifest image of Solidity. In remaking our conception of solidity, we simply emphasize those parts of the manifest image and deemphasize other parts, providing some continuity between Solidity in the manifest image and solidity in the scientific image.

Much the same goes for color. In the manifest image, colors seem to be primitive Color qualities that suffuse the surfaces of objects. The apple is Red, the grass is Green, and the sky is Blue. In the manifest image, these Colors aren't complicated physical properties, nor do they seem to depend on us. In Eden, Colors are simple intrinsic qualities of the external world.

Friedrich Nietzsche once said that there are no beautiful surfaces without terrible depths. After the fall, scientists examined objects in the external world and discovered that they don't have simple intrinsic color qualities in themselves. Instead, they have complicated physical properties that allow them to reflect and transmit light to perceivers. There's a long chain of optical and electrical transmission from the apple to the eye, and from the eye to the brain, before the perceiver experiences the apple as red.

Galileo and other scientists reacted to this by saying that the apple is not really red at all—that colors exist only in the mind and not in external reality. This view is intellectually coherent, but it has never really caught on, except when scientists and philosophers are in a philosophical mood. One reason is that it discards a key tool we have for classifying entities in the world. There's an important difference between apples and bananas marked by saying that apples are red and bananas are yellow. To eliminate color from our picture of the external world would be to lose this tool entirely.

A more popular reaction has been to say that even though we don't live in Eden, objects still have colors. Apples aren't primitively Red, but they're still red. The grass is not Green, but it *is* green. The great British empiricist John Locke provided a crucial tool to make sense of this reaction, arguing that we should conceive of colors not as simple sensory qualities but as *causal powers*. Redness is not a primitive quality; instead, it's the power to elicit certain distinctive sensory experiences in normal perceivers. To a first approximation: An apple is red because it has the power to look red.

In effect, we have moved from *color primitivism*, which understands colors as primitive sensory qualities, to *color functionalism*, which understands colors in terms of the functional role they play. Color is as color does. Redness is as redness does. Here, the chief role of color is causing specific experiences in human perceivers.

Space in the manifest and the scientific image

What is Space in the manifest image? The ordinary experience of space presents it as a three-dimensional container for everything. It is largely Euclidean. It is absolute: If something is square, it's absolutely square and not just square relative to something. Space is fundamental: It's the arena in which everything exists, and nothing underlies space. And just as Colors are intrinsically Colorful, Space is intrinsically Spatial—it has a special Spatial quality that is hard to capture in words.

In our first bite from the Tree of Science, Newton's physics invoked a three-dimensional Euclidean space that was close enough to the Edenic picture that we did not fall far. Einstein's physics led to a much greater fall. According to the general theory of relativity, there's no absolute, Euclidean, three-dimensional space. Instead, space is non-Euclidean, relative to a reference frame, and part of a four-dimensional integrated spacetime. There are no absolute squares. There are just approximate squares relative to certain reference frames.

In relativity theory, spacetime at least remains fundamental, but later theories have questioned that. On some interpretations of quantum mechanics, what is fundamental is a much higher-dimensional space inhabited by a quantum wave function, and our three-dimensional space is merely something derivative. In string theory and other theories that try to reconcile quantum mechanics and relativity, there's increasing speculation that space is not fundamental but emergent. The fundamental laws of many theories don't postulate space at all. Instead, space emerges at a derivative level.

This is another factor in our fall from Eden. There is no Space as it exists in the manifest image. One reaction is to say that genuine space doesn't exist at all. But again, this throws away a central tool we have for making sense of the world. Size, distance, angle, and other spatial measurements are vital to our worldview. It makes more sense to say that even though we don't live in Eden, space still exists. Nothing is primitively and absolutely Square, but there are still square objects. Space is not the fundamental container for everything, but objects still have locations in space.

Spatial functionalism understands space in terms of the roles it plays. Space is as space does. To maintain continuity with the manifest image, we need these to be roles that space plays in the manifest image. What are these roles? Where the main role of solidity is in interaction with other objects, and the main role of color is in perception, space plays at least three major roles.

First, space mediates *motion*. In Eden, things move continuously through space. Second, space mediates *interaction*. In Eden, physical things interact when they're spatially in contact with each other or at least close to each other. There's no action at a distance. Third, space causes our spatial *perception*. In Eden, square things look square, at least in normal conditions.

Spatial functionalism identifies space as whatever plays these roles. That is, space is what mediates motion and interaction, and what causes spatial perception.

A slogan from the Canadian philosopher Brian Cantwell Smith is apt here. Smith takes the old slogan "There's no action at a distance" and inverts it to say "Distance is what there's no action at." Or better, since science seems to support some action at a distance, but with less interaction than there is at close range: "Distance is what there's less action at."

We can also capture the Edenic idea that motion is continuous by saying, "There's no motion at a distance." Then we can invert it and qualify it as before, yielding "Distance is what there's no motion at," or better, "Distance is what there's less motion at." In effect, we understand distance as what plays the distance role, where the key role of distance is mediating motion and interaction.

This way, even if there's no Space in the scientific image, there's still space. Relativistic space still mediates motion and interaction, and it still produces spatial experience. In theories in which space isn't fundamental, we can locate space at a derivative level by finding the quantities that (1) tend to bring about spatial experience and (2) tend to mediate motion and interaction, especially in macroscopic objects. That helps us to explain how space emerges.

Solidity, color, and space in virtual reality

These functionalist analyses show us how objects can be solid, colored, and spatial even in virtual reality or in a simulation. Virtual objects are not Solid, Colored, or Spatial in the Edenic sense. But they can still be solid, colored, and spatial in the functional sense.

How can a virtual object be solid? Like a physical object, a virtual object is (virtually) solid when it is rigid and resists penetration. One virtual object might let other virtual objects pass right through it—so it isn't solid. Another might block other virtual objects from penetrating it; it's solid. In VR as it is today, you can pass your hand through many virtual objects, though they may resist penetration from other virtual objects. As a result, they are at best partially solid. In the future, there might be some kind of resistance so that virtual objects are more fully solid.

How can a virtual object be colored? Like a physical object, a virtual object is (virtually) red when it produces the right sort of reddish experiences in normal conditions for perception. A virtual apple can produce reddish experiences in normal conditions for virtual perception (through a headset). If so, we'll count it as virtually red.

How can a virtual object be spatial? Like physical space, virtual space is whatever produces spatial experience and what mediates motion and interaction. In VR, this is usually a digital property involving an encoded location. Objects often move discontinuously in VR—by teleportation, for example. Objects can interact at a distance, as when one's avatar picks up a ball by pointing to it. Nevertheless, there's usually less motion at a distance and less action at a distance—at least there are standard modes of motion and of interaction that work best when distances are short.

We saw in chapter 10 that spatiality is essential for a virtual world. We can now see why. Space is, in effect, a measure of motion and interaction. Without space, objects cannot move continuously through the universe. Without space, any two objects could interact equally easily. In order for the universe to display some distinction between the local and the global, we need structured motion and interaction so that we interact with a local environment in a distinctive way. Once we have this, we have space.

Virtual space isn't the same as physical space—at least if the simulation hypothesis is false. The same goes for virtual colors and virtual solidity. These are simply analogs of physical space, physical color, and physical solidity. Whereas physical space plays the space role in the physical world, virtual space plays the space role in a virtual world.

If the simulation hypothesis is true, on the other hand, then solidity *is* a sort of virtual solidity, and virtual solidity is what has played the solidity role for us all along. Likewise, colors are virtual colors, and space is a virtual space. In our it-from-bit world, these are what play the color role and the space role.

There's a powerful intuition that if we're in a simulation, nothing is laid out in space as it seems to be. We can now diagnose this as an Edenic intuition. It's true that if we're in a simulation, nothing is laid out in Space as it seems to be. But the same is true of relativity, quantum mechanics, and string theory. Once we reconceive space as whatever plays the space role, there's room for space in those physical theories, and there's also room for space in a simulation.

Is reality an illusion?

Was Vishnu right, after all? In light of the fall from Eden, is ordinary reality an illusion? We seem to be in an Edenic world, but we are not. Things seem to be Red and Square, but nothing in the external world is Red or Square. As Cornel West put it: Is it illusions all the way down?

Yes and no. Insofar as perception says that objects are Red, it's illusory. Nothing is Red. Insofar as perception says that objects are red, it's not illusory. Many things are red. I've argued that for perception to be *perfectly accurate*, things would have to be Red. But they can be *imperfectly accurate*, even if things are merely red. Perception has some illusory elements but nevertheless can be used as an accurate guide to reality.

We think of Eden as a place where we have highly reliable knowledge of the external world. In Eden, we have direct acquaintance with everything. But if I'm right, our Edenic model of the external world also leads to Cartesian skepticism. When our model attributes Space, Time, and Color to the external world, it's too easy for it to be wrong. Once we think of the external world as containing merely space, time, and color, it's much easier for our model to be right.

As we've fallen from Eden, we've moved from a *fragile* to a *robust* picture of the manifest world. The Edenic model of Colors as primitive, intrinsic properties is fragile—easy to doubt or disprove. The post-fall model of colors as functional properties is robust; there's certainly something that plays the color role.

The key point is that in a post-Edenic world, our knowledge of the structure in the external world is more robust than our knowledge of its intrinsic qualities. We perceive the apple as being Red. In Eden, we were directly acquainted with objects in the world so we Perceived the Redness of the apple directly. But after the fall, there's a long causal chain between us and the apple: a Green apple or a Colorless apple could in principle have reflected the same light and caused the same experiences of Redness in us, and they would have looked just the same. So it's hard to see how we can know whether the apple is Red or Green, or whether it has a Color at all. But we can know whether an apple is red or green. These apples ordinarily look red, so they have a property that plays the redness role in causing perception. That's all it takes to be red, according to the structural conception of redness.

Similarly, it's hard to see how we can know that the external world contains genuine Space. A world with the same structure but no Space could have affected our experiences in the same way as a world with Space. But it's much easier to know that something plays the space role. That's all it takes to be space, according to the structural conception of space.

In his recent book *The Case Against Reality: Why Evolution Hid the Truth from Our Eyes*, the cognitive scientist Donald Hoffman makes an evolutionary case for skepticism. He argues that evolution doesn't care about whether our beliefs about the world are true. It just cares about whether we're fit—whether we survive and leave offspring. He also argues that there are many more ways for our beliefs to be massively false than for them to be true, so we should expect most of our beliefs to be false: The world is almost certainly not as it appears.

Hoffman's argument assumes something like an Edenic model of perception. It's true that the Edenic content of our beliefs is likely to be false. We can't know that an apple is Red or a ball is Spherical. But once we move to a structuralist conception of perception and reality, our model of reality is robust. We can be much more confident that an apple is red or that a ball is spherical. We can no longer conclude that reality is not as it appears. I make this case in detail in an online appendix.

That said, I agree with Hoffman where our perception of an Edenic world is concerned. This perception doesn't latch on to reality. There are no Colors and Sizes in the external world. I can even happily endorse Hoffman's idea that Edenic qualities serve as a sort of "interface" in perception. In effect, we're presented with an Edenic world that serves as a useful guide to the structure of the true external world, even though the true external world is not itself Edenic.

I diverge from Hoffman on the idea that perception doesn't tell us anything about the true nature of the external world. It tells us about the colors and sizes of things just fine. Knowing about colors and sizes may not tell us about the intrinsic Colors and Sizes of things, but it still tells us a great deal about the structure of external reality.

Imperfect realism

In a 1996 essay, the Slovenian philosopher Slavoj Žižek said: "the ultimate lesson of virtual reality is the virtualization of the very true reality." We can now see that there's an element of truth to this. Reflection on virtual reality led us to conclude that virtual worlds are as real as an ordinary physical world. But one could equally put the point by saying that our ordinary physical world has been effectively virtualized: that is, it is only as real as a virtual world. Rather than being an Edenic world of Color and Space, it is a world of structures that play the relevant roles.

In light of all this, I think we should reject a naïve realism about the external world that says things are exactly as they seem, but we should also reject a simple illusionism. The right view is a sort of *imperfect realism*. There are no Colors exactly as they seemed to be. But there are still colors, which can play many of the roles of Colors and are still crucial in making sense of the world.

I think imperfect realism is the correct view for many important philosophical topics. We may not have Free Will, by means of which we can perform undetermined actions, but we still have free will. There may be no Right and Wrong, absolute standards of morality entirely independent of human beings, but there is right and wrong.

Some people will say that lowercase free will is not really free will. They'll insist that only uppercase Free Will is genuine free will. There is some danger of getting into an argument over words here, but there is a important issue underneath. That issue is whether free will, or only Free Will, can give us what we care about.

Why do we care about free will? One major reason is that we think that free will is required for moral responsibility. Only if we act freely are we responsible for our actions. Do we need Free Will in order to be responsible for our actions? It's not obvious. If you act merely with free will, performing an action you choose, then even if your choice was determined, it's arguable that you're responsible all the same. If so, free will gives us what we care about where moral responsibility is concerned, and we don't need full-scale Free Will. That said, perhaps there are other reasons to care about Free Will.

Under imperfect realism, is reality an illusion? Yes and no. Edenic reality is an illusion: Edenic Color is not real. Nor are Space, Solidity, or Free Will. If we believe that the apple is Red and Spherical, we're wrong. But color is real, and so are space, solidity, and free will. If we believe that the apple is red and spherical, we're right. According to imperfect realism, the manifest image is partly an illusion—but it has a structural core that isn't an illusion at all.

What about Consciousness? Can we say that Consciousness is an illusion, but lowercase consciousness exists? This is a respectable view with a long history, but it faces some obstacles. When we structuralize color and space, we move Color and Space into the mind. The special redness of red remains as an aspect not of the external world but of Consciousness. To structuralize Consciousness, we need to move these qualities somewhere else or get rid of them entirely.

A structuralist might dismiss the special qualities of Consciousness as an illusion—that is, we think these qualities exist when they do not. It's harder to dismiss these qualities than the qualities in the external world, though. Our introspective experience of the redness of red makes it something of a datum. Jackson's thought experiment of Mary in the black-and-white room looms large. Mary knows a full structural description of the brain, but when she leaves the room, she gains new knowledge of color experience that her structural knowledge could not give her. Unless the structuralist is prepared to dismiss this new knowledge as an illusion, there remain large obstacles in reducing consciousness to structure.

I'm not dismissing imperfect realism about consciousness entirely. I'm certain that many of our ordinary beliefs about consciousness are wrong, and that a fully developed science of consciousness will have many surprises. So there are sure to be differences between Consciousness in the manifest image and consciousness in the scientific image. But as things stand, there are strong reasons to think consciousness goes beyond structure. At least, to fully explain consciousness in terms of pure structure would take a scientific revolution or two. That said, there have been scientific revolutions before.

Plato's Forms and the fall from Eden

For Plato, the most fundamental reality is the world of Forms. The Forms are the special essences of all things in their purest and most perfect nature. Among many others, there is a Form of Largeness, of Squareness, of Solidity, of Beauty, and of Good. The Forms are timeless, absolute, and never-changing.

Plato thought that ordinary physical reality is a mere imitation of the world of Forms. Ordinary square objects are mere shadows of the Form of Squareness. He introduced his allegory of the cave to illustrate the difference between ordinary reality and the world of Forms. He thought that the world of Forms is a far deeper reality than the ordinary world of shadows, and that everyone will prefer to be in it, given the choice.

My world of Eden has some elements in common with Plato's world of Forms. Eden isn't timeless or unchanging; events there unfold in time and space (or at least in Time and Space). But it includes essences such as Squareness, Solidity, and Goodness in a perfect form. People in Eden are directly acquainted with these essences and have a kind of direct access to reality.

By contrast, the structural world of science looks a little like Plato's cave of shadows. There's no Redness; there's just something that plays its role. Objects in ordinary reality look like shadows of objects in Eden.

Is the fall from Eden analogous to a retreat into Plato's cave? In our non-Edenic world, are we missing something crucial to the good life?

I'd say no. Even after the fall from Eden, we have consciousness much as before. Through consciousness, we have acquaintance with Color, Space, Causation, and other pure forms. The objects around us may not be Red, but we experience a world of Redness all the same. We have knowledge of Redness as a universal form, which Plato took to be the most important sort of knowledge of all.

Has the onset of illusion in the fall from Eden made our life worse? In Eden, people perceived Red and Square things directly. We seem to experience a world of Red and Square things, but our world is merely a world of red and square things. But even if this makes our lives imperfect, it's hard to see how it makes our lives much worse.

Eden is a kind of hypothetical ideal. We can think of it as Reality 0.0. Ordinary physical reality, after the fall, is Reality 1.0, at least assuming we're not in a simulation. Virtual reality is Reality 2.0. In both ordinary reality and virtual reality, Eden has been stripped down to its structural core, with consciousness at its center. But virtual reality remains on a par with ordinary reality.

Chapter 24

Are we Boltzmann brains in a dream world?

A *HUMAN, A SIM, AND A BOLTZMANN BRAIN WALK INTO A BAR . . .* I don't know the punchline to this joke, but it probably involves the Boltzmann brain exploding. In a Boltzmann brain, named after the 19th-century Austrian physicist Ludwig Boltzmann, matter assembles randomly into the exact configuration of a human brain for a moment. The vast majority of Boltzmann brains will explode back into chaos a moment later. Boltzmann brains are extraordinarily improbable, but in a big enough universe, improbable things will eventually happen. In fact, some physical theories predict an infinite spacetime in which we can expect Boltzmann brain duplicates of my brain to form an infinite number of times.

For one brief shining moment, these Boltzmann brains may have experiences just like mine. This raises the question: how do I know I'm

Figure 56 The short life of a Boltzmann brain.

not a Boltzmann brain right now? I seem to remember a long past and anticipate a long future, but then so does a Boltzmann brain. I can test the hypothesis by waiting a moment to see if I survive—but seconds later when I find myself intact, for all I know I am a Boltzmann brain who formed at that moment with a false memory of testing the hypothesis seconds beforehand.

If I am a Boltzmann brain, the reality of my external world is threatened. Unlike sims, Boltzmann brains are not surrounded by detailed simulations of an external reality. They are typically surrounded by disordered randomness. Even if I am right that sims have largely true beliefs, almost everything that Boltzmann brains believe about the external world is false. So Boltzmann brains threaten to reinvigorate the skeptical challenge. If I can't know I'm not a Boltzmann brain, I can't know that the world around me is real.

The Boltzmann brain is just one of many alternative skeptical challenges. For most of this book, my focus has been on near-perfect simulations, like the one in *The Matrix*. In the most extreme version of the Matrix simulation, the whole universe is simulated, the laws of physics are simulated faithfully and exactly, and I've been in the simulation my whole life. I've argued that if we're in this sort of Matrix-style simulation, reality is not an illusion. So this version of the simulation hypothesis doesn't threaten our knowledge of the external world.

But what about other skeptical hypotheses? For a start, we can consider *local, temporary,* and *imperfect* simulations. In these simulations, we'll certainly have many false beliefs. Might these threaten our knowledge of the external world? There's also Descartes's dream hypothesis and the evil-demon hypothesis. Might I be in a dream world right now? Are there threats to our knowledge here? Finally, can my knowledge of the external world survive the Boltzmann brain hypothesis?

I'll consider many of these hypotheses in this chapter, starting with local simulations and ending with Boltzmann brains. I'll argue that while there are some important skeptical challenges here, none lead to global skepticism. This will allow us to step back and assess how far we've gotten in responding to Descartes's challenge. Toward the end of the chapter, I'll make a case that we may even have some knowledge of the external world.

Local simulations

In a local simulation, just a part of the universe is simulated. Perhaps only New York City is simulated, and places beyond it are not. As we saw in chapter 2, it's not easy to make a local simulation work. I may live in New York City, but I have memories of Australia. I read articles and see videos about places all over the world. I often travel out of the city. I regularly talk with people who are outside the city. For a simulation of just my life in New York City to work, a lot of what's outside the city will need to be simulated, too.

Compared to a global simulation, a local simulation can cut some corners by not simulating parts of the world that have little impact on my life. Perhaps the simulation can get away with only a relatively simple model of uninhabited parts of the Antarctic, say, or the deep sea floor. Or perhaps we could get away with a detailed simulation of Earth, a less detailed simulation of the Moon and Sun to capture their effects on Earth, and a fairly simple model of the universe beyond.

Suppose we're in this sort of local simulation. What should we say about reality? I would say that ordinary things on Earth are real—tables and chairs, dogs and cats, oceans and deserts. The local simulation hypothesis is equivalent to a local it-from-bit creation hypothesis, where a creator creates Earth by setting bits in motion but pays much less attention to the rest of the universe. As a result, the Earth and everything on it will be real. The Sun and Moon will probably be real, too, although they may not have the detailed physical structure we thought they had. Distant stars will be even more different from what we thought. Unobserved stars may not exist at all.

If this is right, the local simulation hypothesis leads to *local simulation realism*: The local things that are simulated are still real, even if more distant things aren't. Some of our beliefs about the wider world may be false. For example, I may be wrong in believing that the

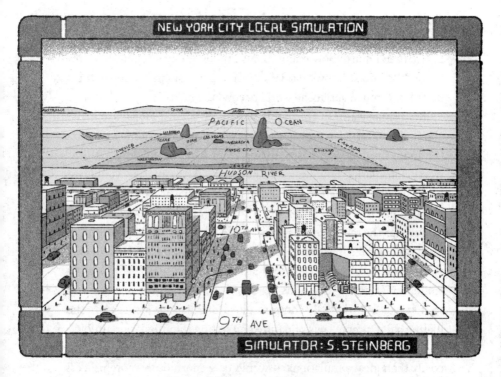

Figure 57 A local simulation of New York City
(with apologies to Saul Steinberg).

Sun is governed by the same physics that governs Earth, or that unobserved stars exist. But most of my ordinary beliefs about the local environment will be true. Even if I'm only in a local simulation, I'm still sitting in a chair at a desk looking out the window at the city beyond.

In a local simulation, our situation is akin to that of Truman in *The Truman Show*, who lives in a bubble where everything is arranged for him. Truman believes many things about the world beyond his bubble, but many of them are false. Ordinary physical things in the bubble, however, such as tables and chairs, are still real, and his beliefs about them are largely true.

Like *The Truman Show*, the local simulation scenario is a local skeptical scenario. In that scenario, some of our beliefs are false, but our core beliefs about the local environment are mostly true. So the local simulation hypothesis can be used to support skepticism about some of our beliefs. If I don't know that I'm not in a local simulation of Earth, then I don't know many things I thought I knew about Mars and beyond. If I don't know that I'm not in a Truman-style bubble, then I don't know much about the world beyond the bubble.

My primary aim in this book is to argue against global skepticism about the external world, and I haven't given an argument against the more limited skepticism arising from local simulations. Still, it's worth considering how we might handle it. To do this, we would need to go beyond simulation realism and argue that we're not in a local simulation.

Perhaps the most obvious objection to the idea that we might be in a local simulation is Bertrand Russell's appeal to simplicity. Like *Truman Show* scenarios, local simulations are much more complex than global simulations because decisions need to be continually made about what to simulate and when. By contrast, global simulations just require simulating a few simple laws of nature and letting the simulation unfold.

On the other hand, global simulations may well be much more costly than local simulations. It would take enormously more processing to simulate a large universe like ours than it would to merely simulate Earth. So there's a reasonable case that, given the costs of global simulations of a universe as large as ours, local computer simulations may well be more common than global simulations. So I'm not sure we can know we're not in a local simulation.

Still, we've seen that local simulations are unlikely to be *too* local. Insofar as there's a sweet spot for local simulations around Earth or the solar system, the resulting skepticism will be skepticism about the very far away: distant stars, the core of Earth, and so on. Perhaps that's a degree of skepticism we can live with.

Temporary simulations

A variant of the local simulation hypothesis is the *temporary* simulation hypothesis, in which only a limited period of time is simulated.

Perhaps only the 21st century has been simulated, or perhaps the simulation just started today.

In one version of the temporary simulation hypothesis, we first existed in a nonsimulated world, and were then transferred to a simulation which is so good that we never notice the difference. Suppose that just yesterday you were kidnapped while sleeping and transferred to a fully convincing simulation. Is the world around you real?

In this scenario, when you see a cat, it's not a real cat, it's a virtual cat. Assuming that we didn't grow up in a simulation, a virtual cat is quite different from a real biological cat, as we saw in chapter 11. Virtual cats are real digital objects with causal powers independent of us, but they're not real cats. So when you believe there's a cat in front of you, you're wrong.

At the same time, many of your beliefs remain true. Your memories of growing up may still be perfectly accurate. Your hometown and your grandparents may be just as you remember them. Your beliefs about what's going on elsewhere in the world may be fine, too, assuming that the world hasn't been destroyed in the interim. It's your beliefs about your current environment that are wrong. So the temporary simulation hypothesis leads at worst to local skepticism about your current and recent environment and not to global skepticism about the external world.

What if the temporary simulation is extended backward in time? If your whole life from birth onward is simulated, I'd say that this is the only reality you've known, so the objects in it are real. The cats around you are real cats—it's just that cats in your reality are made of bits. (Your word "cat" has always picked out virtual cats, as in chapter 20.) Even if you entered the simulation at five years old, it may be that by now you've spent enough time there so that simulated cats count as real cats for you. If you entered two years ago, who knows? Perhaps your word "cat" now applies equally to virtual and nonvirtual cats. Anyway, there's no threat of global skepticism.

Russell once asked: What if God created reality five minutes ago, with our memories and the fossil record intact? We can ask the same question about a simulation: What if it was created five minutes ago out of whole cloth, with simulators programming in our memories? In the standard view of Russell's scenario, we'll have mostly correct beliefs about our present environment, just not about the past. I'm not so sure. How were our memories and the fossil record created? The obvious way would be to run a detailed simulation of the past, so as to capture our memories and the fossil record. But now this begins to look like a situation in which we began life in one simulation and were then transferred to a different one or to a nonsimulated environment. In that situation, I'd say our beliefs about the past are largely accurate. It's just our beliefs about the last five minutes that may be false.

What's the best response to skepticism about the last five minutes or the last year? I'd be inclined, again, to appeal to simplicity. It requires a much more complex setup to bring someone up in a non-simulated world and then transfer them to a simulated world. The same goes with transferring somebody from one simulation to another (especially since considerable differences between them would have to be involved if they aren't to be one continuous environment). Unlike the local simulation scenario, there are no obvious efficiency advantages to doing this. It seems unlikely that many simulations will be like this. So perhaps there is at least a simplicity case against this sort of skepticism.

In an extreme version of temporary simulation, only the present moment is simulated. Sometimes this will be easy. If I am just waking up from a nap in a dark room, simulators may need only to simulate a few thoughts and experiences with a minimal world model. When I am fully awake and attending to the world as I am now, temporary simulation will take more work. The simulators will at least need a model or a basis for everything I am perceiving and that I am thinking about. If I might turn my thought and perception to other things, simulators should be ready to simulate those things, too; but perhaps this simulation is so brief that this is unnecessary.

Could I be in such a momentary simulation right now? I can't rule it out with certainty, though I don't see much reason to think these momentary simulations will be common. In any case, the world model that simulators use for what I am perceiving and thinking about will always serve at least as a local reality, as with the local simulations discussed earlier. Another interesting case is a momentary simulation of my whole brain, but in this case my brain's historical environment will serve as a model (as in the two-brain preprogramming scenarios we'll discuss shortly). In all these cases, the model will serve as the basis for at least my local reality, thereby avoiding global skepticism.

Imperfect simulations

In an imperfect simulation, physics is not simulated perfectly. There are various scenarios here.

In an *approximate* simulation, there's an approximation to physics that cuts corners by approximating physical values only to a few decimal places (as in the work by Beane, Davoudi, and Savage discussed in chapter 2). If we're in an approximate simulation, I'd say that physics is only approximately true. There are still particles, forces, and so on, but they obey physical laws only approximately rather than exactly. Ordinary objects such as tables and chairs will be unthreatened.

In a *loophole* simulation, physics is mostly simulated perfectly, but there are occasional loopholes to physical laws, which allow things like Neo's red pills and other means of communication with the world beyond. If we're in a loophole simulation, I would say, again, that physics is only approximately true. There are occasional exceptions to allow special events to occur, as there might be in a physical universe where God allows occasional miracles. Again, there's nothing here to threaten the reality of physical objects.

In a *macroscopic* simulation, things are mostly simulated at the macroscopic level of ordinary objects, bringing in lower levels such as biology, chemistry, and physics only when necessary. Many macroscopic phenomena depend on physics, so it's far from obvious that there will be an adequate macroscopic simulation that ignores physics most of the time. But it may be possible to cut corners on lower-level phenomena at least some of the time. If we're in a simulation in which physics mostly goes unsimulated, then most objects in our world won't be made of atoms, so many of our beliefs about physics will be false. If the simulation operates directly on simulated chairs and tables, perhaps we should embrace an it-from-bit view in which chairs and tables are composed directly of bits without the intermediate levels of physics and chemistry. But, once again, these ordinary objects will be real. This hypothesis might raise a skeptical question about whether we know physics to be true, but it doesn't make us question ordinary reality.

As with the local simulation hypothesis, there's a reasonable case that approximate and macroscopic simulations will be common since they're cheaper to run. I can't rule out that we're in a simulation like this. So there remains a skeptical question about our knowledge of physics and the microscopic world.

There are some possible defenses to that skepticism. For example, this sort of simulation may work best when the simulators (1) have an antecedent model of the laws of physics and (2) make sure that all observations (or at least all recorded observations) are consistent with those laws, which means (3) simulating the laws of physics and microscopic structure in detail whenever necessary for observation. You then could argue that in this world, observations and observed reality are governed by the laws of physics in a "just-in-time" way: That is, if some unobserved reality becomes relevant to a later observation, it's simulated in detail. You could further argue that this is enough for the apparent laws of physics to be the laws of the universe—or, at least, the laws of the observed universe. Any skepticism left is relatively minor.

More generally: imperfect simulation leads at worst to a relatively limited skepticism about theoretical matters, and not to skepticism about the ordinary macroscopic world.

Preprogrammed simulations

You'll recall from chapter 1 that Robert Nozick's experience machine was a *preprogrammed* simulation, following a script that is fixed in advance. It contrasts with ordinary, open-ended simulations, which leave room for users to make choices, and in which many different histories can be simulated, depending on how things develop.

What if we're in a preprogrammed simulation? Is the world around us still real?

It's not obvious how a preprogrammed simulation can work. What happens if a user performs an unscripted action? The simulation can't handle going off-script. So a preprogrammed simulation must be set up so that the user always sticks to the script. Perhaps the user's brain can be manipulated so that they always do what the script dictates. I'll set this sort of case aside until later. Another possibility is that the scenario is finely tuned to the user. Perhaps the user's brain is analyzed in advance and a script is developed that the user will never deviate from.

An elegant way for this setup to work, suggested to me by Adam Fontenot, is *two-brain preprogramming*. We start with two identical brains, or perhaps two identical simulated brains, and we run them through two versions of the simulation. The first simulation is a "trial run" that isn't preprogrammed. We put the first brain in a simulated environment and see what it does. We record all the inputs the brain receives. For the second, preprogrammed simulation, we feed all those inputs to the second brain. Because it's an identical brain receiving identical inputs, then (assuming determinism) it will behave exactly as the first brain did. If consciousness depends on certain brain activity, the second brain will have exactly the same conscious experiences as its twin. But it isn't interacting with an external world, or even with a simulation. It's simply receiving a preprogrammed set of inputs.

The two-brain preprogramming case is one of the harder challenges to my view of simulation and skepticism that I have encountered. Suppose we're in this sort of preprogrammed simulation: What should we say about external reality? It's tempting to conclude that the external world we perceive doesn't exist, since we're receiving inputs without even simulated objects behind them. And certainly any actions we might take wouldn't make a difference to the external world. Our action systems may well be set up so that our decisions have no effect on anything outside the brain. This threatens us with global skepticism.

I think global skepticism can be avoided. The situation is analogous to other scenarios with one environment and two brains, like the one developed by Daniel Dennett in his story "Where Am I?" (discussed in chapter 14). In Dennett's story, a biological brain and a silicon brain receive the same inputs and are kept in sync until one day they come apart. Greg Egan lays out a similar scenario in his short story "Learning to Be Me," in which everyone is born with a dual system that serves as a backup in case something happens to their brain in later life. In these scenarios, only one system controls external action at a given time. The other system's actions don't affect anything outside the brain.

In the Dennett/Egan two-brain scenario, it is natural to say there are two people and two streams of experience. Both have largely accurate perceptions of the external world and largely true beliefs. However, only one of them is controlling action. The other system will believe that it's kicking a football, but it's wrong. It is perceiving the football and trying to kick it, but the actual kicking is coming from the other brain.

The two-brain preprogramming scenario is much like the Dennett/Egan scenario. Inputs are fed to both brains, both process the inputs and have the experience of performing actions, but only one of them really does so. The only difference is that in the preprogramming scenario, there's a delay in feeding the input to the second brain. I don't think this changes anything essential, however. If Dennett's silicon brain was delayed one second behind the biological brain, I'd say it's still accurately perceiving the world. You might think it's mistaken when it thinks, "The ball is in front of me now" when the ball was there a second ago. However, this turns on an Edenic model of Time. If we embrace temporal functionalism, then when this system applies the word "now" to the physical world, it means something like the time in the world that causes the experience of now (compare: "Red" means the property that brings about the experience of red). So the second brain's belief will still be accurate—and there's no reason why the length of the delay should make a principled difference. Generalizing, the second brain has largely accurate beliefs. When it experiences a football, it is experiencing the same digital football that the first brain interacted with, just via a time delay. Its world is perfectly real. The second brain has false beliefs about its decisions affecting the world, but the same goes for the second brain in the Dennett/Egan scenario.

At worst, these two-brain cases challenge our knowledge that our decisions and other mental states guide our actions and affect the world. The threat is that our minds don't affect the world. That's an important issue in its own right and has been the subject of much debate. Again, there may be various defenses. Perhaps you could make a case that two-brain scenarios are too complex to be common. (On the other hand, you can see a case for simulated creatures having ongoing backups, as in Egan's story.) Perhaps you could argue that there's only one *mind* corresponding to the two brains, and that mind affects the world. In any case, skepticism about the mind's effects on the world is a different issue from those we've been addressing. If I'm right, these two-brain cases don't justify global skepticism about the external world.

Gods and evil demons

What about classical skeptical scenarios involving gods or evil demons? Let's start with a case in which your sensory experiences come from God, who holds a model of the universe in his head. In this case, God is playing the role of the computer. God will presumably have thoughts corresponding to this table and this chair, or at least to the particles that make them up. Instead of the table and chair being constituted by bits in a computer, they're constituted by ideas in God's mind. But they remain perfectly real: They exist, they have causal powers, and they are real tables and chairs. It's true that they're not entirely mind-independent, as they depend on God's mind. But at least they don't depend on the mind of ordinary observers like you and me.

As for the evil demon: I'd say that despite the demon's bad intentions, his case is equivalent to God's. The demon will presumably need some sort of model or internal simulation to keep track of what sensory inputs to send you, and what to do with the outputs it receives. If the demon has a full-scale model of physics throughout the universe, the demon scenario will be akin to the perfect simulation hypothesis. If the demon cuts corners with a limited model, it will be akin to the local or temporary or imperfect simulation hypothesis. Either way, tables and chairs will exist. They will simply be constituted by processes in the evil demon's mind.

Another sort of evil demon messes with your reasoning. Descartes entertained the idea that God could make you go wrong when you added two and three. He thought God wouldn't deceive you like that, but an evil demon might. What if an evil demon is manipulating your mind into believing that two plus three is five when it's really six? This is an extreme skeptical scenario, and I *think* I can rule it out by reasoning, but if an evil demon is manipulating my mind, then my reasoning is not to be trusted. More generally, what if the evil demon is manipulating me into believing assorted things about the external world that I have no evidence for? The simulation need not even be coherent. My *impression* that my world is coherent may itself be a product of the evil demon's manipulation.

I think of this sort of skepticism primarily as *metacognitive* skepticism, or skepticism about one's own reasoning. It raises difficult and fascinating issues somewhat distinct from skepticism about the external world. Still, it may have consequences for external-world skepticism: If I can't rule out that an evil demon may be tampering with my reasoning, do I know anything about the external world at all?

Briefly: I'd argue that as long as we reason well enough about mathematics, then we know. When we reason well enough to prove that two plus three is five, then we can know that two plus three is five. Perhaps we can't rule out that an evil demon is tampering with our mind, but we can still rule out, by good reasoning, that two plus three is six. Admittedly, someone whose reasoning was twisted by the demon might say there was proof that two plus three is six. They're reasoning badly, so they don't know the truth. But when we reason well, we can know. The same goes for knowledge of the external world. *If* good reasoning supports the conclusion that there is a chair in front of me, then the mere possibility of this demon doesn't undermine this reasoning. I'm still able to know about the chair.

Dreams and hallucinations

What about the *dream hypothesis*? Am I dreaming right now? Most dream experiences are unstable and fragmented, and my experiences aren't like that. But let's imagine an unusual dream—one that's stable and unfragmented. In this case, I'd say the dream hypothesis is like the God or evil-demon hypothesis, except that I'm the one running the simulation. My dream experiences are determined by a model of the world somewhere in my own mind.

If I've been dreaming for just a short while, we'll have an analog of the recent simulation hypothesis in which the simulation kicked off only recently. In this case, my beliefs about the world around me may be false, but my memories may be fine.

If the idea is that I've been dreaming my whole life, then the dream environment will be my reality. In the extreme case, in which the dream involves a full-scale model of the whole universe, we'll have an analog of the perfect simulation hypothesis. In that case, everything in my world will be real. The body I perceive is real, even though it is a dream body. I have really lived this life. I may have another body and another life in the world of the dreamer, but that does not make this body and this life unreal. In a less extreme case in which the dream involves a par-tial model, we'll have something more like the local, temporary, and imperfect simulation hypotheses. In those cases, electrons and faraway galaxies may not be real, but tables and chairs and my body will still be dream entities that are real in many respects.

The same goes for Zhuangzi's butterfly dream. When Zhuangzi dreams of the butterfly, there is a real dream butterfly in the dream world, grounded in his mind. When a butterfly dreams of Zhuangzi, there is a real dream Zhuangzi in the dream world. If these are ordinary short and sketchy dreams, the dream butterfly and the dream Zhuangzi may be simple entities whose embodiment is akin to a simple avatar in a video game. In the extreme case of a lifelong dream of a full-scale world, however, then the dream butterfly and the dream Zhuangzi may be as real and as complex as the nonvirtual versions. Perhaps Zhuangzi cannot know that he is not in a butterfly's dream, any more than we can know that we are not in a simulation. He may even be a dreaming butterfly with its own butterfly body in the next world up. But even if he is a dreaming butterfly, he is also Zhuangzi, and his world is real.

The lifelong dream hypothesis falls short of reality in one important respect: in a dream, my reality depends on my own mind. My dreams are constructed by me, although the construction is not something I'm conscious of. (This is why Freud thought that dreams are a key to the unconscious mind.) If I'm in a lifelong dream, my whole reality is constructed by me. This world falls short on the third element of the reality checklist we built in chapter 6—mind-independence. So there's at least one reasonable sense in which dream objects aren't real. Still, if the lifelong dream hypothesis is true, the ordinary objects I perceive still exist, even though they're dream objects inside my mind. Most of my beliefs about them may be true. In principle, dream objects need not be illusions.

What about ordinary dreams? Does my view imply that these are a sort of reality? Ordinary dreams are relatively brief, so they are closer to a temporary VR experience than a lifelong simulation. They are also far more unstable and fragmented than most simulations. Typically, one does not know one is dreaming. As a result, there will be many illusions and false beliefs in dreams: you think you are chasing a dragon but you are not. But if the dream is coherent enough, the object you are chasing may be a real object: a virtual dragon, say, constituted by processes in your mind. And just as you can virtually chase a virtual dragon in VR, you might also virtually chase a virtual dragon in your dream. The virtual dragon will exist and have causal powers as a process in your mind, though it will be mind-dependent, illusory, and not a real dragon. By meeting two of the five criteria of the reality checklist, it perhaps qualifies for a very limited sort of reality.

What about lucid dreams: dreams in which you know you are dream-ing? Lucid dreams are perhaps a mind-dependent analog of ordinary VR, where users typically know they are in VR. I've argued that expert users need not suffer illusions in VR because of their sense of virtuality. Perhaps the same could be true for some expert lucid dreamers. If they experience the objects in their dream as virtual objects rather than as physical objects, the objects need not be illusory. The dragon will still not be a real dragon, and it will still be mind-dependent, but this way lucid dream objects may be one step closer to reality than objects in ordinary dreams.

What about the hallucinations and delusions brought on in men-tal disorders such as schizophrenia? If one experiences a hallucinatory object as a physical object in the surrounding world, this will be akin to the ordinary dream case in some respects. The hallucinatory object will be an illusion and a construction of one's mind, though it may also exist as a virtual object in the mind with causal powers. Some sophisticated subjects with these disorders may experience hallucinatory objects *as* virtual objects or as constructions of their minds. As in a lucid dream, these experiences will not be entirely illusory. But again, the halluci-nated people will not be real people, and they will be constructions of one's mind.

I'm not saying that merely imagining something creates a virtual object or virtual world. If I simply conjure up an image of an elephant, there is not usually an interactive simulation. I don't perform different actions toward the elephant and get different outcomes. If so, there is no genuine virtual object with elephant-like causal powers here. The same goes for calling up memories from the past. If this is more like a script than a simulation, there will not be memory objects with the robust causal powers that virtual objects have. Many hallucinations of the external world may be like this, too. But in special cases in which there is an interactive world model, we will have something like a mind-dependent virtual world.

(Novels, interactive novels, and text-based adventure games raise related issues that I discuss in the notes.)

The chaos hypothesis and Boltzmann brains

So far, none of the skeptical scenarios we've considered have turned out to be *global* skeptical scenarios in which no ordinary objects are real and few ordinary beliefs are true. The reason is that in each case, something is generating our experiences. In order for it to generate our experiences, it has to have a good deal of the structure that we ordinarily attribute to the external world. As a result, many of our ordinary beliefs will be true.

For a genuine global skeptical scenario, we need a case in which our experiences are not generated systematically by the external world at all. One such scenario is the *chaos hypothesis*, which says that there is no external world, just a flow of random experiences. Thanks to a massive coincidence, this flow yields the regular stream of experiences I've had. Or at least, it yields the highly ordered state of consciousness, complete with memory experiences, that I'm having right now.

If the chaos hypothesis is true, I'd say that the external world isn't real. The tables and chairs I seem to perceive don't exist. The perceptions are real, but there's nothing beyond them. If I accept the chaos hypothesis, I should reject most of my beliefs about the external world.

At the same time, the chaos hypothesis is extraordinarily improbable. It would take an extremely unlikely group of coincidences to realize all the regularities in my experience by chance. So I'd argue that we can rule out the chaos hypothesis on probabilistic grounds. If we can do this, then the mere possibility of a chaos hypothesis doesn't support global skepticism.

However, there's a relative of the chaos hypothesis that we need to take seriously. The Boltzmann brain hypothesis says that I am a random fluctuation in matter that happened to produce a physical object that's just like a fully functioning brain. As we saw at the start of this chapter, some physical theories predict that many Boltzmann brains will exist in the history of the universe. When the expanding universe has finally reached a uniform state of thermodynamic equilibrium, random fluctuations from the equilibrium will eventually produce exactly the structure of my brain.

Given an infinite period of time and an infinite amount of space, we should expect that an infinite number of brains like mine will fluctuate into existence. Could I be one of them? The vast majority of Boltzmann brains will immediately decay out of existence. A tiny but still infinite subset of them will continue to function as my brain does, for a period of a few seconds, with genuine cause-and-effect processing among their neurons. We can expect that these brains will have memories and conscious experiences just like mine.

If I'm a Boltzmann brain, are objects in the external world I experience real? For a few Boltzmann brains, they may be. Very occasionally an entire Boltzmann city or Boltzmann planet may come into existence, where Boltzmann brains are surrounded by an ordinary physical environment. But for the vast majority of Boltzmann brains, the world outside the brain won't be included in the package. For these Boltzmann brains, the experience of an external world will be an illusion.

Here's the problem: There will be an infinite number of Boltzmann brains with the structure of my brain in the history of the universe. There's probably, at most, one ordinary non-Boltzmann brain with that structure. Now we can apply statistical reasoning, analogous to our reasoning about the simulation hypothesis in chapter 5. Statistical reasoning leads to the Boltzmann brain hypothesis: I'm almost certainly a Boltzmann brain. That means in turn that my external world is an illusion. Boltzmann brains have led us back to global skepticism!

As the theoretical physicist Sean Carroll has pointed out, however, the thesis that I am almost certainly a Boltzmann brain is "cognitively unstable." If it's true, then I cannot stably endorse it. If I endorse it, I then must endorse that my perception of the external world is almost certainly an illusion. But then I must reject all of my scientific reasoning that's based on my perception of the external world. In particular, I should reject the scientific reasoning that led to the physical theories on which the existence of Boltzmann brains was based in the first place. Those theories are the only reason to take the Boltzmann brain hypothesis seriously. Without those theories as support, we're back to the original situation: The hypothesis that I am a Boltzmann brain is extremely improbable.

You might argue: Even without scientific support for the physical theories in question, isn't there *some* likelihood that we're in a universe of random fluctuations that will produce an infinite number of Boltzmann brains, or at least Boltzmann minds? Let's say that, a priori, there's a one percent chance that we're in such a randomly fluctuating universe. However, in such a universe, most conscious beings would have highly disordered experiences. Only a tiny minority would have highly ordered experiences, apparently experiencing a coherent and regular external world, as I am right now. So my highly ordered experiences are strong evidence against the hypothesis that we're in such a random universe, and strong evidence in favor of the hypothesis that we're in an ordered universe.

Conclusion: No escape from reality

Where does this leave the external world? Or at least, where does this leave Cartesian arguments for global skepticism about the external world?

A Cartesian argument takes a scenario and says the following. First, it says no to the Knowledge Question: *We can't know we're not in the scenario.* Second, it says no to the Reality Question: *If we're in the scenario, nothing is real.* It concludes: *We can't know that anything is real.*

For the Cartesian argument to work, we need a scenario for which the answers to the Knowledge and Reality Questions are both no. We haven't found such a scenario. For all the versions of the simulation scenario we've considered, the answer to the Reality Question is yes: At least some of the things we perceive in the external world are real. The same goes for simulation-like relatives such as the evil-demon and lifelong-dream scenarios. For some nonsimulation scenarios, such as the chaos hypothesis and the Boltzmann brain hypothesis, the answer to the Reality Question is no, but the answer to the Knowledge Question is yes: We can know that we're not Boltzmann brains. So it looks as if there's no scenario that supports a Cartesian argument for skepticism.

Here's a diagnosis. Either there's an explanation for the regularities in our experience (as in the simulation hypothesis and its relatives), or there's no such explanation (as in the chaos hypothesis and its relatives).

If there's no explanation for the regularities in our experience, as in the chaos hypothesis, then the scenario requires massive coincidence and can be ruled out on probabilistic grounds. In this case, the answer to the Knowledge Question is yes: We can know we're not in the scenario.

If there's an explanation for the regularities in our experience, as in the simulation hypothesis, then there's an external world of some sort. Furthermore, to explain all the regularities, the external world must share some of the structure of the external world that we perceive and believe in. Given structuralism, it follows that the answer to the Reality Question is yes: If we're in this scenario, at least some of the things we perceive and believe in are real.

To put things even more briefly: There must be *some* explanation for everything we perceive and believe. If there's an explanation, then this will be an external world with a structure that vindicates much of what we perceive and believe. Explanation yields structure, and structure yields reality!

If I'm right, the classic Cartesian argument for global skepticism about the external world fails.

Still, there are many other sorts of skepticism, and there are many other arguments for skepticism. Simulation realism does not refute those.

In particular, simulation realism doesn't rule out various forms of local skepticism. Skepticism about what I perceive in the present stays on the table, thanks to the hypothesis that I entered a simulation overnight. The same goes for skepticism about the recent past. Skepticism about the far away and the very small stays on the table, thanks to the local and macroscopic simulation hypotheses. Then there's skepticism about other minds, about the mind's role in action, and about reasoning—though these go beyond skepticism about the external world as usually conceived.

In fact, each of my beliefs might be threatened by a different local skeptical hypothesis. I think I know there's an iPhone on my desk. Maybe mirrors are distorting my perception, or I unknowingly entered VR a few minutes ago. I think I know that my partner is Brazilian. Maybe she's a well-trained Russian spy with a cover story. And so on. Could this lead to a piecemeal variant of global skepticism about the external world, where each of my beliefs is threatened for local reasons?

There are limits to this piecemeal skepticism. I can't be too badly wrong about the shape of the world and the shape of my life. For example, I know for sure that I have a body, or at least that I *have* had a body. Perhaps my original body was vaporized and I was uploaded to a simulation with an avatar five minutes ago. But my avatar is a body, so I still have a body. Even if we don't count the avatar as a body, my belief that I *have had* a body is correct. You might try to find a scenario in which I never had a body in the first place. Perhaps I was a brain in a vat the whole time. But if so, my experiences of having a body must have come from somewhere, like a simulation or a model of a body. If so, that's where my body is to be found. You can try to spin a scenario in which I have multiple bodies, in which my body is fragmented, or in which someone replaced all my memories by memories of having had a body. If we work through the reasoning in this book, these will all be scenarios in which I've had a body.

In the same spirit: There must be some explanation of my experience of having a body. If there is, then there's something that plays the role of a body—and that entity will be my body. Once again, explanation yields structure, and structure yields reality.

The same reasoning extends to other broad claims about the shape of the world and my life. I know there are other people, or at least that there *have been* other people. I'm not claiming to know that other people are conscious—the problem of other minds is a problem for another day. But conscious or not, other people exist—I've experienced them. Perhaps you'll try to explain my experiences of them as products of a simulation. Perhaps other people were only recently implanted in my memories. If so, you'll have told me only that other people are digital beings who are part of a simulation that produced my memories. The same goes for other explanations.

You could try to extend this further. I think I know that there is water, or at least that there has been water. The same goes for trees and cats. At least, it's harder than many have thought to come up with a scenario in which there have never been water, trees, or cats. Maybe cats are always played by small dogs wearing a cat suit; if so, then cats are dogs wearing cat suits. Maybe our experience of trees comes from implanted memories of trees; if so, the sources of those memories are trees. Maybe our experiences of water come from a simulation via augmented reality technology in the midst of an otherwise physical world; even so, I think the right verdict is that water exists and is virtual.

This strategy for responding to piecemeal skepticism has limits. I don't claim that I can know, in this way, that a specific person exists. Can I be certain that my colleague Ned Block, who I talk to frequently, has existed? Probably not: I can't rule out that he's been played by a series of actors. Can I be certain that Australia has existed? Probably not. There might be a geographical conspiracy, and all the time I thought I spent in Australia was in fact spent in sound stages around the world. These conspiracies would have to be complex, and I could perhaps rule them out on those grounds, but that sort of anti-skeptical strategy would be a project for another day.

What's happening here? I think structuralism is at work again. The regularities in our experience are a good guide to at least the approximate structure of the world: there are certain entities that cause our experiences and that interact with one another in certain patterns. Experience is by no means a perfect guide to structure; we get the fine-grained structure of the world wrong all the time. But experience, combined with the appeal to simplicity, tells us at least about approximate structure. Even where we find skeptical scenarios, such as local and temporary simulation hypotheses, these share a great deal of structure with our ordinary conception of the world. Any remotely simple skeptical hypothesis will share a great deal of structure. We can think of the structural features that are shared across all reasonable scenarios as approximate structure.

Knowing approximate structure is not enough to tell us everything about the world. It doesn't tell me about specific individuals such as Ned Block (his role might be played by many people). It doesn't tell me that there's a kitten in the room now (it might be a virtual kitten with similar approximate structure). It doesn't tell me that other people are conscious (they might be zombies with a similar structure). But it can at least tell me some very basic things about the world, like the fact that I have a body and the fact that other people exist.

This is a start, and it may be possible to go much further. We have a strategy for responding to skepticism about the external world. From experience, we can infer structure. From structure, we can infer reality. Just how much structure we can infer from experience, and just how much reality we can infer from structure, remain open questions. But we have at least made a small dent in the puzzle of the external world.

There remains much more to say. It's an open question how much we can know about reality. There are objective facts about the distant past that we may never know. If we live in a perfect simulation, there will be facts about the world beyond the simulation that we can never know. We don't know how much of reality is accessible and how much is inaccessible. But the truth is out there, and we can know some of it.

Acknowledgments

I COULDN'T HAVE GOTTEN FAR ON THIS PROJECT WITHOUT A LOT of help.

I owe intellectual debts to the many philosophers, scientists, and technology pioneers discussed in the text. Many science-fiction writers and creators also influenced me, from Stanislaw Lem through the Wachowski sisters. I'd especially like to acknowledge some neglected philosophical contributions in this area: Susanne Langer's work in the 1950s on virtual objects and virtual worlds, O. K. Bouwsma and Jonathan Harrison's work in the 1940s and 1960s on skepticism, and Michael Heim and Philip Zhai's work in the 1990s on the metaphysics of virtual reality.

My parents got me started by giving me an Apple II+ computer decades ago. I was first inspired to think hard about virtual worlds by reading Doug Hofstadter and Dan Dennett's book *The Mind's I* when I was a teenager in the early 1980s. Their influence on this book should be obvious. In graduate school, long arguments with Gregg Rosenberg about brains in vats also played a formative role.

I first spoke about these issues in a talk titled "It's Not So Bad to be a Brain in a Vat," arranged by John Heil at Davidson College in March 2002. A month or two later, fortuitously, Chris Grau invited me to write an article for the *The Matrix* website, which helped draw me in further. Since then, I've given talks on these issues to many audiences and have always received interesting and valuable ideas in response. In 2015 and 2016, lecture series at Brown University, the Institut Jean Nicod, Johns

Hopkins University, and the University of Lisbon allowed me to explore some of these ideas in depth. David Yates and Ricardo Santos edited a subsequent symposium in the journal *Disputatio* that provided valuable critical analysis.

The much-missed Tonietta Walters (a.k.a. Xhyra Graf) got me to try out *Second Life* and to think about it as a venue for philosophy. Jacki Morie, Betty Mohler Tesch, and Bill Warren opened up their VR labs and invited me to use the technology for myself. I had a number of useful conversations with Mel Slater and Mavi Sanchez-Vives. The groups in game studies at NYU (Bennett Foddy, Frank Lantz, Julian Togelius) and Copenhagen (Espen Aarseth, Pawel Grabarczyk, Jesper Juul) were full of interesting ideas. Damien Broderick, Jaron Lanier, Ivan Sutherland, and Robert Wright patiently answered my queries about the early history of "virtual reality" and "virtual worlds."

I used early versions of this book in a series of NYU undergraduate classes on Minds and Machines. I'd like to thank many students in those classes for their feedback. I'd especially like to pay tribute to João Pedro Leilia Correa Eboli, possibly the most enthusiastic undergraduate student I've ever had, whose tragic death cut off a promising career in philosophy.

Kati Balog, Steffen Koch, Kelvin McQueen, Charles Siewert, and Scott Sturgeon used the manuscript in their own classes and gave much-appreciated feedback. Miri Albahari, David James Barnett, Banafsheh Beizaei, Christian Coseru, David Godman, Anja Jauernig, Christoph Limbeck, Béatrice Longuenesse, Jake McNulty, Jessica Moss, Paolo Pecere, Anand Vaidya, and Pete Wolfendale gave valuable input on historical matters. Evan Behrle, Adam Lovett, Aidan Penn, and Patrick Wu were helpful on political philosophy.

The band of virtual philosophers—especially Thomas Hofweber, Kris McDaniel, Neil McDonnell, Laurie Paul, Gillian Russell, Jonathan Schaffer, and Robbie Williams—provided valuable experience and good cheer in numerous VR platforms throughout the pandemic.

I got valuable feedback on some or all of the manuscript from many people: Anthony Aguirre, Zara Anwarzai, Axel Barcelo, David James Barnett, Sam Baron, Umut Baysan, Jiri Benovsky, Artem Besedin, Ned

Block, Ben Blumson, Adam Brown, David Jay Brown, Richard Brown, Cameron Buckner, Joe Campbell, Eric Cavalcanti, Andy Chalom, Eddy Keming Chen, Tony Cheng, Jessica Collins, Vince Conitzer, Marcello Costa, Brian Cutter, Barry Dainton, Ernie Davis, Janelle Derstine, Vilius Dranseika, Matt Duncan, Rami El Ali, Lisa Emerson, David Friedell, Philip Goff, David Miguel Gray, Daniel Gregory, Peli Grietzer, Avram Hiller, Jens Kipper, Neil Levy, Matthew Liao, Isaac Mackey, Corey Maley, Steve Matthews, Angela Mendelovici, Bradley Monton, Jennifer Nagel, Eddy Nahmias, Gary Ostertag, Dan Pallies, David Pearce, Steve Petersen, Gualtiero Piccinini, Angel Pinillos, Martin Pleitz, Paavo Pylkkänen, Brian Rabern, Rick Repetti, Adriana Renero, Anton Reutov, Regina Rini, Damien Rochford, Luke Roelofs, Brad Saad, Sascha Seifert, Eric Schwitzgebel, Ankita Sethi, Kerry Shaw, Carl Shulman, Mark Silcox, Vadim Vasilyev, Khai Wager, Kelly Weirich, Shauna Winram, and Roman Yampolskiy. I owe special thanks to Barry Dainton and Jennifer Nagel for urging me to go further in the final chapter in drawing an anti-skeptical conclusion.

Thanks to Dan Pallies and Dylan Sims for suggesting the title, and to any number of friends online for discussion of the title and many other matters. My brother Michael provided the title of chapter 1. He still thinks *Is This the Real Life?* should have been the title of the book.

My literary agent, John Brockman, provided intellectual community and decades of experience. Katinka Matson and Max Brockman were terrifically helpful, too. My editor at W. W. Norton, Brendan Curry, provided extensive feedback and good advice. Laura Stickney at Penguin Press had many useful thoughts. Sara Lippincott and Kelly Weirich went over the book with a fine-tooth comb, checking facts and improving the prose. Becky Homiski at Norton was helpful and responsive in shepherding the book through production.

Many thanks to Tim Peacock for his amazing illustrations. Tim's illustrations don't just bring complex ideas to life. They're an integral part of the philosophical arguments in this book. Tim's creativity led many ideas in new and surprising directions. Collaborating with him on the illustrations has been one of the most exciting parts of writing this book for me.

My partner, the philosopher and psychologist Claudia Passos Ferreira, has been with me throughout the years of writing this book. We've built a life in New York and ridden through the pandemic together. Claudia prefers nonvirtual reality, but she's given me all sorts of help with the project all the same. This book is dedicated to her, with love.

Glossary

Augmented reality: Technology that enables us to experience virtual objects while also perceiving the physical world. (p. 225)

Boltzmann brains: Systems identical to biological brains that arise through random fluctuations. (p. 440)

Cartesian dualism: The view (associated with Descartes) that mind and body are distinct, with a non-physical mind affecting the physical body, and the body affecting the mind. (p. 258)

Cartesian skepticism: A form of external-world skepticism, the view that we don't know anything substantial about the external world, driven by scenarios (such as the dream and evil-demon scenario) in which we appear to be out of touch with reality. (p. 45)

Consciousness: The subjective experience of mind and world. Includes the conscious experience of perceiving, feeling, thinking, doing, and more. A being is conscious when there is something it is like to be that being. (p. 277)

Cosmos: Everything that exists. (p. 107)

Digital object: A structure of bits, or an object that stands to bits as physical objects stand to atoms. (p. 194)

Dualism: The view that mind and body are entirely distinct. (p. 258)

Epistemology: The study of knowledge. (p. 17)

Externalism: The thesis that the meanings of our words and our thoughts depend on the world around us. (p. 372)

Idealism: The view that reality is fundamentally mental, or all in the mind. Berkeley's version of idealism is associated with the slogan "*esse* is *percipi*," or "to be is to be perceived." (p. 68)

Illusion: When things are not the way they seem. (p. 112) More restrictive philosophical usage: When one perceives a real object, but the object is not as it seems. (p. 206)

Immersiveness: In an immersive environment, we experience the environment as a world all around us, with ourselves present at the center. (p. 189)

Impure simulation: A simulation in which some beings are unsimulated; e.g., a brain in a vat connected to a simulation. (p. 30)

It-from-bit thesis: Physical objects, including the entities in physics, are made of bits. That is, they're grounded in a level of digital physics involving the interaction of bits. (p. 160)

It-from-bit-from-it thesis: The it-from-bit thesis, combined with the thesis that the bits are grounded in a further layer of reality. (p. 162)

Metaverse: A virtual world (or system of virtual worlds) in which everyone can spend time, living day-to-day lives with many forms of social interaction. (Alternative use: the sum of all virtual worlds.) (p. 185)

Perfect simulation: One that precisely simulates (as opposed to approximately simulating) the world it is simulating. (p. 35)

Pure it-from-bit thesis: The it-from-bit thesis, combined with the thesis that the bits are at the fundamental level of reality and aren't grounded in anything further. (p. 165)

Pure simulation: A simulation containing only pure sims (see sim); i.e., simulated beings. (p. 30)

Real: See the five conceptions outlined in chapter 6, where being real is understood in terms of existence, causal powers, mind-independence, non-illusoriness, or genuineness. (p. 108)

Reality: There are at least three meanings. First, reality is everything that exists (the entire cosmos). Second, a reality is a world (physical or virtual), and realities (plural) are worlds. Third, the property of reality is the property of real-ness, or of being real, as defined in the previous entry. (p. 107)

Realize: To make real. Used especially for lower-level entities grounding higher-level entities: atoms realize molecules, molecules realize cells, and so on. (p. 161)

Sim: A being who is in a computer simulation. Pure sims are simulated beings inside simulations. Biosims are biological beings connected to simulations. (p. 30)

Simulation argument: A statistical argument for the simulation hypothesis (Hans Moravec), or for a three-way choice between the simulation hypothesis and two other theses (Nick Bostrom). (p. 83)

Simulation hypothesis: The hypothesis that we're living in a computer simulation—that is, that we are and always have been receiving our inputs from and sending our outputs to an artificially-designed computer simulation of a world. (p. 29)

Simulation realism: The thesis that if we're in a simulation, the objects around us are real and not illusory. (p. 106)

Skepticism: The view that we don't know anything. The central form of skepticism here is external-world skepticism, the view that we don't know anything substantial about the external world. (p. 45)

Structuralism (or structural realism): The thesis that scientific theories are equivalent to structural theories, cast in terms of mathematics plus connections to observation. Epistemic structural realism says that science tells us only the structure of reality (though there may be more to reality than this). Ontic structural realism says that reality itself is entirely structural. (p. 402)

Universe: Same as *world*. (p. 108)

Utilitarianism: The thesis that one should do the greatest good for the greatest number. (p. 335)

Verificationism: The view that a claim is meaningful only if there's some way to verify it. (p. 74)

Virtual digitalism: The thesis that virtual objects are digital objects. (pp. 107, 194)

Virtual fictionalism: The thesis that virtual objects and virtual worlds are fictional. (p. 193)

Virtual-inclusive: A category X or a word "X" is virtual-inclusive when a virtual X counts as a real X. Otherwise it is virtual-exclusive. (p. 200)

Virtual realism: The thesis that virtual reality is a genuine reality, with emphasis on the thesis that virtual objects are real and not illusory. (pp. 105–6)

Virtual reality: An immersive, interactive, computer-generated world. (p. 189)

Virtual world: An interactive, computer-generated world. (pp. 191–92)

World: A complete interconnected space (physical or virtual), including everything in it. (pp. 107–8)

Notes

MORE EXTENSIVE NOTES, INCLUDING PHILOSOPHICAL DISCUSSION and technical and historical details, can be found online at consc.net/reality. This site also includes appendices with extended discussions of topics such as the following: shortcut simulations (chapters 2, 5, and 24), responses to external-world skepticism (chapter 4), objections to the simulation argument (chapter 5), Nick Bostrom on the simulation argument (chapter 5), Michael Heim and Philip Zhai on virtual realism (chapter 6), varieties of information (chapter 8), the history of the expressions "virtual," "virtual reality," and "virtual worlds" (chapters 10–11), free will in the experience machine and virtual reality (chapter 17), Donald Hoffman's case against reality (chapter 23), novels, experience worlds, and other skeptical scenarios (chapter 24), notes on the illustrations (all chapters), and other topics.

Introduction: Adventures in technophilosophy

xviii ***Neurophilosophy and technophilosophy:*** Patricia Churchland, *Neurophilosophy: Toward a Unified Science of the Mind-Brain* (MIT Press, 1986). A classic statement of technophilosophy (without the name) is Aaron Sloman's 1978 book *The Computer Revolution in Philosophy* (Harvester Press, 1978). To date, technophilosophy has been most influential at the nexus between artificial intelligence and the philosophy of mind; pioneers include Daniel Dennett ("Artificial Intelligence as Philosophy and Psychology," in *Brainstorms* [Bradford Books, 1978]), and Hilary Putnam ("Minds and Machines," in *Dimensions of Minds*, ed. Sidney Hook [New York University Press, 1960]).

xviii ***Philosophy of technology:*** For overviews, see Jan Kyrre Berg, Olsen Friis, Stig Andur Pedersen, and Vincent F. Hendricks, eds., *A Companion to the Philosophy* *of Technology* (Wiley-Blackwell, 2012); Joseph Pitt, ed., *The Routledge Companion to the Philosophy of Technology* (Routledge, 2016).

xx   **My views about consciousness:** More precisely, my views about the hard problem of consciousness, zombies, physicalism, dualism, and panpsychism play only a minor role in this book. The main arguments about reality are equally available to materialists and dualists about consciousness. My views about the distribution of consciousness, and especially that machines can be conscious, play a somewhat larger role.

xxii   **Some chapters of the book go over ground I've discussed in academic articles:** The arguments in chapter 9 (and a little of chapters 6, 20, and 24) are based on ideas in my online essay "The Matrix as Metaphysics," thematrix.com, 2003; reprinted in Christopher Grau, ed., *Philosophers Explore the Matrix* (Oxford University Press, 2005), 132–76. Chapters 10 and 11 (and a little of chapter 17) are based on themes from "The Virtual and the Real," *Disputatio* 9, no. 46 (2017): 309–52. The central idea of chapter 14 is based on an old unpublished note on "How Cartesian Dualism Might Have Been True" (online manuscript, consc.net/notes/dualism.html, February 1990). Chapter 15 is largely based on my work on consciousness, especially in *The Conscious Mind* (Oxford University Press, 1996). Chapter 16 is largely based on joint work with Andy Clark, "The Extended Mind," *Analysis* 58 (1998): 7–19. Chapters 21–23 are based on ideas in "On Implementing a Computation," *Minds and Machines* 4 (1994): 391–402; "Structuralism as a Response to Skepticism," *Journal of Philosophy* 115, no. 12 (2018): 625–60; and "Perception and the Fall from Eden," in *Perceptual Experience*, eds. Tamar S. Gendler and John Hawthorne (Oxford University Press, 2006), 49–125, respectively. There are plenty of new ideas in these chapters, and most of the material in the other chapters is new.

xxiii   **I give some possible paths, depending on your interests:** If you want to follow the narrative on Descartes's problem of the external world and my response to it, the central chapters are 1–9 and 20–24. If your main interest is near-term virtual-reality technology, you could read chapters 1 and 10–20. If you're especially interested in the simulation hypothesis, you might read chapters 1–9, 14–15, 18, 20–21, and 24. If you want an introduction to traditional problems in philosophy, perhaps I'd focus on chapters 1, 3–4, 6–8, and 14–23. It's also worth noting that chapter 4 presupposes chapter 3, chapter 9 presupposes chapter 8 (and to some extent 6 and 7), chapter 11 presupposes chapter 10, and chapter 22 presupposes chapter 21. Parts 4–7 can be read in any order, but part 7 presupposes much of parts 2 and 3.

Chapter 1: Is this the real life?

3   **Lead singer Freddie Mercury sings:** The video for *Bohemian Rhapsody* depicts all four members of Queen singing the first few lines, but in fact Freddie Mercury, who wrote the song, sang all of the parts in the opening. It seems apt that in asking whether this is just fantasy, all of the voices belong to the same person.

4   **"Zhuangzi Dreams of Being a Butterfly":** In *The Complete Works of Zhuangzi*, trans. Burton Watson (Columbia University Press, 2013). There the passage is translated using Zhuangzi's informal name "Zhuang Zhou," but I have used "Zhuangzi" for simplicity. For a different translation and interpretation of Zhuangzi's butterfly dream, focusing on the reality of both Zhuangzi and the butterfly rather than on issues about knowledge, see Hans Georg Moeller, *Daoism Explained: From the Dream of the Butterfly to the Fishnet Allegory* (Open Court, 2004).

5 *1999 movie* **The Matrix:** See Adam Elga, "Why Neo Was Too Confident that He Had Left the Matrix," http://www.princeton.edu/~adame/matrix-iap.pdf. I'll discuss issues arising from *The Matrix Resurrections* (released in December 2021 after this book went to press) in an online note.

5 *Ancient Indian philosophers were gripped by issues of illusion and reality:* For an excellent guide to issues about illusion in Indian philosophy, religion, and literature (including Narada's transformation), see Wendy Doniger O'Flaherty, *Dreams, Illusions, and Other Realities* (University of Chicago Press, 1984).

14 *James Gunn's 1954 science fiction story:* James Gunn, "The Unhappy Man" (*Fantastic Universe*, 1954); collected in Gunn's *The Joy Makers* (Bantam, 1961).

15 *In his 1974 book:* Robert Nozick, *Anarchy, State, and Utopia* (Basic Books, 1974).

15 *Life in the experience machine:* In *The Examined Life* (Simon & Schuster, 1989, 105), Nozick himself distinguishes versions of our Knowledge, Reality, and Value Questions about the experience machine: "The question of whether to plug in to this experience machine is a question of value. (It differs from two related questions: an epistemological one—Can you know you are not already plugged in?—and a metaphysical one—Don't the machine experiences themselves constitute a real world?)."

15 *In a 2020 survey:* See http://philsurvey.org/. Here and throughout, when I give PhilPapers Survey results by saying, for example, "13 percent said they would enter the experience machine," this is shorthand for: 13 percent of respondents indicated that they accept or lean toward this view. For broader surveys beyond professional philosophers, see Dan Weijers, "Nozick's Experience Machine Is Dead, Long Live the Experience Machine!," *Philosophical Psychology* 27, no. 4 (2014): 513–35; Frank Hindriks and Igor Douven, "Nozick's Experience Machine: An Empirical Study," *Philosophical Psychology* 31 (2018): 278–98.

16 *In a 2000 article in* **Forbes** *magazine:* Robert Nozick, "The Pursuit of Happiness," *Forbes*, October 2, 2000.

17 *Mind Question:* Online note.

18 *These six further questions each correspond to an area of philosophy:* There are many other areas of philosophy: for example, the philosophy of action, the philosophy of art, the philosophy of gender and race, the philosophy of mathematics, and many areas of the history of philosophy. I touch on all of these areas along the way as well, though not in as much depth as the nine areas I've listed.

18 *Survey of professional philosophers:* For the 2009 PhilPapers Survey of professional philosophers, see David Bourget and David Chalmers, "What Do Philosophers Believe?," *Philosophical Studies* 170 (2014): 465–500. For the 2020 Survey, see http://philsurvey.org/. On progress in philosophy, see David J. Chalmers, "Why Isn't There More Progress in Philosophy?," *Philosophy* 90, no. 1 (2015): 3–31.

18 *Disciplines founded or cofounded by philosophers:* Aside from Newton, I have in mind Adam Smith (economics), Auguste Comte (sociology), Gustav Fechner (psychology), Gottlob Frege (modern logic), and Richard Montague (formal semantics).

Chapter 2: What is the simulation hypothesis?

21 *The Antikythera mechanism is an attempt to simulate the solar system:* See Tony Freeth et al., "A Model of the Cosmos in the ancient Greek Antikythera Mechanism," *Scientific Reports* 11 (2021): 5821.

21 *Mechanical simulation of the San Francisco Bay:* For a philosophical discussion of the San Francisco Bay mechanical simulation, see Michael Weisberg's book *Simulation and Similarity: Using Models to Understand the World* (Oxford University Press, 2013).

21 *Computer simulations are ubiquitous in science and engineering:* There is a large philosophical literature on computer simulations and the role they play in science: Eric Winsberg, *Science in the Age of Computer Simulation* (University of Chicago Press, 2010); Johannes Lenhard, *Calculated Surprises: A Philosophy of Computer Simulation* (Oxford University Press, 2019); and Margaret Morrison, *Reconstructing Reality: Models, Mathematics, and Simulations* (Oxford University Press, 2015).

22 *Computer simulations of human behavior:* Daniel L. Gerlough, "Simulation of Freeway Traffic on a General-Purpose Discrete Variable Computer" (PhD diss., UCLA, 1955); Jill Lepore, *If Then: How the Simulmatics Corporation Invented the Future* (W. W. Norton, 2020).

23 *In his 1981 book:* Jean Baudrillard, *Simulacres et Simulation* (Editions Galilée, 1981), translated as *Simulacra and Simulation* (Sheila Faria Glaser, trans.; University of Michigan Press, 1994).

23 *Baudrillard is talking about cultural symbols and not computer simulations:* Baudrillard's four levels (which map only very loosely onto mine) are: "It is the reflection of a profound reality," "It masks and denatures a profound reality," "It masks the absence of a profound reality," and "It has no relation to any reality whatsoever: It is its own pure simulacrum." At some points, Baudrillard counts only the fourth level as simulation.

24 *The vast cosmos of possible universes:* Online note.

25 *Ursula Le Guin's classic 1969 novel:* Ursula K. Le Guin, *The Left Hand of Darkness* (Ace Books, 1969). The passages on thought experiments and on psychological reality come from Le Guin's introduction to the 1976 edition of the novel. "Is Gender Necessary?" was published in *Aurora: Beyond Equality*, eds, Vonda MacIntyre and Susan Janice Anderson (Fawcett Gold Medal, 1976).

27 *James Gunn's 1955 story:* Remarkably, Gunn's *The Joy Makers* closely anticipates two of the most important thought experiments in recent philosophy: the experience machine and the simulation hypothesis. In a preface to a later edition, he describes how he was inspired by a 1950 *Encyclopædia Britannica* article on the psychology of feeling.

27 *Simulations in science fiction:* Online note.

28 *Invited to write about philosophical ideas for its official website:* "The Matrix as Metaphysics" and many other articles were solicited by Christopher Grau, then a graduate student in philosophy who worked as an editor and producer for RedPill Productions, the production company for *The Matrix*. They were later published in Grau's edited collection, *Philosophers Explore the Matrix* (Oxford University Press, 2005). At least three other edited collections of *Matrix*-themed philosophy have been published: William Irwin's *The Matrix and Philosophy:*

Welcome to the Desert of the Real (Open Court, 2002) and *More Matrix and Philosophy: Revolutions and Reloaded Decoded* (Open Court, 2005); and Glenn Yeffeth's *Taking the Red Pill: Science, Philosophy and Religion in* The Matrix (BenBella Books, 2003).

29 **Bostrom published his important article:** Bostrom's original article on the simulation argument was "Are You Living in a Computer Simulation?," *Philosophical Quarterly* 53, no. 211 (2003): 243–55. His article introducing the label "simulation hypothesis" is "The Simulation Argument: Why the Probability that You Are Living in a Matrix Is Quite High," *Times Higher Education Supplement*, May 16, 2003.

30 **I will use the word sim:** The economist Robin Hanson has introduced the related term *em* for beings constructing by emulating a human brain. Ems and sims are distinct: an impure sim (like Neo) is a sim but not an em, and an emulated human brain in a robot body is an em but not a sim

31 **Local *simulation hypothesis:*** Online note.

32 **Philosophers revel in distinctions:** In "Innocence Lost: Simulation Scenarios: Prospects and Consequences" (2002, https://philarchive.org/archive/DAIILSv1), the British philosopher Barry Dainton makes a number of related distinctions: hard vs. soft simulations, active vs. passive simulations, original-psychology vs. replacement-psychology simulations, communal vs. individual simulations.

32 **Any such evidence could be simulated:** One might try appealing to externalism about evidence, as defended, e.g., by Timothy Williamson in *Knowledge and Its Limits* (Oxford University Press, 2000) to make a case that we can know we're not in a simulation. I discuss externalism about evidence in the online appendix on objections to the simulation argument.

35 **2012 article:** Silas R. Beane, Zohreh Davoudi, and Martin J. Savage, "Constraints on the Universe as a Numerical Simulation," *European Physical Journal A* 50 (2014): 148.

36 **Analog computer:** In this book, this means a computer that uses precise continuous quantities such as real numbers. For alternative conceptions, see Corey J. Maley, "Analog and Digital, Continuous and Discrete," *Philosophical Studies* 115 (2011): 117–31.

36 **Analog quantum computer:** At least in theory, standard quantum computers are analog computers because they use continuous quantities to serve as amplitudes for qubits, but in practice their precision is limited. There is also a theory of quantum computing with continuous variables in place of binary qubits: e.g., Samuel L. Braunstein and Arun K. Pati, eds., *Quantum Information with Continuous Variables* (Kluwer, 2001).

36 **Classical computers cannot efficiently simulate quantum processes:** Zohar Ringel and Dmitry Kovrizhin, "Quantized Gravitational Responses, the Sign Problem, and Quantum Complexity," *Science Advances* 3, no. 9 (September 27, 2017). See also Mike McRae, "Quantum Weirdness Once Again Shows We're Not Living in a Computer Simulation," *ScienceAlert*, September 29, 2017; Cheyenne Macdonald, "Researchers Claim to Have Found Proof We Are NOT Living in a Simulation," *Dailymail.com*, October 2, 2017; and Scott Aaronson, "Because You Asked: The Simulation Hypothesis Has Not Been Falsified; Remains Unfalsifiable," *Shtetl-Optimized*, October 3, 2017.

36 **No universe can contain a perfect simulation of itself:** Online note.

37   *Finite simulation that lags behind reality:* See Mike Innes, "Recursive Self-Simulation," https://mikeinnes.github.io/2017/11/15/turingception.html.
37   *Imperfect simulation hypotheses:* Online note.
39   **Tetris** *and* **Pac-Man** *can be regarded as simulations:* Online note.

Chapter 3: Do we know things?

43   *Philosophers have questioned each of these kinds of knowledge:* Sextus Empiricus: See Michael Frede, "The Skeptic's Beliefs," chap. 10, in his *Essays in Ancient Philosophy* (University of Minnesota Press, 1987); Nāgārjuna: see Ethan Mills, *Three Pillars of Skepticism in Classical India: Nāgārjuna, Jayarāśi, and Śrī Harsa* (Lexington Books, 2018); al-Ghazali: *Deliverance from Error*, and https://www.aub.edu.lb/fas/cvsp/Documents/Al-ghazaliMcCarthytr.pdf; David Hume: *A Treatise of Human Nature* (1739); Eric Schwitzgebel, *Perplexities of Consciousness* (MIT Press, 2011); Grace Helton, "Epistemological Solipsism as a Route to External World Skepticism," *Philosophical Perspectives* (forthcoming); Richard Bett, *Pyrrho: His Antecedents and His Legacy* (Oxford University Press, 2000).

50   *A shade of "dark yellow":* Paul M. Churchland, "Chimerical Colors: Some Phenomenological Predictions from Cognitive Neuroscience," *Philosophical Psychology* 18, no. 5 (2005): 27–60.

51   *Christia Mercer has recently charted:* Christia Mercer, "Descartes' Debt to Teresa of Ávila, or Why We Should Work on Women in the History of Philosophy," *Philosophical Studies* 174, no. 10 (2017): 2539–2555. Teresa of Ávila, *The Interior Castle*, trans. E. Allison Peers (Dover, 2012).

52   *Michel de Montaigne:* Montaigne explores skeptical ideas most deeply in his 1576 essay "Apology for Raymond Sebond" (Montaigne, *The Complete Essays*, M. A. Screech, ed. and trans., Penguin, 1993).

54   *Brain in a vat:* Hilary Putnam, *Reason, Truth and History* (Cambridge University Press, 1981).

55   *Barry Dainton has put it:* Barry Dainton, "Innocence Lost: Simulation Scenarios: Prospects and Consequences," 2002, https://philarchive.org/archive/DAIILSv1.

56   *If you can't know you're not in a simulation:* Online note.

56   *The point of philosophy:* Bertrand Russell, "The Philosophy of Logical Atomism," *The Monist* 28 (1918): 495–527.

58   *Cogito, ergo sum:* As with most new ideas in philosophy, the *cogito* wasn't entirely original to Descartes. In *City of God*, in the 5th century CE, the North African philosopher Saint Augustine of Hippo wrote "I am certain that I am. . . . If I am mistaken, I am. For if one does not exist, he can by no means be mistaken. Therefore I am, if I am mistaken."

58   *Philosophers have interpreted Descartes's celebrated slogan in many different ways:* For an interpretation that denies that the *cogito* is an inference or an argument, see Jaakko Hintikka, "*Cogito ergo sum*: Inference or Performance?," *Philosophical Review* 71 (1962): 3–32.

60   *Descartes explicitly defines thought:* In responding to objections in the second edition of the *Meditations* (1642), Descartes defines *thought* [*cogitatio*] as follows: "I use this term to include everything that is within us in such a way that we are

immediately aware [conscii] of it. Thus all the operations of the will, the intellect, the imagination, and the senses are thoughts." See Paolo Pecere, *Soul, Mind and Brain from Descartes to Cognitive Science* (Springer, 2020).

60   **Consciousness could be an illusion:** See Keith Frankish, ed., *Illusionism as a Theory of Consciousness* (Imprint Academic, 2017).

Chapter 4: Can we prove there is an external world?

62   *A wonderful and long-neglected story:* Jonathan Harrison, "A Philosopher's Nightmare or the Ghost Not Laid," *Proceedings of the Aristotelian Society* 67 (1967): 179–88.

65   *An idea of God as a perfect being:* Descartes's argument about the perfect idea of God wasn't original with him. In the 11th century, Saint Anselm of Canterbury put forward related "ontological" arguments for the existence of God, which we'll discuss in chapter 7. An argument very much like Descartes's perfect idea argument was put forward by the 16th-century Spanish scholar Francisco Suárez.

67   *Idealism:* For more recent discussions of idealism, see Tyron Goldschmidt and Kenneth L. Pearce, eds., *Idealism: New Essays in Metaphysics* (Oxford University Press, 2017) and *The Routledge Handbook of Idealism and Immaterialism*, eds. Joshua Farris and Benedikt Paul Göcke (Routledge & CRC Press, 2021).

69   *If reality is made up of all our minds together:* An intersubjective version of idealism is developed by the German philosopher Edmund Husserl in his 1929 *Cartesian Meditations: An Introduction to Phenomenology.* A more recent version is perhaps suggested by the (non-idealist) neuroscientist Anil Seth's slogan (in *Being You*, Dutton, 2021): "We're all hallucinating all the time. It's just that when we agree about our hallucinations, that's what we call reality." Intersubjective idealism suffers from many of the same problems as regular idealism: not least, we need some reality beyond appearances to explain the regularities in our collective appearances, such as why we all seem to see the same tree.

71   *Why do we need God?:* For a modern version of idealism that uses algorithmic information theory to avoid the need for God or an external world, see Markus Müller, "Law Without Law: From Observer States to Physics via Algorithmic Information Theory," *Quantum* 4 (2020): 301.

71   *I'll argue that some forms of idealism should be taken seriously:* See the discussion of the it-from-bit-from-consciousness thesis in chapters 8 and 22. See also David J. Chalmers "Idealism and the Mind-Body Problem," in T*he Routledge Handbook of Panpsychism*, ed. William Seager (Routledge, 2019); reprinted in *The Routledge Handbook of Idealism and Immaterialism.*

71   *Carnap held that many philosophical problems are meaningless "pseudoproblems":* Rudolf Carnap, *Scheinprobleme in der Philosophie* (Weltkreis, 1928); Rudolf Carnap, *The Logical Structure of the World & Pseudoproblems in Philosophy*, trans. Rolf A. George (Carus, 2003). For an introduction to the Vienna Circle, see David Edmonds, *The Murder of Professor Schlick: The Rise and Fall of the Vienna Circle* (Princeton University Press, 2020).

72   *Skeptical hypotheses are meaningless:* Ludwig Wittgenstein, *Tractatus Logico-Philosophicus* (Kegan Paul, 1921). In *Language, Truth, and Logic* (Victor Gollancz,

1936), A. J. Ayer says "Consequently, anyone who condemns the sensible world as a world of mere appearance as opposed to reality, is saying something which, according to our criterion of significance, is literally nonsensical." In "Empiricism, Semantics, and Ontology" (*Revue Internationale de Philosophie* 4 [1950]: 20–40), Carnap says that the question of "the reality of the thing world" involves a "concept cannot be meaningfully applied to the system itself." None of the Vienna Circle members explicitly discussed the simulation hypothesis, of course.

75   *In his 1981 book:* Hilary Putnam, *Reason, Truth and History* (Cambridge University Press, 1981).

77   ***Bertrand Russell's appeal to* simplicity:** See Bertrand Russell, *The Problems of Philosophy* (Henry Holt, 1912), 22–23; see also Jonathan Vogel, "Cartesian Skepticism and Inference to the Best Explanation," *Journal of Philosophy* 87, no. 11 (1990): 658–66.

79   *Moore said "Here is one hand":* G. E. Moore, "Proof of an External World," *Proceedings of the British Academy* 25, no. 5 (1939): 273–300. Moore's "common sense" approach to philosophy was influenced by the 18th-century Scottish philosopher Thomas Reid, e.g., in his 1764 book *An Inquiry into the Human Mind on the Principles of Common Sense.* See also James Pryor, "What's Wrong with Moore's Argument?", *Philosophical Issues* 14 (2004): 349–78.

80   *Other replies to external-world skeptics:* Online appendix.

Chapter 5: Is it likely that we're in a simulation?

83   *"Pigs in Cyberspace":* Hans Moravec, "Pigs in Cyberspace," in *Thinking Robots, an Aware Internet, and Cyberpunk Librarians*, eds. H. Moravec et al. (Library and Information Technology Association, 1992). Reprinted in *The Transhumanist Reader*, eds. Max More and Natasha Vita-More (Wiley, 2013).

83   *Interview with Hans Moravec:* Charles Platt, "Superhumanism," Wired, October 1, 1995.

83   *Nick Bostrom in his 2003 article:* Nick Bostrom, "Are You Living in a Computer Simulation?" *Philosophical Quarterly* 52 (2003): 243–55.

83   *The entrepreneur Elon Musk:* Elon Musk interview at Code Conference 2016, Rancho Palos Verdes, CA, May 31–June 2, 2016; "Why Elon Musk Says We're Living in a Simulation," *Vox*, August 15, 2016.

85   *I'll simplify by assuming that all populations have the same size:* Online note.

85   *The conclusion that we are probably sims:* Online note.

86   *Math and other complications:* Online note.

86   *Sim blockers:* Online note.

86   *Intelligent sims are impossible:* For arguments that simulating human-level intelligence is impossible (using Gödel's theorem to argue that humans have capacities that go beyond any computer), see J. R. Lucas, "Minds, Machines and Gödel," *Philosophy* 36, no. 137 (1961): 112–27, and Roger Penrose, *The Emperor's New Mind* (Oxford University Press, 1989). For a response to Penrose, see my "Minds, Machines, and Mathematics," *Psyche* 2 (1995): 11–20.

87   *Quantum gravity computers:* For one conception of these, see Lucien Hardy, "Quantum Gravity Computers: On the Theory of Computation with

Indefinite Causal Structure," in Wayne Myrvold and Joyce Christian, eds., *Quantum Reality, Relativistic Causality, and Closing the Epistemic Circle* (Springer, 2009).

87 *On current estimate, the brain's computing speed is around 10 petaflops:* Online note.

87 *The universe has enormous unused capacity for computing:* Richard Feynman, "There's Plenty of Room at the Bottom," *Engineering & Science* 23, no. 5 (1960): 22–36; Seth Lloyd, "Ultimate Physical Limits to Computation," *Nature* 406 (2000): 1047–54; Frank Tipler, *The Physics of Immortality* (Doubleday, 1994), 81.

88 *Computronium:* The name "computronium" was introduced for the idea of programmable matter by Tommaso Toffoli and Norman Margolus; see their "Programmable Matter: Concepts and Realization," *Physica D*, 47, no. 1–2 (1991): 263–72; and Ivan Amato, "Speculating in Precious Computronium," *Science* 253, no. 5022 (1991): 856–57. The now-common usage for maximally efficient programmed matter was popularized in science-fiction works such as Charles Stross's *Accelerando* (Penguin Random House, Ace reprint, 2006), in which much of the solar system is turned into computronium.

88 *If we're in a simulation, evidence about our computer power may be misleading:* For versions of this objection, see Fabien Besnard, "Refutations of the Simulation Argument," http://fabien.besnard.pagesperso-orange.fr/pdfrefut.pdf, 2004; and Jonathan Birch, "On the 'Simulation Argument' and Selective Scepticism," *Erkenntnis* 78 (2013): 95–107. See the online appendix on further objections to the simulation argument for more discussion.

88 *Existential risks:* Toby Ord, *The Precipice: Existential Risk and the Future of Humanity* (Hachette, 2020).

89 *Nonsims will die out before creating sims:* Online note.

90 *Simulate the decision first and see how things go:* Online note.

91 *We could be nanoscale nonsims:* Online note.

91 *Sim sign:* In his "The PNP Hypothesis and a New Theory of Free Will" (*Scientia Salon*, 2015), Marcus Arvan argues that a version of the simulation hypothesis is the best explanation of free will and of various features of quantum mechanics, suggesting in effect that these phenomena are sim signs.

92 *Interestingness is a sim sign:* Robin Hanson, "How to Live in a Simulation," *Journal of Evolution and Technology* 7 (2001).

93 *Our position early in the universe is a sim sign:* Online note.

93 *Sims can't be conscious!:* See John Searle, *Minds, Brains, and Science* (Harvard University Press, 1986).

94 *Simulators will avoid creating conscious sims:* Thanks to Barry Dainton, Grace Helton, and Brad Saad for versions of this suggestion. In her "Epistemological Solipsism as a Route to External World Skepticism" (*Philosophical Perspectives*, forthcoming), Helton argues that ethical simulators may well create simulations in which only one being is conscious—in which case, any conscious being should take seriously the solipsistic thesis that they are the only conscious being in the universe.

94 *Sims running on serial von Neumann architectures will not be conscious:* See Christof Koch in *The Feeling of Life Itself: Why Consciousness Is Widespread But Can't Be Computed* (MIT Press, 2019) and Giulio Tononi and Christof Koch, "Consciousness: Here, There, and Everywhere?" *Philosophical Transactions of the Royal Society B* (2015).

95   *Sims won't experience large universes:* Online note.
95   *Simulation that takes shortcuts:* Online appendix.
98   *Major sim signs:* Online note.
98   *I don't think Bostrom's formula or his conclusions are quite correct as they stand:* Online appendix on Bostrom on the simulation argument.
98   *If there are no sim blockers, we are probably sims:* Online note.
99   *Both premises now require only relatively small assumptions:* Online note.
102   *If it came up heads, she connected me to a perfect simulation:* Online note.

Chapter 6: What is reality?

105   *Virtual realism:* Other authors whose work contains elements of virtual realism include David Deutsch and Philip Zhai (discussed in chapter 6) and Philip Brey (discussed in chapter 10). Elements of simulation realism are endorsed by Douglas Hofstadter (discussed in chapter 20) as well as in the articles by Andy Clark and Hubert Dreyfus in *Philosophers Explore the Matrix*. In addition, O. K. Bouwsma (chapter 6) and Hilary Putnam (chapter 20) explore views akin to simulation realism without explicitly discussing simulations per se.

109   *What is it to exist?:* For contrasting perspectives on existence, see W. V. Quine, "On What There Is," *Review of Metaphysics* 2 (1948): 21–38; Rudolf Carnap, "Empiricism, Semantics, and Ontology," *Revue Internationale de Philosophie* 4 (1950): 20–40; and the articles in D. J. Chalmers, D. Manley, and R. Wasserman, eds., Metametaphysics: *New Essays in the Foundations of Ontology* (Oxford University Press, 2009).

110   *Eleatic dictum:* Online note.

111   *Philip K. Dick's dictum:* Philip K. Dick, "I Hope I Shall Arrive Soon." First published as "Frozen Journey" in Playboy, December 1980. Reprinted in Dick, *I Hope I Shall Arrive Soon* (Doubleday, 1985).

111   *Dumbledore's dictum:* J. K. Rowling, *Harry Potter and the Deathly Hallows* (Scholastic, 2007), p. 723.

113   *Austin's lectures:* J. L. Austin, *Sense and Sensibilia* (Oxford University Press, 1962).

115   *There are other strands we could have added:* Other strands include *Reality as observability. Reality as measurability. Reality as theoretical utility.* (These are related to the causal power strand.) *Reality as authenticity. Reality as naturalness. Reality as originality. Reality as fundamentality.* (These are related to the genuineness strand.) Then there are the senses of "really"—what we mean when we say that something is *really* the case. Here the strands include *Reality as truth. Reality as actuality. Reality as factuality.* (These are related to the nonillusoriness strand.) *Reality as objectivity. Reality as intersubjectivity. Reality as evidence-independence.* (These are related to the mind-independence strand.) Each of these senses of "really" arguably yields a corresponding sense of "real" by translating "X is real" into "X really exists." (I set aside the strands at play in "real number" and "real estate"—though it's worth noting that the terminology of real and imaginary numbers comes from Descartes!) Of these many strands, perhaps those that most threaten the status of simulated objects as real are some of those in the genuineness strand, such as *Reality as originality* and *Reality as funda-*

mentality, which I discuss in the text. For further discussions of the many senses of "real," "really," and "reality," see Jonathan Bennett, "Real," *Mind* 75 (1966): 501–15; and Steven L. Reynolds, "Realism and the Meaning of 'Real,'" *Noûs* 40 (2006): 468–94.

117   ***Theoretical physicist David Deutsch:*** David Deutsch, *The Fabric of Reality* (Viking, 1997).

120   ***Striking how uncommon this view has been:*** Online note.

120   ***Bouwsma's article:*** O. K. Bouwsma, "Descartes' Evil Genius," *Philosophical Review* 58, no. 2 (1949): 141–51.

122   ***Philip Zhai in his important 1998 book:*** Philip Zhai, *Get Real: A Philosophical Adventure in Virtual Reality* (Rowman and Littlefield, 1998).

Chapter 7: Is God a hacker in the next universe up?

125   ***Most interesting argument for the existence of God in a long time:*** https://www.simulation-argument.com/.

132   ***The fine-tuning argument is controversial:*** In a 2020 PhilPapers Survey question about what explains fine-tuning, 17 percent said design explains it, 15 percent said a multiverse explains it, 32 percent said it's a brute fact, and 22 percent said there's no fine-tuning.

135   ***Naturalism:*** Online note.

136   ***Simulation theology:*** Other sources for simulation theology are Bostrom's "Are You Living in a Computer Simulation?" (*Philosophical Quarterly* 53, no. 211 [2003]: 243–55), which talks about "naturalist theogeny," and Eric Steinhart's "Theological Implications of the Simulation Argument," *Ars Disputandi* 10, no. 1 (2010): 23–37.

138   ***Simulation and decision-making:*** Online note.

138   ***Simulations set up for entertainment will be terminated:*** Preston Greene, "The Termination Risks of Simulation Science," *Erkenntnis* 85, no. 2 (2020): 489–509.

141   ***The "end of history":*** G. W. F. Hegel, *Lectures on the Philosophy of History*, 1837. The Hegelian idea is echoed in the epigraph to Douglas Adams' *The Restaurant at the End of the Universe* (Pan Books, 1980): "There is a theory which states that if ever anyone discovers exactly what the Universe is for and why it is here, it will instantly disappear and be replaced by something even more bizarre and inexplicable. There is another theory which states that this has already happened."

141   ***Simulation afterlife:*** For an optimistic perspective, see Eric Steinhart's *Your Digital Afterlives: Computational Theories of Life after Death* (Palgrave Macmillan, 2014).

141   ***Hard to keep it contained:*** Eliezer Yudkowsky, "The AI-Box Experiment," https://www.yudkowsky.net/singularity/aibox; David J. Chalmers, "The singularity: A Philosophical Analysis," *Journal of Consciousness Studies* 17 (2010): 9–10.

142   ***"Turtles all the way down":*** The story (which is often told with Bertrand Russell or others in place of William James) appears to be apocryphal. James himself alludes to the "old story" involving "rocks all the way down" in his "Rationality, Activity, and Faith" (Princeton Review, 1882). The 18th-century philosophers Johann Gottlieb Fichte (in *Concerning the Conception of the Science of Knowledge*

Generally, 1794) and David Hume (in *Dialogues Concerning Natural Religion*, 1779) both allude to versions of the story involving elephants or tortoises.

142   **Jonathan Schaffer has argued:** Jonathan Schaffer, "Is There a Fundamental Level?" *Nous* 37 (2003): 498–517. See also Ross P. Cameron, "Turtles All the Way Down: Regress, Priority, and Fundamentality," *Philosophical Quarterly* 58 (2008): 1–14.

Chapter 8: Is the universe made of information?

145   **Leibniz invented the bit:** Gottfried Wilhelm Leibniz, "De Progressione Dyadica" (manuscript, March 15, 1679); "Explication de l'arithmétique binaire," *Memoires de l'Academie Royale des Sciences* (1703). It is sometimes said that the *I Ching* inspired Leibniz's discovery. In fact, he formulated binary arithmetic some years before Joachim Bouvet introduced him to the *I Ching* and pointed out the resemblance, after which Leibniz built it into his exposition. There is also a case for Thomas Hariot inventing the bit a century before Leibniz: See John W. Shirley, "Binary Numeration before Leibniz" (*American Journal of Physics* 19, no. 8 [1951]: 452–54). The 20th-century American mathematician Claude Shannon, who coin-troduced the label "bit," is sometimes called the "inventor of the bit." As we'll see, what Shannon invented was an information-theoretic measure and not the binary digit.

147   **You can try out the Game of Life:** playgameoflife.com. The default starting point is a glider, but you can try many other arrangements, including a glider gun: playgameoflife.com/lexicon/Gosper_glider_gun.

148   **Many indigenous cultures have their own metaphysical systems:** Robert Lawlor, *Voices of the First Day: Awakening in the Aboriginal Dreamtime* (Inner Traditions, 1991); James Maffie, *Aztec Philosophy, Understanding a World in Motion* (University Press of Colorado, 2014); Anne Waters, ed., *American Indian Thought* (Blackwell, 2004).

149   **Metaphysical theorizing:** For metaphysical systems across different cultures, see Gary Rosenkrantz and Joshua Hoffman, *Historical Dictionary of Metaphysics* (Scarecrow Press, 2011); Jay Garfield and William Edelglass, eds., *The Oxford Handbook of World Philosophy* (Oxford University Press, 2014); Julian Baggini, *How the World Thinks* (Granta Books, 2018); A. Pablo Iannone, *Dictionary of World Philosophy* (Routledge, 2001).

151   **An oscillation among materialism, dualism, and idealism:** In the 2020 Phil-Papers Survey, 52 percent accept physicalism about the mind while 22 percent reject it. In a question about consciousness, 22 percent accept dualism and 8 percent accept panpsychism (33 percent accept functionalism, 13 percent accept the mind-brain identity theory, and 5 percent accept eliminativism, which we haven't discussed here). In a question about the external world, 7 percent accept idealism (5 percent accept skepticism and 80 percent accept non-skeptical realism).

152   **Semantic information:** See Rudolf Carnap and Yehoshua Bar-Hillel, "An Outline of a Theory of Semantic Information," Technical Report No. 247, MIT Research Laboratory of Electronics (1952), reprinted in Bar-Hillel, *Language and Information* (Reading, MA: Addison-Wesley, 1964); Luciano Floridi, "Semantic Conceptions of Information" in *Stanford Encyclopedia of Philosophy* (2005).

152   **Structural, semantic, and symbolic information:** See an online appendix for more in-depth discussion. This is my own way of dividing up the territory, but related distinctions have been made many times before. There are many different taxonomies of information. See, for example, Mark Burgin, *Theory of Information: Fundamentality, Diversification and Unification* (World Scientific, 2010); Luciano Floridi, *The Philosophy of Information* (Oxford University Press, 2011); and Tom Stonier, *Information and Meaning: An Evolutionary Perspective* (Springer-Verlag, 1997).

155   *These three measures of structural information:* Online note.

155   *Analog computation:* George Dyson, *Analogia: The Emergence of Technology beyond Programmable Control* (Farrar, Straus & Giroux, 2020); Lenore Blum, Mike Shub, and Steve Smale, "On a Theory of Computation and Complexity over the Real Numbers," *Bulletin of the American Mathematical Society* 21, no. 1 (1989): 1–46; Aryan Saed et al., "Arithmetic Circuits for Analog Digits," *Proceedings of the 29th IEEE International Symposium on Multiple-Valued Logic,* May 1999; Hava T. Siegelmann, *Neural Networks and Analog Computation: Beyond the Turing Limit* (Birkhäuser, 1999); David B. Kirk, "Accurate and Precise Computation Using Analog VLSI, with Applications to Computer Graphics and Neural Networks" (PhD diss., Caltech, 1993).

156   *Continuous digits:* The terms "continuous-valued digit" and "analog digit" are sometimes used in the literature (e.g., Saed et al., "Arithmetic Circuits for Analog Digits"), but as far as I know, there's no standard abbreviation. *Ant* and *cont* are unlovely terms, so I'm reluctant using *real* despite its imperfect connotations. For example, while it suggests purely mathematical real numbers, physically realized reals (as with bits) are more crucial for our purposes. (Also, reals shouldn't be confused with *real* in the sense of reality; and continuous quantities are often complex numbers rather than real numbers.) Whereas a bit is physically embodied as a binary state in a physical system, a real is physically embodied as a real-valued state in a physical system (where both are individuated in a substrate-neutral way). Note that there isn't really a measure of the quantity of continuous information analogous to Shannon-style bit measures, in part because multiple reals can be recoded as a single real and vice versa.

157   *Structural information can be physically embodied:* Online note.

157   *Physical information:* Online note.

157   *A difference that makes a difference:* Gregory Bateson, *Steps to an Ecology of Mind* (Chandler, 1972). Bateson gives credit to Donald Mackay, who said, "Information is a distinction that makes a difference."

158   *Anatoly Dneprov published a short story:* Anatoly Dneprov, "The Game," *Knowledge-Power* 5 (1961): 39–41. English translation by A. Rudenko at http://q-bits.org/images/Dneprov.pdf. Dneprov's "Portuguese stadium" can be seen as an antecedent of John Searle's well-known "Chinese room" argument ("Minds, Brains, and Programs," *Behavioral and Brain Sciences* 3 (1980): 417–57), with Dneprov's system translating a sentence where Searle's carries out a conversation.

159   *Information is physical:* this slogan was put forward by the physicist Rolf Landauer in "Information Is Physical," *Physics Today* 44, no. 5 (1991): 23–29.

160   *Digital physics:* Konrad Zuse, *Calculating Space* (MIT Press, 1970); Edward Fredkin, "Digital Mechanics: An Information Process Based on Reversible Universal Cellular Automata," *Physica D* 45 (1990): 254–70; Stephen Wolfram, *A New Kind of Science* (Wolfram Media, 2002).

160   *Wheeler's now-familiar slogan: it from bit:* John Archibald Wheeler, "Information, Physics, Quantum: The Search for Links," *Proceedings of the 3rd International Symposium on the Foundations of Quantum Mechanics* (Tokyo, 1989), 354–68.

162   ***Space and time emerge from some sort of underlying digital physics:*** I discuss this idea in "Finding Space in a Nonspatial World," in *Philosophy beyond Spacetime*, eds. Christian Wüthrich, Baptiste Le Bihan, and Nick Huggett, which contains many other discussions of emergent spacetime (Oxford University Press, 2021).

163   ***It from qubit:*** David Deutsch, "It from qubit," in *Science and Ultimate Reality: Quantum Theory, Cosmology, and Complexity*, eds. John Barrow et al. (Cambridge University Press, 2004); Seth Lloyd, *Programming the Universe: A Quantum Computer Scientist Takes on the Cosmos* (Alfred A. Knopf, 2006); P. A. Zizzi, "Quantum Computation Toward Quantum Gravity," 13th International Congress on Mathematical Physics, London, 2000, arXiv:gr-qc/0008049v3.

163   ***It-from-bit-from-it:*** For related discussion, see Anthony Aguirre, Brendan Foster, and Zeeya Merali, eds., *It from Bit or Bit from It? On Physics and Information* (Springer, 2015); and Paul Davies and Niels Henrik Gregersen, *Information and the Nature of Reality* (Cambridge University Press, 2010).

165   ***It-from-bit-from-consciousness:*** See Gregg Rosenberg, *A Place for Consciousness: Probing the Deep Structure of the Natural World* (Oxford University Press, 2004).

165   ***Pure it-from-bit thesis:*** see Aguirre et al., *It from Bit or Bit from It*; Eric Steinhart, "Digital Metaphysics," in *The Digital Phoenix*, eds. T. Bynum and J. Moor, (Blackwell, 1998). For critical analyses, see Luciano Floridi, "Against Digital Ontology," *Synthese* 168 (2009): 151–78; Nir Fresco and Philip J. Staines, "A Revised Attack on Computational Ontology," *Minds and Machines* 24 (2014): 101–22; and Gualtiero Piccinini and Neal Anderson, "Ontic Pancomputationalism," in *Physical Perspectives on Computation, Computational Perspectives on Physics*, eds. M. E. Cuffaro and S. E. Fletcher (Cambridge University Press, 2018).

166   ***Reality is grounded in continuous information:*** Online note.

Chapter 9:   Did simulation create its from bits?

171   *I need only establish that the simulation hypothesis leads to the it-from-bit creation hypothesis:* Online note.

171   *The arguments generalize to simulations run on a quantum computer:* For discussions of simulated worlds in the context of quantum computing and the it-from-qubit hypothesis, see Seth Lloyd, *Programming the Universe* (Knopf, 2006) and Leonard Susskind, "Dear Qubitzers, GR=QM" (2017, arXiv:1708.03040 [hep-th]).

172   *The bits that our simulator is creating aren't* **fundamental:** The issue of whether the simulation hypothesis is compatible with fundamental bits or non-programmable processes raises potential objections to the reverse (inessential) claim that the it-from-bit creation hypothesis entails the simulation hypothesis. I discuss them in an online note.

175   *If digital physics realizes standard physics, then photons are real:* Some scientists and philosophers deny that individual photons are real even in standard physics. Perhaps there are just quantum waves or electromagnetic fields, for exam-

ple. If that's correct, then digital physics will instead support these waves or fields, or whatever else is physically real.

175 *If the structure of atomic physics is present, then atomic physics is true, and atoms exist:* There are some forms of structuralism that deny the existence of entities such as atoms (James Ladyman and Don Ross, *Every Thing Must Go*, Oxford University Press, 2009). If we accept this radical structuralism without any "its," then as in the previous note, the reality of the physical domain if we're in a simulation will still be on a par with its reality in a nonsimulated world.

177 *I'm not saying that necessarily, any simulation of physics makes physics real:* In the philosophical terms of Saul Kripke's *Naming and Necessity* (Harvard University Press, 1980), the simulation hypothesis is not necessarily equivalent to the it-from-bit hypothesis, but it is (roughly) a priori equivalent: if the simulation hypothesis is true *in the actual world*, so is the it-from-bit hypothesis. Similarly, we fix reference to photons as those things that actually play the photon role, even if photons themselves have a further metaphysical nature (whether unsimulated or simulated) that goes beyond this role.

Chapter 10: Do virtual reality headsets create reality?

185 *Snow Crash:* Neal Stephenson, *Snow Crash* (Bantam, 1992).
186 *Attempts at a Metaverse:* The leading platforms for social VR at the time of writing in 2021 included *VRChat, Rec Room, Altspace VR, Bigscreen,* and *Horizon.*
186 *Or one giant Metaverse:* Online note.
187 *Definitions:* Ludwig Wittgenstein, *Philosophical Investigations*, 4th edition (Blackwell, 2009); Eleanor Rosch, "Natural Categories," *Cognitive Psychology* 4, no. 3 (1973): 328–50.
187 *Charles Sanders Peirce enshrined this definition:* C. S. Peirce, "Virtual," in *Dictionary of Philosophy and Psychology*, ed. James Mark Baldwin (Macmillan, 1902). Peirce goes on to say that this meaning of "virtual" as *in effect* should be distinguished from the sense where it means *potential*, as with an embryo that is a potential person. An embryo does not have the power of a person, so it is not a virtual person in the "in-effect" sense, but it has the power to become a person, so it is a virtual person in the *potential* sense. The idea of virtuality as potentiality is no longer central in ordinary uses of the word, but it has led to an important philosophical tradition associated with the French philosophers Henri Bergson (in his 1896 book *Matter and Memory*) and Gilles Deleuze (in his 1966 book *Bergsonism* and other works). As Deleuze puts it: "virtual" (in his sense) is opposed not to "real" but to "actual," where "actual" is understood in the sense of *actualization*. The virtual is not yet actualized (like an embryo, or a possible path in Borges's Garden of Forking Paths), or is in the process of being actualized (like a decision to take one path), or was once actualized (like a memory). For a guide to the many senses of virtuality, see Rob Shields, *The Virtual* (Routledge, 2002).
188 *La réalité virtuelle:* Strictly speaking, Artaud's first published use was "la realidad virtual." "The Alchemical Theater" was first published in a Spanish translation, "El Teatro Alquímico," in the Argentinian journal *Sur* in 1932. The French version was published in 1938 as "Le Théâtre Alchimique," in *Le théâtre et son double*

(Gallimard). The English translation (by Mary Caroline Richards) was published in *The Theatre and Its Double* (Grove Press, 1958).

188   **Constitutes the virtual reality:** Antonin Artaud, *The Theatre and Its Double*, 49.

188   **Early uses of "virtual reality" and "virtual world":** Online appendix.

192   **We call it a merely virtual object:** Susanne K. Langer, *Feeling and Form: A Theory of Art* (Charles Scribner's Sons, 1953), 49.

193   **Virtual fictionalism:** Varieties of virtual fictionalism are expounded by Jesper Juul, *Half-Real: Videogames between Real Rules and Fictional Worlds* (MIT Press, 2005); Grant Tavinor, *The Art of Videogames* (Blackwell, 2009); Chris Bateman, *Imaginary Games* (Zero Books, 2011); Aaron Meskin and Jon Robson, "Fiction and Fictional Worlds in Videogames" in *The Philosophy of Computer Games*, eds. John Richard Sageng et al. (Springer, 2012); David Velleman, "Virtual Selves," in his *Foundations for Moral Relativism* (Open Book, 2013); Jon Cogburn and Mark Silcox, "Against Brain-in-a-Vatism: On the Value of Virtual Reality," *Philosophy & Technology* 27, no. 4 (2014): 561–79; Neil McDonnell and Nathan Wildman, "Virtual Reality: Digital or Fictional," *Disputatio* 11, no. 55 (2020): 371–97. The first four of these theorists are making claims about video-game worlds, so it's not always clear that they would endorse fictionalism about virtual worlds more generally. Some of these fictionalists also distinguish special respects in which virtual realities are real: For example, virtual worlds involve real rules (Juul) or agents who perform fictional actions with fictional bodies (Velleman). Espen Aarseth, in "Doors and Perception: Fiction vs. Simulation in Games," *Intermedialities* 9 (2007): 35–44, denies that virtual worlds are fictional while nevertheless holding that they're not real: They have the same sort of status as dream worlds and thought experiments, which he also understands as not fictional and not real.

195   **Made of atoms:** Philosophers have understood the sense in which physical objects are "made of" atoms in many different ways. The currently most popular way is in terms of *grounding* (Jonathan Schaffer, "On What Grounds What," in *Metametaphysics: New Essays on the Foundations of Ontology*, eds. David J. Chalmers, David Manley, and Ryan Wasserman (Oxford University Press, 2009); Kit Fine, "The Pure Logic of Ground," *Review of Symbolic Logic* 5, no. 1 (2012): 1–25. Physical objects are grounded in atoms; by analogy, digital objects are grounded in bits. In "The Virtual as the Digital" (*Disputatio* 11, no. 55 [2019]: 453–86), I suggest calling structures of bits *narrowly digital objects* and objects grounded in structures of bits and mental states *broadly digital objects*.

196   **Why should we accept virtual digitalism over virtual fictionalism?:** For defenses of virtual fictionalism against some of these arguments, see Claus Beisbart, "Virtual Realism: Really Realism or Only Virtually So? A Comment on D. J. Chalmers's Petrus Hispanus Lectures," *Disputatio* 11, no. 55 (2019): 297–331; Jesper Juul, "Virtual Reality: Fictional all the Way Down (and That's OK)," *Disputatio* 11, no. 55 (2019): 333–43; and McDonnell and Wildman, "Virtual Reality: Digital or Fictional?" For further discussion of virtual digitalism, see also Peter Ludlow, "The Social Furniture of Virtual Worlds," *Disputatio* 11, no. 55 (2019): 345–69. I reply in "The Virtual as the Digital."

196   **As Philip Brey puts it:** Philip Brey, "The Social Ontology of Virtual Environments," *The American Journal of Economics and Sociology* 62, no. 1 (2003): 269–82. See also Philip Brey, "The Physical and Social Reality of Virtual Worlds," in *The*

Oxford Handbook of Virtuality, ed. Mark Grimshaw (Oxford University Press, 2014).

200   **When is a virtual X a real X?:** More precisely I'd say that X is virtual-inclusive (a virtual X is a real X) as long as X is a causal/mental invariant: something that depends only on the abstract causal organization and the mental properties of a situation (see "The Matrix as Metaphysics" and "The Virtual and the Real"). Philip Brey (see previous note) suggests that a virtual X is an X if and only if X is an institutional kind (such as money), one that is constituted by collective social agreements in the right way. I think that the "only if" claim is too strong as many causal/mental invariants are not institutional (e.g., virtual calculators are calculators) but many institutional kinds are plausibly causal/mental invariants, so Brey's "if" claim is more plausible.

201   **Conceptual engineering:** See Herman Cappelen, *Fixing Language: An Essay on Conceptual Engineering* (Oxford University Press, 2018); Alexis Burgess, Herman Cappelen, and David Plunkett, eds., *Conceptual Engineering and Conceptual Ethics* (Oxford University Press, 2020). On inclusiveness and gender concepts, see Katharine Jenkins, "Amelioration and Inclusion: Gender Identity and the Concept of Woman", *Ethics* 126 (2016): 394–421.

Chapter 11: Are virtual reality devices illusion machines?

203   **Jaron Lanier wrote:** Jaron Lanier, *Dawn of the New Everything: Encounters with Reality and Virtual Reality* (Henry Holt, 2017).

203   **His 1956 novel:** Arthur C. Clarke, *The City and the Stars* (Amereon, 1999).

204   **Psychologist Mel Slater:** Mel Slater, "A Note on Presence Terminology," *Presence Connect* 3, no. 3 (2003): 1–5; Mel Slater, "Place Illusion and Plausibility Can Lead to Realistic Behaviour in Immersive Virtual Environments," *Philosophical Transactions of the Royal Society of London B* 364, no. 1535 (2009): 3549–57.

204   **Plausibility Illusion:** A philosopher might have called this the Event Illusion or the Happening Illusion because it centers on the sense that certain events are really happening.

204   **Body Ownership Illusion:** Olaf Blanke and Thomas Metzinger, "Full-Body Illusions and Minimal Phenomenal Selfhood," *Trends in Cognitive Sciences* 13, no. 1 (2009): 7–13; Mel Slater, Daniel Perez-Marcos, H. Henrik Ehrsson, and Maria V. Sanchez-Vives, "Inducing Illusory Ownership of a Virtual Body," *Frontiers in Neuroscience* 3, no. 2 (2009): 214–20; Antonella Maselli and Mel Slater, "The Building Blocks of the Full Body Ownership Illusion," *Frontiers in Human Neuroscience* 7 (March 2013): 83.

207   **VR is not an illusion:** Philip Zhai also argues against the illusion view of VR in his 1998 book *Get Real* (Rowman & Littlefield), discussed in chapter 6 and in the online appendix on Heim and Zhai on virtual realism.

210   **Virtual shape and size:** For arguments against the simple view of physical and virtual space laid out here, see E. J. Green and Gabriel Rabin, "Use Your Illusion: Spatial Functionalism, Vision Science, and the Case against Global Skepticism," *Analytic Philosophy* 61, no. 4 (2020): 345–78; and Alyssa Ney, "On Phenomenal Functionalism about the Properties of Virtual and Non-Virtual Objects," *Disputatio* 11, no. 55

(2019): 399–410. I reply in "The Virtual as the Digital," *Disputatio* 11, no. 55 (2019): 453–86.

211   *Illusion and no-illusion view of mirrors:* I develop these arguments in "The Virtual and the Real," *Disputatio* 9, no. 46 (2017): 309–52. Maarten Steenhagen independently argues that mirror perception need not be illusory in "False Reflections," *Philosophical Studies* 5 (2017): 1227–42. For related philosophical discussion of mirrors, see Roberto Casati, "Illusions and Epistemic Innocence," in *Perceptual Illusion: Philosophical and Psychological Essays*, ed. C. Calabi (Palgrave Macmillan, 2012) and Clare Mac Cumhaill, "Specular Space," *Proceedings of the Aristotelian Society* 111 (2011): 487–95.

213   *Cognitive penetration:* Zenon W. Pylyshyn, *Computation and Cognition: Toward a Foundation for Cognitive Science* (MIT Press, 1984); Susanna Siegel, "Cognitive Penetrability and Perceptual Justification," *Noûs* 46, no. 2 (2012): 201–22; John Zeimbekis and Athanassios Raftopoulos, eds., *The Cognitive Penetrability of Perception: New Philosophical Perspectives* (Oxford University Press, 2015); Chaz Firestone and Brian J. Scholl, "Cognition Does Not Affect Perception: Evaluating the Evidence for 'Top-Down' Effects," *Behavioral & Brain Sciences* 39 (2016): 1–77.

216   *Phenomenology of virtuality:* For other phenomenological analyses of virtuality, see Sarah Heidt, "Floating, Flying, Falling: A Philosophical Investigation of Virtual Reality Technology," *Inquiry: Critical Thinking Across the Disciplines* 18 (1999): 77–98; Thomas Metzinger, "Why Is Virtual Reality Interesting for Philosophers?," *Frontiers in Robotics and AI* (September 13, 2018); Erik Malcolm Champion, ed., *The Phenomenology of Real and Virtual Places* (Routledge, 2018). For a "postphenomenological" approach, see Stefano Gualeni, *Virtual Worlds as Philosophical Tools: How to Philosophize with a Digital Hammer* (Palgrave Macmillan, 2015).

218   *Sense of reality:* Albert Michotte, "Causalité, permanence et réalité phénoménales," Publications Universitaires (1962), translated as "Phenomenal Reality" in *Michotte's Experimental Phenomenology of Perception*, eds. Georges Thinès, Alan Costall, and George Butterworth (Routledge, 1991); Anton Aggernaes, "Reality Testing in Schizophrenia," *Nordic Journal of Psychiatry* 48 (1994): 47–54; Matthew Ratcliffe, *Feelings of Being: Phenomenology, Psychiatry and the Sense of Reality* (Oxford University Press, 2008); Katalin Farkas, "A Sense of Reality," in *Hallucinations*, eds. Fiona MacPherson and Dimitris Platchias (MIT Press, 2014).

218   *The sense of reality and unreality also arises in VR:* Gad Drori, Paz Bar-Tal, Yonatan Stern, Yair Zvilichovsky, and Roy Salomon, "Unreal? Investigating the Sense of Reality and Psychotic Symptoms with Virtual Reality," *Journal of Clinical Medicine* 9, no. 6 (2020): 1627, DOI:10.3390/jcm9061627.

222   *Experiencing a real virtual body:* Avatars in VR appear to significantly affect people's behavior. For example, people with taller avatars are more likely to behave self-confidently. Psychologists Nick Yee and Jeremy Bailenson call this the *Proteus effect* after the shape-shifting Greek god Proteus. See Yee and Bailenson, "The Proteus Effect: The Effect of Transformed Self-Representation on Behavior," *Human Communication Research*, 33 (2007): 271–90; Jim Blascovich and Jeremy Bailenson, *Infinite Reality* (HarperCollins, 2011).

Chapter 12:   Does augmented reality lead to alternative facts?

231   *Alternative facts gained notoriety:* "Conway: Trump White House offered 'alternative facts' on crowd size" (CNN, January 22, 2017), https://www.cnn.com/2017/01/22/politics/kellyanne-conway-alternative-facts/index.html.

231   *Relativism is a deeply controversial idea:* For an overview, see Maria Baghramian and Annalisa Coliva, *Relativism* (Routledge, 2020). For a contemporary defense of a moderate form of relativism using tools from the philosophy of language, see John MacFarlane, *Assessment Sensitivity: Relative Truth and Its Applications* (Oxford University Press, 2014).

236   *Reality-virtuality continuum:* Paul Milgram, H. Takemura, A. Utsumi, and F. Kishino (1994). "Augmented Reality: A Class of Displays on the Reality-Virtuality Continuum," *Proceedings of the SPIE—The International Society for Optical Engineering* 2351 (1995), https://doi.org/10.1117/12.197321.

Chapter 13:   Can we avoid being deceived by deepfakes?

238   *Henry Shevlin posted an interview online:* Available at https://www.facebook.com/howard.wiseman.9/posts/4489589021058960 and http://henryshevlin.com/wp-content/uploads/2021/06/chalmers-gpt3.pdf. Thanks to Henry Shevlin for permission to use this piece.

240   *Deepfakes can be found in contexts:* Sally Adee, "What Are Deepfakes and How Are They Created?," *IEEE Spectrum* (April 29, 2020).

241   *Knowledge Question for deepfakes:* The Knowledge Question for deepfakes is raised by Don Fallis, "The Epistemic Threat of Deepfakes," *Philosophy & Technology* (August 6, 2020): 1–21; and *Philosophers' Imprint* 20, no. 24 (2020): 1–16.

241   *Knowledge Question for fake news:* For more on the Knowledge Question for fake news, see Regina Rini, "Fake News and Partisan Epistemology," *Kennedy Institute of Ethics Journal* 27, no. 2 (2017): 43–64; M. R. X. Dentith, "The Problem of Fake News," *Public Reason* 8, no. 1–2 (2016): 65–79; and Christopher Blake-Turner, "Fake News, Relevant Alternatives, and the Degradation of Our Epistemic Environment," *Inquiry* (2020).

245   *As the philosopher Regina Rini has observed:* Regina Rini, "Deepfakes and the Epistemic Backstop," *Philosophers' Imprint* 20, no. 24 (2020): 1–16.

248   *The term "fake news" itself has now become controversial:* See Josh Habgood-Coote, "Stop Talking about Fake News!," *Inquiry* 62, no. 9–10 (2019): 1033–65; and Jessica Pepp, Eliot Michaelson, and Rachel Sterken, "Why We Should Keep Talking about Fake News," *Inquiry* (2019).

248   *Fake news isn't the same thing as false or inaccurate news:* On the definition of fake news, see Axel Gelfert, "Fake News: A Definition," *Informal Logic* 38, no. 1 (2018): 84–117; Nikil Mukerji, "What is Fake News?," *Ergo* 5 (2018): 923–46; Romy Jaster and David Lanius, "What is Fake News?," *Versus* 2, no. 127 (2018): 207–27; and Don Fallis and Kay Mathiesen, "Fake News Is Counterfeit News," *Inquiry* (2019).

250   *Interconnected in webs of mutual endorsement:* For a network analysis of fake news and other misinformation, see Cailin O'Connor and James Owen Weatherall, *The Misinformation Age: How False Beliefs Spread* (Yale University Press, 2019).

251 *The Origins of Totalitarianism:* Hannah Arendt, *The Origins of Totalitarianism* (Schocken Books, 1951).

251 *Manufacturing Consent:* Edward S. Herman and Noam Chomsky, *Manufacturing Consent: The Political Economy of the Mass Media* (Pantheon Books, 1987).

Chapter 14: How do mind and body interact in a virtual world?

255 *Second-ever conference on artificial life:* Christopher G. Langton, Charles Taylor, J. Doyne Farmer, and Steen Rasmussen, eds., *Artificial Life II* (Santa Fe Institute, 1993).

255 *Kay's "Vivarium":* Larry Yaeger, "The Vivarium Program," http://shinyverse.org/larryy/VivHist.html.

256 *It struck me that these creatures would almost certainly become dualists about the mind:* David J. Chalmers, "How Cartesian Dualism Might Have Been True," February 1990, https://philpapers.org/rec/CHAHCD.

258 *Dualism can be found in many different cultures:* Kwame Gyekye, "The Akan Concept of a Person," *International Philosophical Quarterly* 18 (1978): 277–87, reprinted in *Philosophy of Mind: Classical and Contemporary Readings*, 2nd edition, ed. D. J. Chalmers (Oxford University Press, 2021); Avicenna (Ibn Sina), *The Cure*, ca. 1027, excerpted as "The Floating Man" in *Philosophy of Mind*, ed. Chalmers.

258 *Descartes articulated the classic form of dualism:* René Descartes, *Meditations on First Philosophy* (Meditations 2 and 6, 1641) and *Passions of the Soul* (1649), both excerpted in *Philosophy of Mind*, ed. Chalmers.

260 *Posed by Princess Elisabeth of Bohemia:* Lisa Shapiro, ed. and trans., *The Correspondence between Princess Elisabeth of Bohemia and René Descartes* (University of Chicago Press, 2007). Excerpted in Chalmers, ed., *Philosophy of Mind*.

262 *Speculated that the mind could play a role in quantum mechanics:* Eugene Wigner, "Remarks on the Mind-Body Question," in *The Scientist Speculates*, ed. I. J. Good (Heinemann, 1961); David J. Chalmers and Kelvin J. McQueen, "Consciousness and the Collapse of the Wave Function," in *Consciousness and Quantum Mechanics*, ed. Shan Gao (Oxford University Press, 2022).

268 *Animism:* Graham Harvey, *The Handbook of Contemporary Animism* (Routledge, 2013). For a contemporary animism with roots in indigenous animism, see Val Plumwood, "Nature in the Active Voice," *Australian Humanities Review* 46 (2009): 113–29.

269 *Biological and virtual brains synchronized:* This is a little reminiscent of Leibniz's theory where there is preestablished harmony between mind and body, although Leibniz's picture avoided any causal interaction between the two.

269 *Daniel Dennett's story "Where Am I?":* Daniel C. Dennett, "Where Am I?," in *Brainstorms* (MIT Press, 1978).

Chapter 15: Can there be consciousness in a digital world?

276   *Mind uploading:* Russell Blackford and Damien Broderick, eds., *Intelligence Unbound: The Future of Uploaded and Machine Minds* (Wiley-Blackwell, 2014).

278   *My first book:* David J. Chalmers, *The Conscious Mind: In Search of a Fundamental Theory* (Oxford University Press, 1996).

279   *The hard problem:* My original article on the hard problem is "Facing Up to the Problem of Consciousness," *Journal of Consciousness Studies* 2, no. 3 (1995): 200–219. It was reprinted with 26 replies and my reply in turn in Jonathan Shear, ed., *Explaining Consciousness: The Hard Problem* (MIT Press, 1997).

279   *None of this was because the idea was radical or original:* It's arguable that Leibniz gave one of the first clear statements of the hard problem in his parable of the brain as a "mill" in his 1714 *Monadology* ("We should only find on visiting it, pieces which push one against another, but never anything by which to explain a perception"). Thomas Huxley gave an even clearer statement in his 1866 *Lessons in Elementary Physiology:* "How it is that anything so remarkable as a state of consciousness comes about as a result of irritating nerve tissue, is just as unaccountable as the appearance of the Djin when Aladdin rubbed his lamp." I review the history of the problem in "Is the Hard Problem of Consciousness Universal?" *Journal of Consciousness Studies* 27 (2020): 227–57.

280   *Nagel famously defined consciousness:* Thomas Nagel, "What Is It Like to Be a Bat?," *The Philosophical Review* 83, no. 4 (1974): 435–50.

281   *Mary is a neuroscientist:* Frank Jackson, "Epiphenomenal Qualia," *The Philosophical Quarterly* 32, no. 127 (1982): 127–36. See also Peter Ludlow, Y. Nagasawa, and D. Stoljar, eds., *There's Something about Mary: Essays on Phenomenal Consciousness and Frank Jackson's Knowledge Argument* (MIT Press, 2004).

282   *Knut Nordby:* Knut Nordby, "Vision in a Complete Achromat: A Personal Account," in *Night Vision: Basic, Clinical and Applied Aspects,* eds. R. F. Hess, L. T. Sharpe, and K. Nordby (Cambridge University Press, 1990). Knut Nordby, "What Is This Thing You Call Color? Can a Totally Color-Blind Person Know about Color?," in *Phenomenal Concepts and Phenomenal Knowledge: New Essays on Consciousness and Physicalism,* eds. Torin Alter and Sven Walter (Oxford University Press, 2007).

283   *Panpsychism:* See Godehard Brüntrup and Ludwig Jaskolla, eds., *Panpsychism: Contemporary Perspectives* (Oxford University Press, 2017); Philip Goff, *Galileo's Error: Foundations for a New Science of Consciousness* (Pantheon, 2020); David Skrbina, *Panpsychism in the West* (MIT Press, 2007). Historical panpsychists include Margaret Cavendish (*The Blazing World,* 1666), Gottfried Wilhelm Leibniz (*Monadology,* 1714), and Baruch Spinoza (*Ethics,* 1677), among many others.

283   *Illusionism:* See Keith Frankish, ed., *Illusionism as a Theory of Consciousness* (Imprint Academic, 2017). Historically, there are elements of illusionism in materialist philosophers from Thomas Hobbes (*De Corpore,* 1655) to David Armstrong (*A Materialist Theory of the Mind,* 1968).

284   *Zhuangzi observes some jumping fish:* In *The Complete Works of Zhuangzi,* trans. Burton Watson (Columbia University Press, 2013).

284   *Other minds:* In the 2020 PhilPapers Survey, 89 percent of respondents say cats are conscious, 35 percent say flies are conscious, 84 percent say newborn babies are conscious, 3 percent say current AI systems are conscious, and 39 percent say

future AI systems can be conscious (while 27 percent deny this, and the rest adopt various forms of neutrality).

288   *We might* **become** *the machine:* Online note.

288   *Gradual uploading:* For more on uploading and machine consciousness, see my "Mind Uploading: A Philosophical Analysis," in *Intelligence Unbound: The Future of Uploaded and Machine Minds*, eds. Russell Blackford and Damien Broderick (Wiley-Blackwell, 2014). The argument here is elaborated in more detail in chapter 7 of *The Conscious Mind*. For related analyses of personal identity, see Derek Parfit, *Reasons and Persons* (Oxford University Press, 1984).

289   *2019 book* **Artificial You:** Susan Schneider, *Artificial You: AI and the Future of Your Mind* (Princeton University Press, 2019).

293   *Many people accept that the original person dies:* In a 2020 PhilPapers Survey question on "Mind uploading (brain replaced by digital emulation)," 27 percent of professional philosophers said this is a form of survival and 54 percent said it's a form of death.

Chapter 16: Does augmented reality extend the mind?

294   *Charles Stross's 2005 science-fiction novel:* Charles Stross, *Accelerando* (Penguin Random House, Ace reprint, 2006).

296   *The extended mind:* Andy Clark and David Chalmers, "The Extended Mind," *Analysis* 58 (1998): 7–19. We certainly weren't the first to propose that external processes can be akin to cognitive processes. Versions of the idea appear in Daniel Dennett, *Kinds of Minds* (Basic Books, 1996); John Haugeland, "Mind Embodied and Embedded," in Y. Houng and J. Ho, eds., *Mind and Cognition* (Academica Sinica, 1995); Susan Hurley, "Vehicles, Contents, Conceptual Structure and Externalism," Analysis 58 (1998): 1–6; Edwin Hutchins, *Cognition in the Wild* (MIT Press, 1995); Ron McClamrock, *Existential Cognition* (University of Chicago Press, 1995); Carol Rovane, "Self-Reference: The Radicalization of Locke," *Journal of Philosophy* 90 (1993): 73–97; Francisco Varela, Evan Thompson, and Eleanor Rosch, *The Embodied Mind* (MIT Press, 1995); Robert Wilson, "Wide Computationalism," *Mind* 103 (1994): 351–72; and many others. Earlier predecessors include the British anthropologist Gregory Bateson, the American philosopher John Dewey, the European phenomenologists Martin Heidegger and Maurice Merleau-Ponty, the Canadian media theorist Marshall McLuhan, and the Russian psychologist Lev Vygotsky.

296   *The Extended Phenotype:* Richard Dawkins, *The Extended Phenotype* (Oxford University Press, 1982).

296   *A number of books:* Robert D. Rupert, *Cognitive Systems and the Extended Mind* (Oxford University Press, 2009); Frederick Adams and Kenneth Aizawa, *The Bounds of Cognition* (Wiley-Blackwell, 2008); Richard Menary, ed., *The Extended Mind* (MIT Press, 2010); Annie Murphy Paul, *The Extended Mind: The Power of Thinking Outside the Brain* (Houghton Mifflin Harcourt, 2021).

298   *The webcomic* **xkcd** *published a strip titled "Extended Mind":* xkcd: "A Webcomic of Romance, Sarcasm, Math, and Language," https://xkcd.com/903/.

298   *Pioneers of the computer age:* J. C. R. Licklider, "Man-Computer Symbiosis," *IRE Transactions on Human Factors in Electronics* HFE-1 (March 1960): 4–11;

W. Ross Ashby, *An Introduction to Cybernetics* (William Clowes & Sons, 1956). See also Douglas Engelbart, "Augmenting Human Intellect: A Conceptual Framework," Summary Report AFOSR-3233, Stanford Research Institute, October 1962.

299   *Martin Heidegger observed:* Martin Heidegger, *Being and Time*, trans. J. Macquarrie and E. Robinson (Blackwell, 1962), 98.

305   *2008 cover story by Nicholas Carr in* **The Atlantic:** Nicholas Carr, "Is Google Making Us Stupid?," *The Atlantic* (July–August 2008).

306   *Google-knowing:* Michael Patrick Lynch, *The Internet of Us* (W. W. Norton, 2016), xvi–xvii.

307   *Brain activity is lower:* Amir-Homayoun Javadi et al., "Hippocampal and Prefrontal Processing of Network Topology to Simulate the Future," *Nature Communications* 8 (2017): 14652.

Chapter 17: Can you lead a good life in a virtual world?

312   *Nozick's 1974 fable of the experience machine:* Online note.

313   *Would you plug in?:* Robert Nozick, *Anarchy, State, and Utopia* (Basic Books, 1974), 44–45.

313   *Jennifer Nagel has suggested:* Email, January 5, 2021.

314   *Experience machine is unlike standard VR:* Barry Dainton, "Innocence Lost: Simulation Scenarios: Prospects and Consequences," 2002, https://philarchive .org/archive/DAIILSv1; Jon Cogburn and Mark Silcox, "Against Brain-in-a-Vatism: On the Value of Virtual Reality," *Philosophy & Technology* 27, no. 4 (2014): 561–79.

314   *Nozick's 2000* **Forbes** *article:* Robert Nozick, "The Pursuit of Happiness," *Forbes,* October 2, 2000.

315   *VR is not preprogrammed:* Online note.

317   *A philosophy fit for swine:* Thomas Carlyle, 1840/1993, *On Heroes, Hero-Worship, and the Heroic in History* (University of California Press, 1993).

317   *"Wireheading":* Online note.

320   *As Nozick puts it in a 1989 discussion:* *The Examined Life: Philosophical Meditations* (Simon & Schuster, 1989).

321   *Ways in which VR may be better:* In *A Defense of Simulated Experience* (Routledge, 2019), Mark Silcox argues that simulated experience—which includes but is not limited to experience in virtual worlds—may be a "unique source of a unique type of human well-being" (p. 81), because of its special role in a social and political context.

Chapter 18: Do simulated lives matter?

331   *Each of these philosophers:* G. E. M. Anscombe, *Intention* (Basil Blackwell, 1957). Mary Midgley, *Beast and Man: The Roots of Human Nature* (Routledge & Kegan Paul, 1978); Iris Murdoch, *The Sovereignty of Good* (Routledge & Kegan Paul, 1970).

331   *Thought-experiment devised by Philippa Foot:* Philippa Foot, "The Problem of Abortion and the Doctrine of Double Effect," *Oxford Review* 5 (1967): 5–15.

332   ***Thomson's version goes like this:*** Judith Jarvis Thomson, "Killing, Letting Die, and the Trolley Problem," *The Monist* 59, no. 2 (April 1976): 204–17.

335   ***Euthyphro's dilemma:*** Online note.

338   ***Classic 1958 article:*** G. E. M. Anscombe, "Modern Moral Philosophy," *Philosophy* 33, no. 124 (January 1958): 1–19.

338   ***Chinese philosophers in the New Confucian movement:*** Yu Jiyuan and Lei Yongqiang, "The Manifesto of New-Confucianism and the Revival of Virtue Ethics," *Frontiers of Philosophy in China* 3 (2008): 317–34.

338   ***Virtue ethics has recently had a resurgence:*** See Nancy E. Snow, ed., *The Oxford Handbook of Virtue* (Oxford University Press, 2018). In the 2020 PhilPapers Survey, 32 percent of respondents endorsed deontology, 31 percent endorsed consequentialism, and 37 percent endorsed virtue ethics. In the 2009 PhilPapers Survey, virtue ethics was in last place. The figures are not exactly comparable, but a longitudinal analysis shows that virtue ethics has increased in popularity relative to the other options.

339   ***A being has moral status:*** For a review of general issues about moral status, see Agnieszka Jaworska and Julie Tannenbaum, "The Grounds of Moral Status," *Stanford Encyclopedia of Philosophy* (Spring 2021), https://plato.stanford.edu/entries/grounds-moral-status/. On issues about the moral status of AI systems, see Matthew Liao, "The Moral Status and Rights of Artificial Intelligence," in *The Ethics of Artificial Intelligence*, ed. Matthew Liao (Oxford University Press, 2020) and Eric Schwitzgebel and Mara Garza, "Designing AI with Rights, Consciousness, Self-Respect, and Freedom," in *Ethics of Artificial Intelligence*, ed. Matthew Liao.

342   ***Sentience is what matters for moral status:*** Peter Singer, *Animal Liberation* (Harper & Row, 1975).

343   ***Reported cases of humans who do not experience pain, fear, or anxiety:*** "The woman who doesn't feel pain," BBC Scotland News, March 28, 2019. There are other partial-Vulcan syndromes such as pain asymbolia, where people report feeling pain that doesn't hurt, and anhedonia, where people can't experience pleasure. Peter Carruthers ("Sympathy and Subjectivity," *Australasian Journal of Philosophy* 77 [1999]: 465–82) has argued that a partial Vulcan he calls "Phenumb" has moral status. Phenumb never experiences affect when desires are satisfied or frustrated, but still experiences pain and other sorts of affect.

346   ***Any reproduction is immoral:*** See David Benatar, *Better Never to Have Been: The Harm of Coming into Existence* (Oxford University Press, 2006).

347   ***Simulation theodicy:*** The first simulation-based solution to the problem of evil that I know of was given by Barry Dainton in "Innocence Lost: Simulation Scenarios: Prospects and Consequences," 2002, https://philarchive.org/archive/DAIILSv1; see also Dainton's "Natural Evil: The Simulation Solution" (*Religious Studies* 56, no. 2 (2020): 209–30, DOI:10.1017/S0034412518000392). For a discussion of Dainton's idea, see David Kyle Johnson, "Natural Evil and the Simulation Hypothesis," *Philo* 14, no. 2 (2011): 161–75; and Dustin Crummett, "The Real Advantages of the Simulation Solution to the Problem of Natural Evil," *Religious Studies* (2020): 1–16. On simulation theodicies, see Brendan Shea, "The Problem of Evil in Virtual Worlds," in *Experience Machines: The Philosophy of Virtual Worlds*, ed. Mark Silcox (Rowman & Littlefield, 2017).

Chapter 19: How should we build a virtual society?

350   *Julian Dibbell reported a conversation:* Julian Dibbell, "A Rape in Cyberspace," *Village Voice*, December 21, 1993. Reprinted in his *My Tiny Life: Crime and Passion in a Virtual World* (Henry Holt, 1999).

351   *"Avatar attachment":* Jessica Wolfendale, "My Avatar, My Self: Virtual Harm and Attachment," *Ethics and Information Technology* 9 (2007): 111–19.

352   *"The Gamer's Dilemma":* Morgan Luck, "The Gamer's Dilemma: An Analysis of the Arguments for the Moral Distinction between Virtual Murder and Virtual Paedophilia," *Ethics and Information Technology* 11, no. 1 (2009): 31–36.

353   *Virtual theft:* Nathan Wildman and Neil McDonnell, "The Puzzle of Virtual Theft," *Analysis* 80, no. 3 (2020): 493–99. They cite a decision by the Supreme Court of the Netherlands saying, "[V]irtual items can be regarded as goods and can therefore be the subjects of such property offences." See Hein Wolswijk, "Theft: Taking a Virtual Object in RuneScape: Judgment of 31 January 2012, case no. 10/00101 J," *The Journal of Criminal Law* 76, no. 6 (2012): 459–62.

354   **Grand Theft Auto:** Ren Reynolds, "Playing a 'Good' Game: A Philosophical Approach to Understanding the Morality of Games," *International Game Developers Association*, 2002, http://www.igda.org/articles/rreynoldsethics.php.

355   *Monique Wonderly:* Monique Wonderly, "Video Games and Ethics," in *Spaces for the Future: A Companion to Philosophy of Technology*, eds. Joseph C. Pitt and Ashley Shew (Routledge, 2018), 29–41.

355   *Virtual reality as a superhero:* Robin S. Rosenberg, Shawnee L. Baughman, and Jeremy N. Bailenson, "Virtual Superheroes: Using Superpowers in Virtual Reality to Encourage Prosocial Behavior," *PLOS ONE*, (January 30, 2013) DOI:10.1371/journal.pone.0055003.

355   *VR analog of Milgram's experiment:* Mel Slater, Angus Antley, Adam Davison, David Swapp, Christoph Guger, Chris Barker, Nancy Pistrang, and Maria V. Sanchez-Vives, "A Virtual Reprise of the Stanley Milgram Obedience Experiments," *PLOS ONE*, https://doi.org/10.1371/journal.pone.0000039.

355   *Equivalence Principle:* Erick Jose Ramirez and Scott LaBarge, "Real Moral Problems in the Use of Virtual Reality," *Ethics and Information Technology* 4 (2018): 249–63.

356   *Ethical guidelines for researchers:* Michael Madary and Thomas K. Metzinger, "Real Virtuality: A Code of Ethical Conduct," *Frontiers in Robotics and AI* 3 (2016): 1–23.

356   *Chinese philosopher Mozi:* "Identification with the Superior I," Chinese Text Project, https://ctext.org/mozi/identification-with-the-superior-i/ens.

357   *"Nasty, brutish, and short":* Thomas Hobbes, *Leviathan* i. xiii. 9.

359   *The Alphaville Herald:* See Peter Ludlow and Mark Wallace, *The Second Life Herald: The Virtual Tabloid that Witnessed the Dawn of the Metaverse* (MIT Press, 2007). On governance in virtual worlds, see Peter Ludlow, ed., *Crypto Anarchy, Cyberstates, and Pirate Utopias* (MIT Press, 2001).

359   **EVE Online:** Pétur Jóhannes Óskarsson, "The Council of Stellar Management: Implementation of Deliberative, Democratically Elected, Council in EVE," https://www.nytimes.com/packages/pdf/arts/PlayerCouncil.pdf.   See   also   Nicholas O'Brien, "The Real Politics of a Virtual Society," *The Atlantic*, March 10, 2015.

360   *Vast range of virtual worlds:* This scenario bears some similarity to Robert

Nozick's conception of utopia (in *Anarchy, State, and Utopia*) as a "meta-utopia" of countless different societies organized in different ways. For more on digital and virtual meta-utopias, see "Could Robert Nozick's Utopian Framework Be Created on the Internet?" (*Polyblog*, September 9, 2011, https://polyology.wordpress.com/2011/09/09/the-internet-and-the-framework-for-utopia; Mark Silcox, *A Defense of Simulated Experience: New Noble Lies* (Routledge, 2019); and John Danahaer, *Automation and Utopia: Human Flourishing in a World without Work* (Harvard University Press, 2019). For a philosophical analysis of Nozick's meta-utopia, see Ralf M. Bader, "The Framework for Utopia," in *The Cambridge Companion to Nozick's "Anarchy, State, and Utopia*, eds. Ralf M. Bader and John Meadowcroft (Cambridge University Press, 2011).

362   *Recent article in* **Wired** *magazine:* Matthew Gault, "Billionaires See VR as a Way to Avoid Radical Social Change," *Wired*, February 15, 2021. The John Carmack quotation is from the *Joe Rogan Experience*, episode 1342, 2020.

362   *Artificial scarcity:* An extreme form of artificial scarcity arises with nonfungible tokens (NFTs) attached to digital artworks and other digital objects through blockchain technology. Some people pay large amounts of money for an NFT even though it brings no obvious utility over and above being identified as the owner of the NFT. Here it appears that scarcity is being valued for its own sake. This form of artificial scarcity with no functional utility almost by definition applies only to luxury goods. However, less extreme forms of artificial scarcity for useful goods are only to be expected in a market system.

362   *How will unemployed people pay?:* For more on economic and philosophical issues arising from technological unemployment, see Erik Brynjolffson and Andrew McAfee, *The Second Machine Age* (W. W. Norton, 2014); Danaher, *Automation and Utopia*; Aaron James, Planning for Mass Unemployment: Precautionary Basic Income," in *Ethics of Artificial Intelligence*, ed. Matthew Liao (Oxford University Press, 2020).

363   *Her important 1999 article:* Elizabeth Anderson, "What Is the Point of Equality?," *Ethics* 109, no. 2 (1999): 287–337. For related work in this recent *relational egalitarian* tradition, see Samuel Scheffler, "The Practice of Equality," in *Social Equality: On What it Means to be Equals*, eds. C. Fourie, F. Schuppert, and I. Wallimann-Helmer (Oxford University Press, 2015); Daniel Viehoff, "Democratic Equality and Political Authority," *Philosophy and Public Affairs* 42 (2014): 337–75; and Niko Kolodny, *The Pecking Order* (Harvard University Press, forthcoming). For a related conception of freedom as non-domination, see Philip Pettit, *Republicanism: A Theory of Freedom and Government* (Oxford University Press, 1997).

364   *Coined the term* **intersectionality:** Kimberlé Crenshaw, "Mapping the Margins: Intersectionality, Identity Politics, and Violence against Women of Color," *Stanford Law Review* 44 (1991): 1241–99. See also Angela Davis, *Women, Race, and Class* (Knopf Doubleday, 1983); and Patricia Hill Collins, *Black Feminist Thought: Knowledge, Consciousness and the Politics of Empowerment* (Hyman, 1990).

Chapter 20: What do our words mean in virtual worlds?

367   *Coffeehouse Conversation:* Douglas R. Hofstadter, "A Coffeehouse Conversation on the Turing Test," *Scientific American*, May 1981. Reprinted in *The Mind's I: Fantasies and Reflections on Self and Soul*, eds. Daniel C. Dennett and Douglas R. Hofstadter (Basic Books, 1981). Hofstadter develops this simulation realism further in a discussion of "SimTown" and "SimBowl" in *Le Ton beau de Marot* (Basic Books, 1997), 312–17. He also expresses a sort of virtual realism in discussing a virtual world of blocks on a table used by the AI program SHRDLU (p. 510): "However, whether the table was substantial or ethereal was of little import, since what really mattered was the *patterns* of objects in the situations, and those patterns were not in the least affected by their tangible physical existence or lack thereof."

370   *Analytic and continental philosophy:* For an overview of continental philosophy, see Richard Kearney and Mara Rainwater, eds., *The Continental Philosophy Reader* (Routledge, 1996). For a history of analytic philosophy, see Scott Soames, *The Analytic Tradition in Philosophy*, vols. 1 and 2 (Princeton University Press, 2014, 2017).

370   *Gottlob Frege:* See Michael Beaney's *The Frege Reader* (Blackwell, 1997).

371   *"On Sense and Reference":* Gottlob Frege, "Über Sinn und Bedeutung" (in *Zeitschrift für Philosophie und philosophische Kritik* 100 (1892): 25–50. Translated as "On Sense and Reference" (in Beaney's *The Frege Reader*).

372   *Russell's theory of names and descriptions:* Bertrand Russell, "On Denoting," *Mind* 14, no. 56 (1905): 479–93; and "Knowledge by Acquaintance and Knowledge by Description," *Proceedings of the Aristotelian Society* 11 (1910–11): 108–27.

372   *Small revolution:* Saul Kripke, *Naming and Necessity* (Harvard University Press, 1980); Hilary Putnam, "The Meaning of Meaning," in *Language, Mind, and Knowledge*, ed. Keith Gunderson (University of Minnesota Press, 1975), 131–93; Ruth Barcan Marcus, *Modalities: Philosophical Essays* (Oxford University Press, 1993).

374   *"Water" picks out whatever plays the water role:* This understanding of externalism is controversial as it may retain some of the descriptivism that Kripke argues against, but I defend a weak version of it (where the water role is not explicit, can vary between speakers, and can involve effects on use of the word 'water') in *Constructing the World* and elsewhere. This view goes well with the structuralism discussed in chapters 9 and 22, especially if we think of the roles as ultimately specifiable in structural terms.

374   *Limits to externalism:* Tyler Burge has argued (in "Individualism and the Mental," *Midwest Studies in Philosophy*, 4 [1979]: 73–122) that the meaning of any term—even "seven" or "computer"—can be "outside the head" of many speakers, when those speakers defer to others in their linguistic community. I'm setting aside this sort of social externalism by assuming the speakers are experts and don't defer to others with regard to meaning. For an externalist (or Twin-Earthable) word in my sense, there must be possible twins who don't defer but use their words to refer to different things. "Water" is an externalist word in this sense but "seven" is not.

374   *Internalist words:* In *Constructing the World* (especially online excursus 21) I call these words *non-Twin-Earthable* and give a more careful definition. It's a deep question which words (if any) are internalist, but my view is that internalist

words correspond roughly to causal/mental invariant categories (discussed in the final note to chapter 10), or better, to categories characterizable in structural (logical, mathematical, causal) and mental terms (see chapter 8 of *Constructing the World*).

374   **Two-dimensional *view of meaning:*** David J. Chalmers, "Two-Dimensional Semantics," in *The Oxford Handbook of the Philosophy of Language*, eds. Ernest Lepore and Barry C. Smith (Oxford University Press, 2006).

375   ***Language in virtual worlds:*** Astrin Ensslin, *The Language of Gaming* (Palgrave Macmillan, 2012); Astrid Ensslin and Isabel Balteiro, eds., *Approaches to Videogame Discourse* (Bloomsbury, 2019); Ronald W. Langacker, "Virtual Reality," *Studies in the Linguistic Sciences* 29, no. 2 (1999): 77–103; Gretchen McCulloch, *Because Internet: Understanding the New Rules of Language* (Riverhead Books, 2019).

382   ***If Sim Putnam says "I'm in a computer simulation":*** In "Skepticism Revisited: Chalmers on *The Matrix* and Brains-in-Vats," *Cognitive Systems Research* 41 (2017): 93–98, Richard Hanley suggests that if beliefs like "I'm not in a simulation" are false in a simulation, simulations may be skeptical scenarios after all. My response is that (as acknowledged in chapter 6) we may have some false theoretical beliefs about matters like this in a simulation, but that this does not call our ordinary beliefs into question.

382   ***Paragraph in* Reason, Truth and History:*** Hilary Putnam, *Reason, Truth and History* (Cambridge University Press, 1981), 14.

383   ***Donald Davidson and Richard Rorty:*** Donald Davidson, "A Coherence Theory of Truth and Knowledge," in *Truth and Interpretation: Perspectives on the Philosophy of Donald Davidson*, ed. Ernest Lepore (Blackwell, 1986); Richard Rorty, "Davidson versus Descartes," in *Dialogues with Davidson: Acting, Interpreting, Understanding*, ed. Jeff Malpas (MIT Press, 2011).

Chapter 21: Do dust clouds run computer programs?

385   ***1994 science-fiction novel:*** Greg Egan, *Permutation City* (Orion/Millennium, 1994).

387   ***The dust theory raises any number of questions:*** For philosophical discussion of the dust theory, including a suggestion that the dust may lack causation between its states, see Eric Schwitzgebel, "The Dust Hypothesis," *The Splintered Mind* weblog (January 21, 2009).

388   ***Babbage and Lovelace:*** Doron Swade, *The Difference Engine: Charles Babbage and the Quest to Build the First Computer* (Viking Adult, 2001); Christopher Hollings, Ursula Martin, and Adrian Rice, *Ada Lovelace: The Making of a Computer Scientist* (Bodleian Library, 2018).

388   ***First programmable electronic computer:*** Online note.

389   ***Putnam and Searle:*** Hilary Putnam, *Representation and Reality* (MIT Press, 1988); John Searle, *The Rediscovery of the Mind* (MIT Press, 1992).

390   ***I ended up publishing two articles:*** David J. Chalmers, "On Implementing a Computation," *Minds and Machines* 4 (1994): 391–402; David J. Chalmers, "Does a Rock Implement Every Finite-State Automaton?," *Synthese* 108, no. 3 (1996): 309–33.

392   *In the absence of time:* Online note.
393   *For a million generations or so:* Online note.
395   *The right pattern of counterfactuals:* Online note.
395   *Minority view among philosophers:* In the 2020 PhilPapers Survey, 54 percent of philosophers accepted or leaned toward a non-Humean view of laws of nature, which holds that laws (such as the law of gravity) involve more than regularity; 31 percent accepted or leaned toward a Humean view where laws are a matter of regularities. It's plausible that the distribution of views about causation would be similar.
396   *This isn't the end of the story:* Philosophers arguing that it may still be too easy to meet my stronger constraints on implementing a computation include Curtis Brown, "Combinatorial-State Automata and Models of Computation," *Journal of Cognitive Science* 13, no. 1 (2012): 51–73; Peter Godfrey-Smith, "Triviality Arguments against Functionalism," *Philosophical Studies* 145 (2009): 273–95; Matthias Scheutz, "What It Is Not to Implement a Computation: A Critical Analysis of Chalmers' Notion of Computation," *Journal of Cognitive Science* 13 (2012): 75–106; Mark Sprevak, "Three Challenges to Chalmers on Computational Implementation," *Journal of Cognitive Science* 13 (2012): 107–43; and Gualtiero Piccinini, *Physical Computation: A Mechanistic Account* (Oxford University Press, 2015). I reply to some of them in "The Varieties of Computation," *Journal of Cognitive Science* 13 (2012): 211–48.

Chapter 22: Is reality a mathematical structure?

399   *Carnap's magnum opus:* Rudolf Carnap, *Der Logische Aufbau der Welt* (Felix Meiner Verlag, 1928). Translated as *The Logical Structure of the World* (University of California Press, 1967). For accessible histories of the Vienna Circle, see David Edmonds, *The Murder of Professor Schlick: The Rise and Fall of the Vienna Circle* (Princeton University Press, 2020); and Karl Sigmund, *Exact Thinking in Demented Times: The Vienna Circle and the Epic Quest for the Foundations of Science* (Basic Books, 2017). My initial sentence is inspired by Anders Wedberg, "How Carnap Built the World in 1928," *Synthese* 25 (1973): 337–41.
400   *1932 article:* "Die physikalische Sprache als Universalsprache der Wissenschaft," *Erkenntnis* 2 (1931): 432–65. Translated as "The Physical Language as the Universal Language of Science" in *Readings in Twentieth-Century Philosophy*, eds. William P. Alston and George Nakhnikian (Free Press, 1963), 393–424.
402   *Structuralism about culture:* Claude Lévi-Strauss, *The Elementary Structures of Kinship* (Presses Universitaires de France, 1949;); influenced by Ferdinand de Saussure, *Course in General Linguistics* (Payot, 1916); influenced Louis Althusser, Michel Foucault, Jacques Lacan, and many others.
402   *Scientific realism:* See Juha Saatsi, ed., *The Routledge Handbook of Scientific Realism* (Routledge, 2020); Ernst Mach, *The Science of Mechanics*, 1893; J. J. C. Smart, *Philosophy and Scientific Realism* (Routledge, 1963); Hilary Putnam, "What Is Mathematical Truth?" in *Mathematics, Matter, and Method* (Cambridge University Press, 1975).
404   *Structural realism:* Carnap, *The Logical Structure of the World;* Bertrand Russell, *The Analysis of Matter* (Kegan Paul, 1927); John Worrall, "Structural Realism:

The Best of Both Worlds?" *DIalectica* 43 (1989): 99–124; James Ladyman, "What Is Structural Realism?" *Studies in History and Philosophy of Science Part A* 29 (1998): 409–24.

405 **Technique for structuralizing theories:** Frank Ramsey, "Theories," in *The Foundations of Mathematics and Other Logical Essays* (Kegan Paul, Trench, Trubner, 1931), 212–36.

405 **Remarkable British philosopher:** Cheryl Misak, *Frank Ramsey: A Sheer Excess of Powers* (Oxford University Press, 2020).

405 **There exist seven properties:** "Property" might sound like a nonlogical and nonmathematical term, but just as "there exists an object" can be be formalized into logic with an existential quantifier ∃, "there exists a property" can be similarly formalized in second-order logic.

407 **Mathematical Universe Hypothesis:** Max Tegmark, *Our Mathematical Universe* (Vintage Books, 2014).

412 **The view that physical theories specify purely mathematical structure for reality has to be qualified:** I discuss various possible packages for the structuralist in *Constructing the World*, chapter 8.

412 **It sometimes happens that one physical theory makes another true:** I'm not offering a general analysis of when one physical theory makes another true, which depends on many subtle issues about the precise structural contents of these theories. One puzzle case arises from the so-called "holographic principle" and the associated AdS/CFT correspondence (see Leonard Susskind and James Lindesay, *An Introduction to Black Holes, Information and the String Theory Revolution: The Holographic Universe* [World Scientific, 2005]) in which certain higher-dimensional string theories (e.g., on the three-dimensional interior of a sphere) appear to be mathematically isomorphic to certain lower-dimensional quantum theories (e.g., on the sphere's two-dimensional surface). I discuss the holographic principle and its connection to the simulation hypothesis in an online note.

413 **Argument that starts with structuralism and ends with simulation realism:** I give a more detailed version of this argument in "Structuralism as a Response to Skepticism," *Journal of Philosophy* 115 (2018): 625–60. Where the use of structuralism to respond to external-world skepticism is concerned, I've found one predecessor: a paragraph by the philosopher of physics Lawrence Sklar in his 1982 article "Saving the Noumena" (*Philosophical Topics* 13, no. 1). Sklar entertains the idea that "the brain-in-a-vat account of the world is really equivalent to the ordinary material object world account, so long as the brain-in-a-vat account is suitably formally structured" (p. 98), but immediately dismisses the idea as being too close to instrumentalism.

416 **It does not address the problem of other minds:** Grace Helton, in "Epistemological Solipsism as a Route to External-World Skepticism" (*Philosophical Perspectives*, forthcoming), and in other work on structuralism and skepticism, argues that if others don't have minds, then many ordinary physical objects don't exist, including social entities such as cities, churches, and clubs that depend on minds for their existence. If so, a structuralist anti-skeptical strategy that does not establish the existence of other minds does not establish the existence of social entities, and skepticism about the social realm remains open. Still, I think it is plausible that atoms, cells, trees, planets, and other physical objects do not depend

on other minds for their existence. If so, skepticism about other minds does not lead to skepticism about the ordinary physical world.

420 *Inability to know about things in themselves:* Kant's view is interestingly analogous to the famous first line of Laozi's *Dao De Jing* (or *Tao Te Ching*), one of the founding works of Daoist philosophy: *The Dao that can be named is not the eternal Dao.* Kant says: *The thing that can be known [and named] is not the thing in itself.* For more on this theme, see Martin Schönfeld, "Kant's Thing in Itself or the Tao of Königsberg," *Florida Philosophical Review* 3 (2003): 5–32.

420 *Reminiscent of Kant's transcendental idealism:* In my "The Matrix as Metaphysics" (2003): "One might say that if we are in a matrix, the Kantian ding-an-sich (thing in itself) is part of a computer-an-sich!" Barry Dainton's "Innocence Lost: Simulation Scenarios: Prospects and Consequences" (2002, https://philarchive.org/archive/DAIILSv1) also suggests connections between the simulation hypothesis and transcendental idealism ("In Kantian terms, virtual worlds of the communal variety are *empirically real*, even if *transcendentally ideal*"), as does Eric Schwitzgebel in "Kant Meets Cyberpunk," *Disputatio* 11, no. 55 (2019): 411–35.

420 *Australian philosopher Rae Langton:* Rae Langton, *Kantian Humility: Our Ignorance of Things in Themselves* (Oxford University Press, 1998).

421 *It-from-structure-from-X view:* For a closely related version of the it-from-structure-from-X view, approached in terms of Ramsey sentences, see David Lewis, "Ramseyan Humility," in *Conceptual Analysis and Philosophical Naturalism*, eds. David Braddon-Mitchell and Robert Nola (MIT Press, 2008). In his 1818 book *The World as Will and Representation*, the German philosopher Arthur Schopenhauer is often interpreted as replacing Kant's unknowable X with something knowable and experienceable, namely the will. We might interpret Schopenhauer as holding an *it-from-structure-from-will* view.

Chapter 23: Have we fallen from the Garden of Eden?

425 *The manifest and scientific images:* Wilfrid Sellars, "Philosophy and the Scientific Image of Man," in *Frontiers of Science and Philosophy*, ed. Robert Colodny (University of Pittsburgh Press, 1962), 35–78.

425 *Sellars himself argued that consciousness is real:* Wilfrid Sellars, "Is Consciousness Physical?," *The Monist* 64 (1981): 66–90.

425 *The two images:* Online note.

425 *Patricia and Paul Churchland:* Patricia S. Churchland, *Neurophilosophy* (MIT Press, 1987). Paul M. Churchland, *A Neurocomputational Perspective* (MIT Press, 1989).

429 *Friedrich Nietzsche once said:* Friedrich Nietzsche, Nachgelassene Fragmente (1871) in *Kritische Studienausgabe*, vol. 7 (De Gruyter, 1980), 352.

429 *Colors exist only in the mind:* In *Il Saggitore* (*The Assayer*, 1623), Galileo writes, "[T]hese tastes, odors, colors, etc., so far as their objective existence is concerned, are nothing but mere names for something which resides exclusively in our sensitive body (*corpo sensitivo*), so that if the perceiving creatures were removed, all of these qualities would be annihilated and abolished from existence," in *Introduction* to *Contemporary Civilization in the West*, 2nd edition, vol. 1, trans. A. C. Danto (Columbia University Press, 1954), 719–24.

431   **Spatial functionalism:** I introduced spatial functionalism in chapter 7 of *Constructing the World* (Oxford University Press, 2012) and developed it further in "Three Puzzles about Spatial Experience" (in *Blockheads: Essays on Ned Block's Philosophy of Minds and Consciousness*, eds. Adam Pautz and Daniel Stoljar [MIT Press, 2017]) and "Finding Space in a Nonspatial World," in *Philosophy beyond Spacetime*, eds. Christian Wüthrich, Baptiste Le Bihan, and Nick Huggett (Oxford University Press, 2021). For related discussion of spacetime functionalism in physics, see Eleanor Knox, "Physical Relativity from a Functionalist Perspective," *Studies in History and Philosophy of Modern Physics* 67 (2019): 118–24 and other articles in the *Philosophy beyond Spacetime* volume.

433   **Nothing is laid out in space as it seems to be:** For some questions about my structuralist/functionalist analysis of space, as applied to simulations and skeptical scenarios, see Jonathan Vogel, "Space, Structuralism, and Skepticism," in *Oxford Studies in Epistemology*, vol. 6 (2019); Christopher Peacocke, "Phenomenal Content, Space, and the Subject of Consciousness," *Analysis* 73 (2013): 320–29; and also Alyssa Ney, "On Phenomenal Functionalism about the Properties of Virtual and Non-Virtual Objects," *Disputatio* 11, no. 55 (2019): 399–410; and E. J. Green and Gabriel Rabin, "Use Your Illusion: Spatial Functionalism, Vision Science, and the Case against Global Skepticism," *Analytic Philosophy* 61, no. 4 (2020): 345–78.

435   **Hoffman's Case Against Reality:** Online appendix.

436   **Free will:** See Robert Kane, ed., *The Oxford Handbook of Free Will*, 2nd ed. (Oxford University Press, 2011); John Martin Fischer, Robert Kane, Derk Pereboom, and Manuel Vargas, *Four Views on Free Will* (Blackwell, 2007); Daniel Dennett, *Elbow Room: The Varieties of Free Will Worth Wanting* (MIT Press, 1984).

436   **Slavoj Žižek said:** Slavoj Žižek, "From Virtual Reality to the Virtualization of Reality" in *Electronic Culture: Technology and Visual Representation*, ed. Tim Druckrey (Aperture, 1996), 29095.

438   **Scientific revolutions:** Thomas Kuhn, *The Structure of Scientific Revolutions* (University of Chicago Press, 1962).

438   **The world of Forms:** Sources for these Forms in Plato's dialogues: Large: *Phaedo* 100b and elsewhere. Square: *Republic* 6 510d. Solid: Implied in *Meno* 76a. Beauty: *Republic* V 475e-476d and elsewhere. Good: *Republic* V 476a and elsewhere.

Chapter 24: Are we Boltzmann brains in a dream world?

445   **What if God created reality five minutes ago:** Bertrand Russell, *The Analysis of Mind* (George Allen & Unwin, 1921), 159–60.

445   **Temporary simulation skepticism:** Online note.

446   **Appeal to simplicity:** One limitation of this appeal: while it may justify us in having low confidence in a complex hypothesis, it's not obvious how it allows us to know the hypothesis is false. I should have low confidence that a coin will come up heads 20 times in a row, but I arguably can't know that it won't. In responding to the skeptic, however, I'd settle for being justified in having high confidence in my beliefs, whether or not this strictly counts as knowledge.

450 *Born with a dual system:* Greg Egan, "Learning to Be Me," *Interzone* 37, July 1990.

451 *God is playing the role of the computer:* Peter B. Lloyd ("A Review of David Chalmers' Essay, 'The Matrix as Metaphysics,'" 2003, DOI:10.13140/RG.2.2.11797.99049), who responds to my analysis from a Berkeleyan idealist perspective, suggests that even Berkeley's God might be running some sort of shortcut (just-in-time) simulation in order to be more economical.

453 *When Zhuangzi dreams of the butterfly, there is a real dream butterfly:* Zhuangzi's own discussion has an element of virtual realism, insofar as it stresses the reality of both the butterfly and Zhuangzi (though unlike my analysis, Zhuangzi's analysis also stresses the distinction between the butterfly and Zhuangzi). See Robert Allinson, *Chuang-Tzu for Spiritual Transformation: An Analysis of the Inner Chapters* (SUNY Press, 1989).

455 *Novels and other fictions:* Online appendix.

456 *Experiences not generated by the external world:* Online appendix.

457 *As Sean Carroll has pointed out:* Sean M. Carroll, "Why Boltzmann Brains Are Bad," arXiv:1702.00850v1 [hep-th], 2017. For a related point about self-undermining belief in Boltzmsnn brains, see Bradley Monton, "Atheistic Induction by Boltzmann Brains," in J. Wall and T. Dougherty, eds., *Two Dozen (or So) Arguments for God: The Plantinga Project* (Oxford University Press, 2018).

458 *Only a tiny minority would have ordered experiences:* Online note.

Index